KINETICS of
PHASE TRANSITIONS

KINETICS of PHASE TRANSITIONS

Edited by
Sanjay Puri and Vinod Wadhawan

CRC Press
Taylor & Francis Group
Boca Raton London New York

CRC Press is an imprint of the
Taylor & Francis Group, an **informa** business

CRC Press
Taylor & Francis Group
6000 Broken Sound Parkway NW, Suite 300
Boca Raton, FL 33487-2742

First issued in paperback 2019

© 2009 by Taylor & Francis Group, LLC
CRC Press is an imprint of Taylor & Francis Group, an Informa business

No claim to original U.S. Government works

ISBN-13: 978-0-8493-9065-4 (hbk)
ISBN-13: 978-0-367-38585-9 (pbk)

Library of Congress Cataloging-in-Publication Data

Kinetics of phase transitions / [edited by] Sanjay Puri, Vinod Wadhawan.
    p. cm.
Includes bibliographical references and index.
ISBN 978-0-8493-9065-4 (hardcover : alk. paper)
    1. Phase transformations (Statistical physics) 2. Phase rule and equilibrium. 3. Materials--Thermal properties. I. Puri, Sanjay, 1962- II. Wadhawan, Vinod K. III. Title.

QC175.16.P5K56 2009
530.4'74--dc22   2008054120

Visit the Taylor & Francis Web site at
http://www.taylorandfrancis.com

and the CRC Press Web site at
http://www.crcpress.com

Contents

Preface

This book focuses on the kinetics of phase transitions, that is, the evolution of a system from an unstable or metastable state to its preferred equilibrium state. A system may become thermodynamically unstable due to a sudden change in external parameters like temperature, pressure, magnetic field, and so on. The subsequent dynamics of the far-from-equilibrium system is usually nonlinear and is characterized by complex spatiotemporal pattern formation. Typically, the system evolves toward its new equilibrium state via the emergence and growth of domains enriched in the preferred state. This process is usually referred to as phase-ordering dynamics or domain growth or coarsening. There has been intense research interest in this field over the past few decades, as the underlying physical processes are of great scientific and technological importance. Problems in this field arise from diverse disciplines such as physics, chemistry, metallurgy, materials science, and biology. As a result of this research activity, our understanding of phase-ordering dynamics has reached a high level of sophistication. At the same time, many challenging problems continue to arise in different contexts. It is now clear that the paradigms and concepts of phase-ordering dynamics are of much wider applicability than was initially thought.

In the context of the above developments, we believed that there was a strong need for a book that summarizes our current understanding of domain growth. Furthermore, we believed that this book should be written at a level accessible to the advanced undergraduate; that is, it should be a textbook rather than an advanced research monograph. With this in mind, we wrote to various leaders in this field with a request to each to contribute a chapter. Their responses were very positive, and this book is an outcome of the collective efforts of various colleagues. On our part, we have edited and homogenized the various chapters so that this book reads as a seamless "multiple-author book" rather than as the usual disjointed "edited book."

Let us provide an overview of the various chapters. The first chapter (written by Sanjay Puri) provides an overview of studies of domain growth in simple systems. This chapter develops the theoretical tools and methodology that are used in subsequent chapters. The second chapter (written by Kurt Binder) focuses on the distinction between spinodal decomposition and nucleation and growth, which are common scenarios for domain growth problems. This issue has been discussed extensively in the literature, but there remains considerable confusion over the interpretation of various experiments and simulations. Kurt Binder addresses this issue in great detail, emphasizing that there is no sharp boundary between spinodal decomposition and nucleation.

Chapters 3 and 4 are dedicated to a discussion of simulation techniques in this field. In Chapter 3, Gerard Barkema describes Monte Carlo simulations of kinetic Ising models. In Chapter 4, Giuseppe Gonnella and Julia Yeomans discuss lattice

Boltzmann simulations, which have proved very useful in understanding the late stages of phase separation in fluid mixtures. Numerical simulations have played a crucial role in developing our understanding of phase-ordering dynamics. The methodology described in Chapters 3 and 4 will prove very useful for a researcher entering this field.

In Chapter 5, Marco Zannetti discusses slow relaxation and aging in phase-ordering systems. These phenomena are well known in the context of structural glasses and spin glasses. Recent studies indicate that these concepts are also highly relevant in domain growth problems—Zannetti provides an overview of these studies.

Recent interest in this area has focused on incorporating various experimentally relevant features in studies of phase-ordering systems. In this context, Chapter 6 (by Rajesh Khanna, Narendra Kumar Agnihotri, and Ashutosh Sharma) describes the kinetics of dewetting of liquid films on surfaces. In Chapter 7, Takao Ohta reviews studies of phase separation in diblock copolymers. In these systems, the segregating polymers are jointed, so that the system can only undergo phase separation on micro-scales.

Finally, in Chapter 8 (written by Akira Onuki, Akihiko Minami, and Akira Furukawa), there is a discussion of phase separation in solids. Strain fields play an important role in the segregation kinetics of alloys. Onuki et al. discuss how elastic fields can be incorporated into the description of segregation in solid mixtures.

Before we conclude, it would be appropriate to thank those who have contributed to this project. First, we are grateful to the authors, who have made the effort to write pedagogical reviews of various research problems. Second, we wish to thank our colleagues and collaborators, who have contributed so much to our understanding and appreciation of this fascinating field of research. Finally, we are grateful to the editorial and production staff at CRC Press/Taylor & Francis for their assistance in getting this book into its final form.

Sanjay Puri
New Delhi
Vinod Wadhawan
Mumbai

Editors

Professor Sanjay Puri is an expert in the fields of statistical physics and nonlinear dynamics. He has made important contributions to these fields and has published two books and more than 125 papers. His publications have been extensively cited, and he is an established authority in the kinetics of phase transitions. Dr. Puri received an MS degree in physics in 1982 from the Indian Institute of Technology, Delhi, and a PhD degree in physics in 1987 from the University of Illinois at Urbana-Champaign. In 1987, he joined the School of Physical Sciences at Jawaharlal Nehru University, New Delhi, and has been there ever since. He has received many awards and honors for his research achievements. For example, in 2006 he was elected a fellow of the Indian Academy of Sciences, Bangalore. In 2006, he also received the prestigious Shanti Swarup Bhatnagar Prize from the Government of India.

Dr. Vinod Wadhawan is a condensed-matter physicist with special interest in ferroic materials, phase transitions, and the utilitarian role of symmetry considerations in materials science. He introduced the important notion of "latent symmetry" in composite systems. He has edited/coedited 10 volumes on various aspects of phase transitions and has written two books: one on ferroic materials (the first definitive book on the subject) and the other on smart structures. Dr. Wadhawan received his MSc degree in physics from the University of Delhi in 1967, and a PhD degree from the University of Bombay in 1976. He has been with the Department of Atomic Energy, Government of India, since then—he currently holds the prestigious Raja Ramanna Fellowship at Bhabha Atomic Research Centre, Mumbai. Dr. Wadhawan is an associate editor of *Phase Transitions* (Taylor & Francis), a journal with which he has been associated since 1985. He is also a recipient of the Materials Research Society of India medal.

Contributors

Narendra Kumar Agnihotri
Department of Chemical Engineering
Indian Institute of Technology Delhi
New Delhi, India

Gerard T. Barkema
Institute for Theoretical Physics
Utrecht University
Utrecht, The Netherlands

Kurt Binder
Institute of Physics
Johannes Gutenberg University
    of Mainz
Mainz, Germany

Akira Furukawa
Institute of Industrial Science
University of Tokyo
Tokyo, Japan

Giuseppe Gonnella
Department of Physics
University of Bari

and

National Institute of
    Nuclear Physics
Bari, Italy

Rajesh Khanna
Department of Chemical Engineering
Indian Institute of Technology Delhi
New Delhi, India

Akihiko Minami
Department of Physics
Kyoto University
Kyoto, Japan

Takao Ohta
Department of Physics
Kyoto University
Kyoto, Japan

Akira Onuki
Department of Physics
Kyoto University
Kyoto, Japan

Sanjay Puri
School of Physical Sciences
Jawaharlal Nehru University
New Delhi, India

Ashutosh Sharma
Department of Chemical Engineering
Indian Institute of Technology Kanpur
Kanpur, India

Julia M. Yeomans
The Rudolf Peierls Centre
    for Theoretical Physics
University of Oxford
Oxford, United Kingdom

Marco Zannetti
Department of Mathematics and
    Computer Science
University of Salerno
Fisciano, Italy

1 Kinetics of Phase Transitions

Sanjay Puri

CONTENTS

1

1.1   INTRODUCTION

Many systems exist in multiple phases, depending on the values of external parameters, for example, temperature (T), pressure (P), and so on. In this context, consider a fluid (e.g., water), which can exist in three phases, viz., liquid, solid, and gas. The phase diagram of this fluid in the (T, P)-plane is shown in Figure 1.1. The chosen phase at a particular (T, P)-value is the one with lowest Gibbs potential $G(T, P)$. This phase diagram is characterized by a range of fascinating features, for example, lines of first-order phase transitions, a second-order critical point, a triple point, and so on. The correct understanding of these features is of great scientific and technological importance. We have gained a thorough understanding of the equilibrium aspects of phase transitions (and phase diagrams) through many important works, starting with the seminal contribution of Van der Waals [1,2].

There is also a fascinating class of problems involving the *kinetics of phase transitions*, that is, the evolution dynamics of a system that is rendered thermodynamically unstable by a rapid change of parameters. In the context of Figure 1.1, consider a situation in which the fluid in the solid phase is rapidly heated to a temperature where the preferred equilibrium state is the liquid phase. Clearly, the solid will convert to liquid on some timescale, so the initial and final states of the system are well understood. However, we have less knowledge about the dynamical processes that occur as the solid converts to liquid. These processes play a crucial role in our everyday life. Over the years, our understanding of the kinetics of phase transitions has improved greatly [3–6]. This book provides an overview of developments in this area.

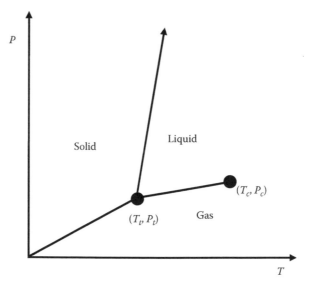

FIGURE 1.1   Phase diagram of a fluid in the (T, P)-plane. The system can exist in either of three phases—liquid, gas, or solid. The solid lines denote lines of first-order phase transitions. At the triple point (T_t, P_t), all three phases coexist. The point labeled (T_c, P_c) is the critical point of the system.

Before we proceed, it is relevant to develop the appropriate terminology first. One is often interested in the evolution of systems whose parameters have been drastically changed. Such systems are referred to as *far-from-equilibrium systems*, and their evolution is characterized by *nonlinear evolution equations* and *spatiotemporal pattern formation*. In most cases, we are unable to obtain exact solutions for the time-dependent evolution of the system. However, the presence of *domain boundaries* or *defects* in these systems provides a convenient analytical tool to understand the resultant pattern dynamics.

Let us consider two other problems in this context. These will serve as paradigms for understanding the kinetics of phase transitions. First, consider a ferromagnet whose phase diagram is shown in Figure 1.2. Focus on the case with zero magnetic field ($h = 0$). At high temperatures, the magnet is in a disordered or paramagnetic state. If the temperature is suddenly quenched to $T < T_c$, this system now prefers to be in the magnetized state with spins pointing in the "up" or "down" directions. The evolution of the system is characterized by the emergence and growth of domains enriched in either up or down spins. As time $t \to \infty$, the system approaches a spontaneously magnetized state.

Second, consider a binary (AB) mixture whose phase diagram is shown in Figure 1.3. The system is mixed or homogeneous at high temperatures. At time $t = 0$, the mixture is suddenly quenched below the *coexistence curve* or *miscibility gap*. This system now prefers to be in the phase-separated state and proceeds to its equilibrium state via the growth of domains that are either A-rich or B-rich. The nonequilibrium dynamics of the magnet or binary mixture is usually referred to as *domain growth* or *coarsening* or *phase-ordering kinetics*.

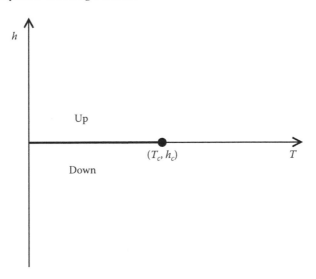

FIGURE 1.2 Phase diagram of a ferromagnet. The system parameters are the temperature (T) and the magnetic field (h). The point ($T_c, h_c = 0$) is a second-order critical point. The line ($T < T_c, h = 0$) corresponds to a line of first-order transitions. At low temperatures ($T < T_c$), the system can be in either of two phases, up or down, depending on the orientation of the magnetic spins.

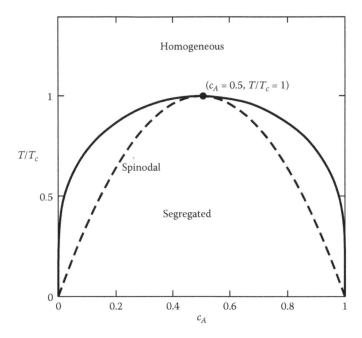

FIGURE 1.3 Phase diagram of a binary (AB) mixture. The system parameters are the concentration of A ($c_A = 1 - c_B$) and the temperature (T). The point ($c_A = 0.5, T/T_c = 1$) corresponds to a second-order critical point. Above the coexistence curve (solid line), the system is in a homogeneous or disordered state. Below the coexistence curve, the system is in a segregated or phase-separated state, characterized by A-rich and B-rich regions. The dashed lines denote spinodal curves. The homogeneous system is metastable between the coexistence and spinodal curves and unstable below the spinodal lines.

There have been many studies of the kinetics of phase transitions. Problems in this area arise in diverse contexts, ranging from *clustering dynamics in the early universe* to the *growth of nanostructures*. This book is a pedagogical exposition of developments in this area and is organized as follows. This chapter reviews the framework of phase-ordering kinetics and develops the tools and terminology used in later chapters. The subsequent chapters are written by leading experts in this area and focus on problems of special interest in the context of phase-ordering dynamics. All the chapters are written in textbook style and are accessible at the level of the advanced undergraduate student. At this point, we should stress that our understanding of this area has been greatly facilitated by numerical simulations of appropriate models. Therefore, two chapters of this book are dedicated to tutorial-level discussions of numerical simulations in this field. The first of these is written by Barkema (Chapter 3)—this chapter focuses on Monte Carlo simulations of *kinetic Ising models*. The second of these is written by Gonnella and Yeomans (Chapter 4) and describes the application of *lattice Boltzmann algorithms* to study phase-ordering systems.

This chapter is organized as follows. In Section 1.2, we introduce the Ising model for two-component mixtures and study its equilibrium properties in the mean-field (MF) approximation. This will enable us to obtain the phase diagrams shown in

Figures 1.2 and 1.3. In Section 1.3, we study kinetic versions of the Ising model. In Section 1.4, we discuss domain growth with a *nonconserved order parameter*, for example, ordering dynamics of a ferromagnet into up and down phases. In this section, we separately examine cases with scalar and vector order parameters. In Section 1.5, we discuss domain growth with a *conserved order parameter*, for example, kinetics of phase separation of an AB mixture. We will separately focus on segregation in binary alloys that is driven by diffusion, and segregation in binary fluids where flow fields drastically modify the asymptotic behavior. Finally, Section 1.6 concludes this chapter with a summary and discussion.

1.2 PHASE DIAGRAMS OF TWO-COMPONENT MIXTURES

1.2.1 ISING MODEL AND ITS APPLICATIONS

The simplest model of an interacting many-body system is the Ising model [7], which was first introduced as a model for phase transitions in magnetic systems. However, with suitable generalizations, it has wide applications to diverse problems in condensed matter physics.

Consider a set of N spins $\{S_i\}$, which are fixed on the sites $\{i\}$ of a lattice. The two-state (spin-1/2) Ising Hamiltonian has the following form:

$$H = -J \sum_{\langle ij \rangle} S_i S_j, \quad S_i = \pm 1, \tag{1.1}$$

where J is the strength of the exchange interaction between spins. We consider the case with nearest-neighbor interactions only, denoted by the subscript $\langle ij \rangle$ in Equation 1.1.

Although the Hamiltonian in Equation 1.1 is formulated for a magnetic system, it is clear that a similar description applies for any interacting two-state system, as the two states can be mapped onto $S = +1$ or -1. A well-known example is the lattice gas or binary (AB) mixture [7]. We can describe this system in terms of occupation-number variables $n_i^\alpha = 1$ or 0, depending on whether or not a site i is occupied by species α (A or B). Clearly, $n_i^A + n_i^B = 1$ for all sites. A more convenient description is obtained in terms of spin variables $S_i = 2n_i^A - 1 = 1 - 2n_i^B$. We associate an interaction energy $-\epsilon_{\alpha\beta}$ between species α and β, located at neighboring sites i and j, respectively. The corresponding Hamiltonian is

$$H = - \sum_{\langle ij \rangle} \left[\epsilon_{AA} n_i^A n_j^A + \epsilon_{BB} n_i^B n_j^B + \epsilon_{AB} \left(n_i^A n_j^B + n_i^B n_j^A \right) \right]$$

$$= - \left(\frac{\epsilon_{AA} + \epsilon_{BB} - 2\epsilon_{AB}}{4} \right) \sum_{\langle ij \rangle} S_i S_j - \frac{q(\epsilon_{AA} - \epsilon_{BB})}{4} \sum_{i=1}^{N} S_i$$

$$- \frac{Nq}{8} (\epsilon_{AA} + \epsilon_{BB} + 2\epsilon_{AB}). \tag{1.2}$$

In Equation 1.2, q denotes the coordination number of a lattice site. The second term on the right-hand side (RHS) is constant because $\sum_i S_i = N_A - N_B$, where N_α is the number of α-atoms in the system. Further, the third term on the RHS is also a constant. The Hamiltonian in Equation 1.2 is analogous to that in Equation 1.1 if we identify

$$J = \frac{\epsilon_{AA} + \epsilon_{BB} - 2\epsilon_{AB}}{4}. \tag{1.3}$$

The Ising model and its variants are not restricted to two-state systems and can be easily generalized to the case of multiple-state systems. Thus, three-state systems can be mapped onto a spin-1 Hamiltonian; four-state systems onto a spin-3/2 Hamiltonian; and so on. In general, higher-spin models have a larger number of possible interaction terms (and parameters) in the Hamiltonian.

We can obtain phase diagrams for magnets (cf. Figure 1.2) and binary mixtures (cf. Figure 1.3) by studying the Ising model in the mean-field (MF) approximation, as described below.

1.2.2 PHASE DIAGRAMS IN THE MEAN-FIELD APPROXIMATION

The equilibrium properties of the Ising model in Equation 1.1 are described in the MF approximation by the Bragg–Williams (BW) form of the Gibbs free energy [7]. This is obtained as follows. Consider a homogeneous state with spatially uniform magnetization $\langle S_i \rangle = \psi$. We approximate the energy as

$$E(\psi) \simeq -J \sum_{\langle ij \rangle} \langle S_i \rangle \langle S_j \rangle = -\frac{NqJ}{2}\psi^2. \tag{1.4}$$

The corresponding probabilities for a site to have up ($\uparrow$) or down ($\downarrow$) spins are

$$p_\uparrow = \frac{1 + \psi}{2}, \\ p_\downarrow = \frac{1 - \psi}{2}. \tag{1.5}$$

Therefore, the entropy for a lattice with N sites is

$$S(\psi) = -Nk_B \left[\left(\frac{1 + \psi}{2} \right) \ln \left(\frac{1 + \psi}{2} \right) + \left(\frac{1 - \psi}{2} \right) \ln \left(\frac{1 - \psi}{2} \right) \right], \tag{1.6}$$

where k_B is the Boltzmann constant.

Then, the Gibbs free energy is obtained as

$$G(\psi) = E(\psi) - hM - TS(\psi), \tag{1.7}$$

where h is the magnetic field, and M ($=N\psi$) is the overall magnetization.

This yields the free energy per spin as

$$g(T, h, \psi) = \frac{G(T, h, \psi)}{N}$$

$$= -\frac{1}{2} q J \psi^2 - h\psi$$

$$+ k_B T \left[\left(\frac{1+\psi}{2} \right) \ln \left(\frac{1+\psi}{2} \right) + \left(\frac{1-\psi}{2} \right) \ln \left(\frac{1-\psi}{2} \right) \right].$$

(1.8)

The RHS of Equation 1.8 is a variational function of the magnetization $\psi = \langle S_i \rangle$. If we Taylor-expand the entropy term in Equation 1.8, the Gibbs free energy assumes the customary ψ^4-form:

$$g(T, h, \psi) = \frac{1}{2} \left(k_B T - q J \right) \psi^2 - h\psi + \frac{k_B T}{12} \psi^4 + O(\psi^6) - k_B T \ln 2. \qquad (1.9)$$

The order parameter ψ in Equation 1.8 or Equation 1.9 can describe both ferromagnetic and antiferromagnetic order, with $J < 0$ in the latter case. Furthermore, in the antiferromagnetic case, ψ refers to the *sublattice magnetization* or *staggered magnetization* [7].

The equilibrium value of ψ at fixed (T, h) is obtained from Equation 1.8 by minimizing the Gibbs free energy:

$$\left. \frac{\partial g}{\partial \psi} \right|_{\psi = \psi_0} = 0. \qquad (1.10)$$

This yields the well-known transcendental equation $[\beta = (k_B T)^{-1}]$:

$$\psi_0 = \tanh(\beta q J \psi_0 + \beta h). \qquad (1.11)$$

For $h = 0$, we identify the MF critical temperature

$$T_c = \frac{q J}{k_B}. \qquad (1.12)$$

For $T > T_c$ and $h = 0$, the transcendental equation has only one solution $\psi_0 = 0$, which corresponds to the paramagnetic state. For $T < T_c$, Equation 1.11 has three solutions $\psi_0 = 0, \pm\psi(T)$. The state with $\psi_0 = 0$ has a higher free energy than do the equivalent states $+\psi(T)$ and $-\psi(T)$. Further, $\psi(T) \to 1$ as $T \to 0$, and $\psi(T) \to 0$ as $T \to T_c^-$. The relevant phase diagram in the (T, h)-plane is shown in Figure 1.2.

Next, let us consider the case of the binary mixture (or lattice gas) with N_A ($= c_A N$) atoms of species A and N_B ($= c_B N$) atoms of species B ($N = N_A + N_B$). The appropriate order parameter in this case is the local density difference, $\psi = \langle n_i^A \rangle - \langle n_i^B \rangle$. The above analysis has to be modified because the appropriate ensemble for a binary mixture is characterized by a fixed *magnetization* rather than a fixed *magnetic field*. The relevant free energy to be minimized is the Helmholtz potential

$$F(T, \psi) = E(\psi) - TS(\psi). \tag{1.13}$$

For the BW free energy, we have the expression

$$f(T, \psi) = \frac{F(T, \psi)}{N}$$

$$= -\frac{1}{2} qJ\psi^2 + k_B T \left[\left(\frac{1+\psi}{2} \right) \ln \left(\frac{1+\psi}{2} \right) + \left(\frac{1-\psi}{2} \right) \ln \left(\frac{1-\psi}{2} \right) \right].$$

$$\tag{1.14}$$

For a system that undergoes phase separation, there are two possibilities:

(a) We can have a *homogeneous* (or one-phase) state with order parameter $\psi_h = c_A - c_B$.

(b) We can have a *phase-separated* state where the system segregates into two regions having order parameter ψ_1 (with fraction x) and ψ_2 [with fraction $(1 - x)$]. The quantity x is determined from the lever rule

$$\psi_h = x\psi_1 + (1 - x)\psi_2. \tag{1.15}$$

Let us minimize the Helmholtz potential $\bar{f}$ for the phase-separated state. (The homogeneous state is the limit $\psi_1 = \psi_2$.) The quantity $\bar{f}$ is obtained as

$$\bar{f} = xf(\psi_1) + (1 - x)f(\psi_2). \tag{1.16}$$

This has to be minimized subject to the constraint in Equation 1.15. We implement this constraint by introducing the Lagrange multiplier λ and minimizing the quantity

$$A = xf(\psi_1) + (1 - x)f(\psi_2) - \lambda[x\psi_1 + (1 - x)\psi_2 - \psi_h]. \tag{1.17}$$

This yields the equations

$$\frac{\partial A}{\partial x} = f(\psi_1) - f(\psi_2) - \lambda(\psi_1 - \psi_2) = 0,$$

$$\frac{\partial A}{\partial \psi_1} = xf'(\psi_1) - \lambda x = 0,$$

$$\frac{\partial A}{\partial \psi_2} = (1 - x)f'(\psi_2) - \lambda(1 - x) = 0, \tag{1.18}$$

$$\frac{\partial A}{\partial \lambda} = x\psi_1 + (1 - x)\psi_2 - \psi_h = 0.$$

The first three equations yield

$$\lambda = \frac{f(\psi_1) - f(\psi_2)}{\psi_1 - \psi_2} = f'(\psi_1) = f'(\psi_2), \tag{1.19}$$

which is referred to as Maxwell's double-tangent construction. This is valid for arbitrary functional forms of the Helmholtz free energy.

The specific form for $f(T, \psi)$ in Equation 1.14 is an even function of ψ with $f(T, -\psi) = f(T, \psi)$. Further, $f(T, \psi)$ has a single minimum at $\psi = 0$ for $T > T_c = qJ$. Thus, the only solution of the double-tangent construction is $\psi_1 = \psi_2$. In that case, the constraint in Equation 1.15 yields $\psi_1 = \psi_h$, corresponding to the homogeneous state.

For $T < T_c$, $f(T, \psi)$ has a symmetric double-well structure with extrema at $f'(T, \psi_0) = 0$, that is,

$$\psi_0 = \tanh(\beta qJ \psi_0). \tag{1.20}$$

The states with non-zero ψ_0 correspond to lower free energy than the state with $\psi_0 = 0$. Thus, a possible solution to the double-tangent construction is

$$\psi_1 = -\psi_0, \quad \psi_2 = +\psi_0, \tag{1.21}$$

where ψ_0 is the positive solution of Equation 1.20. However, this is only an acceptable solution if the lever rule can be satisfied, that is, $-\psi_0 < \psi_h < \psi_0$. Thus, phase separation occurs at $T < T_c$ only if $|\psi_h| < \psi_0$. When phase separation does occur, the segregated states have the composition $-\psi_0$ (B-rich) and $+\psi_0$ (A-rich), respectively. The resultant phase diagram in the $(c_A, T/T_c)$-plane is shown in Figure 1.3.

The phase diagrams in Figures 1.2 and 1.3 will provide the basis for our subsequent discussion of phase-ordering dynamics.

1.3 KINETIC ISING MODELS

1.3.1 INTRODUCTION

The above discussion has clarified the utility of Ising-like models in a wide range of problems. We next consider the issue of *kinetics of Ising models*. For simplicity, we restrict our discussion to the spin-1/2 model described by Equation 1.1. The generalization to higher-spin models is straightforward. The Ising spin variables do not have intrinsic dynamics, as is seen by constructing the relevant Poisson bracket. In order to associate kinetics with the Ising model, we assume that it is placed in contact with a heat bath that generates stochastic spin-flips ($S_i \rightarrow -S_i$) in the system [6]. The heat bath can be interpreted as consisting of phonons that induce spin-flips via a spin-lattice coupling. The resultant *kinetic Ising model* is referred to as the spin-flip or Glauber model [8] and is appropriate for describing the nonconserved kinetics of the paramagnetic $\rightarrow$ ferromagnetic transition. The probability of a jump depends on the configuration of all other spins and the heat-bath temperature, in general.

Next, consider the case where the Ising model describes a lattice gas or a binary (AB) mixture. The appropriate microscopic kinetics involves the diffusion of atoms, for example, atomic jumps to vacant sites in the lattice gas, or $A \leftrightarrow B$ interchanges in the binary mixture. Thus, the heat bath causes spin-exchanges rather than spin-flips, that is, S_i jumps from $+1 \rightarrow -1$ while a neighbor S_j *simultaneously* jumps from $-1 \rightarrow +1$. This process mimics phonon-induced atomic jumps. The resultant model is referred to as the spin-exchange or Kawasaki model [9,10].

It should be emphasized that transition probabilities in both the Glauber and Kawasaki models must satisfy the *detailed-balance condition* [11], which will be discussed shortly. Thus, although the two models describe different time-dependent behavior, the equilibrium state is *unique*. As $t \rightarrow \infty$, we recover properties calculable from the *equilibrium* statistical mechanics of the Ising model in an appropriate ensemble.

1.3.2 THE SPIN-FLIP GLAUBER MODEL

In the Glauber model, the heat bath induces fluctuations in the system in the form of single-spin-flip processes [8]. The Glauber model describes nonconserved kinetics because the spin-flip processes make the total magnetization $M = \sum_{i=1}^{N} S_i$ time-dependent. Let us examine the evolution of the probability distribution for the spin configuration $\{S_i\}$ of a system with N spins. In this context, we introduce the conditional probability $P(\{S_i^0\}, 0 | \{S_i\}, t)$, which is the probability that the ith spin is in state S_i ($i = 1 \rightarrow N$) at time t, given that it was in state S_i^0 ($i = 1 \rightarrow N$) at time $t = 0$. The evolution of P is described by the master equation [11]:

$$\frac{d}{dt} P(\{S_i\}, t) = - \sum_{j=1}^{N} W(S_1, \ldots S_j, \ldots S_N | S_1, \ldots -S_j, \ldots S_N) P(\{S_i\}, t)$$

$$+ \sum_{j=1}^{N} W(S_1, \ldots -S_j, \ldots S_N | S_1, \ldots S_j, \ldots S_N) P(\{S_i'\}, t), \quad (1.22)$$

where we suppress the argument $(\{S_i^0\}, 0|$, for compactness. The first term on the RHS of Equation 1.22 corresponds to the *loss of probability* for the state $\{S_i\}$ due to the spin-flip $S_j \rightarrow -S_j$. The second term on the RHS denotes the *gain of probability* for the state $\{S_i\}$ due to a spin-flip $S_j' \rightarrow -S_j'$ in a state $\{S_i'\}$ with

$$S_i' = S_i \quad \text{for } i \neq j,$$
$$S_j' = -S_j. \quad (1.23)$$

Equation 1.22 assumes that the underlying stochastic process is Markovian. The essential physical input is provided by the modeling of the *transition matrix* $W(\{S_i\}|\{S_i'\})$ for the change $\{S_i\}$ to $\{S_i'\}$). The choice of W must be such that the ensemble approaches the equilibrium distribution $P_{eq}(\{S_i\})$ as $t \rightarrow \infty$:

$$P_{eq}(\{S_i\}) = \frac{1}{Z(T,h,N)} \exp[-\beta(H - hM)]. \tag{1.24}$$

Here, Z is the partition function, which is defined as

$$Z(T,h,N) = \sum_{\{S_i\}} \exp[-\beta(H - hM)]. \tag{1.25}$$

To ensure this, the transition probability $W(\{S_i\}|\{S_i'\})$ should obey the *detailed-balance condition* [11]:

$$W(\{S_i\}|\{S_i'\})P_{eq}(\{S_i\}) = W(\{S_i'\}|\{S_i\})P_{eq}(\{S_i'\}). \tag{1.26}$$

Clearly, in the equilibrium ensemble, this guarantees that the number of systems making the transition from $\{S_i\} \rightarrow \{S_i'\}$ is balanced by the number of systems making the reverse transition $\{S_i'\} \rightarrow \{S_i\}$. Thus, the probability distribution P_{eq} is independent of time, as expected. Further, an arbitrary distribution $P(\{S_i\}, t) \rightarrow P_{eq}(\{S_i\})$ as $t \rightarrow \infty$ under Equation 1.22, provided that W obeys the detailed-balance condition. For the proof of this, we refer the reader to the book by Van Kampen [11].

It is evident that there are many choices of W that satisfy the condition in Equation 1.26. We choose the Suzuki–Kubo form [12]:

$$W(\{S_i\}|\{S_i'\}) = \frac{\lambda}{2}\left\{1 - \tanh\left[\frac{\beta\Delta(H - hM)}{2}\right]\right\}, \tag{1.27}$$

where λ^{-1} sets the timescale of the nonequilibrium process. Here, $\Delta(H - hM)$ denotes the enthalpy difference between the final state $\{S_i'\}$ and the initial state $\{S_i\}$. It is straightforward to confirm that this form of W satisfies the detailed-balance condition.

For the spin-flip Ising model, the states $\{S_i'\}$ and $\{S_i\}$ differ only in one spin, that is, $S_j' = -S_j$. Then

$$(H - hM)_{\text{initial}} = -JS_j \sum_{L_j} S_{L_j} - hS_j + \text{other terms},$$

$$(H - hM)_{\text{final}} = JS_j \sum_{L_j} S_{L_j} + hS_j + \text{other terms}, \tag{1.28}$$

where L_j denotes the nearest neighbors (nn) of j. Thus

$$\Delta(H - hM) = 2JS_j \sum_{L_j} S_{L_j} + 2hS_j, \tag{1.29}$$

and

$$
\begin{aligned}
W\left(\{S_i\}|\{S_i'\}\right) &= \frac{\lambda}{2}\left[1 - \tanh\left(\beta J S_j \sum_{L_j} S_{L_j} + \beta h S_j\right)\right] \\
&= \frac{\lambda}{2}\left[1 - S_j \tanh\left(\beta J \sum_{L_j} S_{L_j} + \beta h\right)\right].
\end{aligned} \tag{1.30}
$$

In Equation 1.30, we can bring S_j outside the argument of the tanh-function because it only takes the values $+1$ or -1. We replace the form of W from Equation 1.30 in Equation 1.22 to obtain the explicit form of the master equation:

$$
\begin{aligned}
\frac{d}{dt}P(\{S_i\},t) = &-\frac{\lambda}{2}\sum_{j=1}^{N}\left[1 - S_j \tanh\left(\beta J \sum_{L_j} S_{L_j} + \beta h\right)\right]P(\{S_i\},t) \\
&+ \frac{\lambda}{2}\sum_{j=1}^{N}\left[1 + S_j \tanh\left(\beta J \sum_{L_j} S_{L_j} + \beta h\right)\right]P(\{S_i'\},t).
\end{aligned}
$$

$$\tag{1.31}$$

We can use this master equation to obtain the evolution of the magnetization:

$$
\langle S_k \rangle = \sum_{\{S_i\}} S_k P(\{S_i\},t). \tag{1.32}
$$

We multiply both sides of Equation 1.31 by S_k and sum over all configurations to obtain

$$
\begin{aligned}
\frac{d}{dt}\langle S_k \rangle = &-\frac{\lambda}{2}\sum_{j=1}^{N}\sum_{\{S_i\}} S_k\left[1 - S_j \tanh\left(\beta J \sum_{L_j} S_{L_j} + \beta h\right)\right]P(\{S_i\},t) \\
&+ \frac{\lambda}{2}\sum_{j=1}^{N}\sum_{\{S_i\}} S_k\left[1 + S_j \tanh\left(\beta J \sum_{L_j} S_{L_j} + \beta h\right)\right]P(\{S_i'\},t) \\
&\equiv A + B.
\end{aligned} \tag{1.33}
$$

In the second term on the RHS of Equation 1.33, we redefine $S_j = -\overline{S_j}$. Clearly, the sum $\sum_{S_j=\pm1}$ is equivalent to the sum $\sum_{\overline{S_j}=\pm1}$. Therefore, the terms in A and B cancel with each other, except for the case $j = k$. This yields the following evolution equation for the magnetization:

$$
\lambda^{-1} \frac{d}{dt} \langle S_k \rangle = - \sum_{\{S_i\}} S_k \left[1 - S_k \tanh \left(\beta J \sum_{L_k} S_{L_k} + \beta h \right) \right] P(\{S_i\}, t)
$$

$$
= - \langle S_k \rangle + \left\langle \tanh \left(\beta J \sum_{L_k} S_{L_k} + \beta h \right) \right\rangle, \tag{1.34}
$$

where we have used $S_k^2 = 1$.

1.3.2.1   Mean-Field Approximation

Unfortunately, the exact time-dependent Equation 1.34 is analytically intractable in $d \geq 2$. (For the $d = 1$ solution, see the work of Glauber [8].) The main obstacle is that the second term on the RHS of Equation 1.34 yields a set of higher-order correlation functions, as can be seen by expanding the tanh-function. These dynamical equations can be rendered tractable by invoking the MF approximation, which truncates the hierarchy by neglecting correlations between different sites, that is, the average of the product of spin operators is replaced by the product of their averages. The result of such a random-phase decoupling is that the angular brackets denoting the statistical average can be taken inside the argument of the tanh-function [13,14]. Thus, we obtain

$$
\lambda^{-1} \frac{d}{dt} \langle S_k \rangle = - \langle S_k \rangle + \tanh \left(\beta J \sum_{L_k} \langle S_{L_k} \rangle + \beta h \right). \tag{1.35}
$$

For time-independent effects in equilibrium, the LHS of Equation 1.35 is identically zero. Thus, we have (as $t \to \infty$)

$$
\langle S_k \rangle^{\mathrm{eq}} = \tanh \left(\beta J \sum_{L_k} \langle S_{L_k} \rangle^{\mathrm{eq}} + \beta h \right). \tag{1.36}
$$

Notice that Equation 1.35 is nonlinear because of the presence of the tanh-function and is only tractable numerically. These equations are often referred to as *mean-field dynamical models* in the literature [15–19]. A further simplification can be effected by expanding the tanh-function and retaining only leading terms. For simplicity, we consider the case of zero magnetic field, that is, $h = 0$. We can then expand various terms on the RHS of Equation 1.35 as follows:

$$\sum_{L_k} \langle S_{L_k} \rangle \simeq q\psi(\vec{r}_k, t) + a^2 \nabla_k^2 \psi(\vec{r}_k, t) + \text{higher-order terms}, \qquad (1.37)$$

where a is the lattice spacing. Further,

$$\tanh\left(\beta J \sum_{L_k} \langle S_{L_k} \rangle\right) \simeq \beta J \sum_{L_k} \langle S_{L_k} \rangle - \frac{1}{3}\left(\beta J \sum_{L_k} \langle S_{L_k} \rangle\right)^3 + \text{higher-order terms}$$

$$\simeq \frac{T_c}{T}\psi(\vec{r}_k, t) - \frac{1}{3}\left(\frac{T_c}{T}\right)^3 \psi(\vec{r}_k, t)^3 + \frac{T_c}{qT}a^2\nabla_k^2\psi(\vec{r}_k, t)$$

$$+ \text{other terms}, \qquad (1.38)$$

where we have used Equation 1.37 to obtain the second expression. Therefore, the order-parameter equation for the Glauber–Ising model simplifies as

$$\lambda^{-1}\frac{\partial}{\partial t}\psi(\vec{r}, t) = \left(\frac{T_c}{T} - 1\right)\psi - \frac{1}{3}\left(\frac{T_c}{T}\right)^3\psi^3 + \frac{T_c}{qT}a^2\nabla^2\psi + \text{other terms},$$

$$(1.39)$$

where we have dropped the subscript k for the position variable.

At this stage, a few remarks are in order. Firstly, Equation 1.39 is referred to as the *time-dependent Ginzburg–Landau* (TDGL) equation. We will discuss the general formulation of the TDGL equation in Section 1.4.1. Secondly, the approximation of neglecting the higher-order terms in Equation 1.39 is justifiable only for $T \simeq T_c$, where the order parameter is small. However, it is generally believed that the TDGL equation is valid even for deep quenches ($T \ll T_c$), at least in terms of containing the correct physics.

1.3.3 THE SPIN-EXCHANGE KAWASAKI MODEL

We mentioned earlier that the Glauber model, which assumes single-spin-flip processes, is appropriate for nonconserved kinetics. On the other hand, when the Ising model describes either phase separation ($J > 0$) or order-disorder ($J < 0$) transitions in an AB mixture [1,7,20,21], the Glauber model is not applicable. For a binary mixture, the Ising spin variable models the presence of an A- or B-atom on a lattice site. Thus, the appropriate microscopic dynamics should involve random exchanges of A- and B-atoms at neighboring sites, with their individual numbers being constant. In practice, these jumps are actually mediated by vacancies [22–25], and the system should be described as a ternary (ABV) mixture [18,19,26,27]. However, when the vacancy concentration is small, it is reasonable to ignore vacancies and assume that the underlying stochastic process is a spin-exchange. As stated earlier, this corresponds to the Kawasaki model, which is based on a stationary Markov process involving a spin-exchange mechanism [9,10]. The resultant master equation is as follows:

$$\frac{d}{dt}P(\{S_i\},t) = -\sum_{j=1}^{N}\sum_{k\in L_j}W(S_1,\ldots S_j,S_k,\ldots S_N|S_1,\ldots S_k,S_j,\ldots S_N)P(\{S_i\},t)$$

$$+\sum_{j=1}^{N}\sum_{k\in L_j}W(S_1,\ldots S_k,S_j,\ldots S_N|S_1,\ldots S_j,S_k,\ldots S_N)P(\{S_i'\},t).$$

$$(1.40)$$

The first term on the RHS is the loss of probability for the state $\{S_i\}$ due to the spin-exchange $S_j \leftrightarrow S_k$. We consider only nearest-neighbor exchanges, where site $k \in L_j$, that is, the nearest-neighbors of j. The second term on the RHS corresponds to the gain of probability for the state $\{S_i\}$ due to an exchange $S_j' \leftrightarrow S_k'$ in a state $\{S_i'\}$. The state $\{S_i'\}$ differs from the state $\{S_i\}$ in only two spins:

$$S_i' = S_i \quad \text{for} \quad i \neq j,k,$$
$$S_j' = S_k, \quad\quad\quad\quad\quad\quad (1.41)$$
$$S_k' = S_j.$$

As in the Glauber case, the transition probability $W(\{S_i\}|\{S_i'\})$ must obey the detailed-balance condition. As we have seen in Section 1.2.2, the binary mixture is described by an ensemble with fixed (T,M,N), where the "magnetization" $M = \sum_{i=1}^{N} S_i = N_A - N_B$. The corresponding equilibrium distribution is

$$P_{eq}(\{S_i\}) = \frac{1}{Z(T,M,N)} \exp(-\beta H)\delta_{\sum_i S_i,M}, \quad\quad (1.42)$$

where the Kronecker delta confines the distribution to configurations with $\sum_{i=1}^{N} S_i = M$. The appropriate partition function is

$$Z(T,M,N) = \sum_{\{S_i\}} \exp(-\beta H)\delta_{\sum_i S_i,M}. \quad\quad (1.43)$$

Again, we choose the Suzuki–Kubo form for the transition probability in Equation 1.40:

$$W(\{S_i\}|\{S_i'\}) = \frac{\lambda}{2}\left[1 - \tanh\left(\frac{\beta\Delta H}{2}\right)\right], \quad\quad (1.44)$$

where ΔH is the change in energy due to the spin-exchange $S_j \leftrightarrow S_k$. For the Ising model,

$$H_{\text{initial}} = -JS_j \sum_{L_j \neq k} S_{L_j} - JS_k \sum_{L_k \neq j} S_{L_k} - JS_j S_k + \text{other terms},$$

$$H_{\text{final}} = -JS_k \sum_{L_j \neq k} S_{L_j} - JS_j \sum_{L_k \neq j} S_{L_k} - JS_j S_k + \text{other terms}.$$

(1.45)

Thus, the energy change resulting from the spin exchange is

$$\Delta H = J(S_j - S_k) \sum_{L_j \neq k} S_{L_j} - J(S_j - S_k) \sum_{L_k \neq j} S_{L_k},$$ (1.46)

and

$$W\left(\{S_i\}|\{S_i'\}\right) = \frac{\lambda}{2}\left\{1 - \tanh\left[\frac{\beta J}{2}(S_j - S_k)\sum_{L_j \neq k} S_{L_j} - \frac{\beta J}{2}(S_j - S_k)\sum_{L_k \neq j} S_{L_k}\right]\right\}$$

$$= \frac{\lambda}{2}\left\{1 - \frac{S_j - S_k}{2}\tanh\left[\beta J\left(\sum_{L_j \neq k} S_{L_j} - \sum_{L_k \neq j} S_{L_k}\right)\right]\right\}.$$ (1.47)

In Equation 1.47, we have used the fact that $(S_j - S_k)/2 = 0, \pm 1$ to factor it out of the argument of the tanh-function. Therefore, the master equation has the form

$$\frac{d}{dt}P(\{S_i\}, t) = -\frac{\lambda}{2}\sum_{j=1}^{N}\sum_{k \in L_j}\left\{1 - \frac{S_j - S_k}{2}\tanh\left[\beta J\left(\sum_{L_j \neq k} S_{L_j} - \sum_{L_k \neq j} S_{L_k}\right)\right]\right\}P(\{S_i\}, t)$$

$$+ \frac{\lambda}{2}\sum_{j=1}^{N}\sum_{k \in L_j}\left\{1 + \frac{S_j - S_k}{2}\tanh\left[\beta J\left(\sum_{L_j \neq k} S_{L_j} - \sum_{L_k \neq j} S_{L_k}\right)\right]\right\}P\left(\{S_i'\}, t\right).$$

(1.48)

We can obtain an evolution equation for the order parameter by multiplying both sides of Equation 1.48 with S_n and summing over all configurations:

$$
\frac{d}{dt} \langle S_n \rangle = -\frac{\lambda}{2} \sum_{\{S_i\}} \sum_{j=1}^{N} \sum_{k \in L_j} S_n \left\{ 1 - \frac{S_j - S_k}{2} \tanh \left[\beta J \left(\sum_{L_j \neq k} S_{L_j} - \sum_{L_k \neq j} S_{L_k} \right) \right] \right\} P(\{S_i\}, t)
$$
$$
+ \frac{\lambda}{2} \sum_{\{S_i\}} \sum_{j=1}^{N} \sum_{k \in L_j} S_n \left\{ 1 + \frac{S_j - S_k}{2} \tanh \left[\beta J \left(\sum_{L_j \neq k} S_{L_j} - \sum_{L_k \neq j} S_{L_k} \right) \right] \right\} P(\{S_i'\}, t).
$$

$$(1.49)$$

In the second term on the RHS of Equation 1.49, we redesignate $S_j = \overline{S}_k$ and $S_k = \overline{S}_j$. This leads to a large-scale cancellation between the first and second terms. The only remaining terms are

$$
\frac{d}{dt} \langle S_n \rangle = -\frac{\lambda}{2} \sum_{\{S_i\}} \sum_{k \in L_n} S_n \left\{ 1 - \frac{S_n - S_k}{2} \tanh \left[\beta J \left(\sum_{L_n \neq k} S_{L_n} - \sum_{L_k \neq n} S_{L_k} \right) \right] \right\} P(\{S_i\}, t)
$$
$$
+ \frac{\lambda}{2} \sum_{\{S_i\}} \sum_{k \in L_n} S_k \left\{ 1 + \frac{S_k - S_n}{2} \tanh \left[\beta J \left(\sum_{L_n \neq k} S_{L_n} - \sum_{L_k \neq n} S_{L_k} \right) \right] \right\} P(\{S_i\}, t)
$$
$$
= -\frac{\lambda}{2} \left\langle \sum_{k \in L_n} (S_n - S_k) \left\{ 1 - \frac{S_n - S_k}{2} \tanh \left[\beta J \left(\sum_{L_n \neq k} S_{L_n} - \sum_{L_k \neq n} S_{L_k} \right) \right] \right\} \right\rangle.
$$

$$(1.50)$$

Some algebra yields the exact evolution equation

$$
2\lambda^{-1} \frac{d}{dt} \langle S_n \rangle = -q \langle S_n \rangle + \sum_{L_n} \langle S_{L_n} \rangle
$$
$$
+ \sum_{k \in L_n} \left\langle (1 - S_n S_k) \tanh \left[\beta J \left(\sum_{L_n \neq k} S_{L_n} - \sum_{L_k \neq n} S_{L_k} \right) \right] \right\rangle.
$$

$$(1.51)$$

This equation is analogous to Equation 1.34, obtained in the context of Glauber kinetics.

Although the Kawasaki model is usually associated with conserved kinetics, we should make a clarifying remark. In the context of binary mixtures, a *ferromagnetic* interaction ($J > 0$) results in phase separation, that is, the equilibrium system consists of domains of A-rich and B-rich phases. The appropriate order parameter is the difference in densities of A and B and is locally conserved by Kawasaki kinetics. The length scale over which the order parameter is conserved increases if we allow *long-ranged* exchanges rather than only *nearest-neighbor* exchanges. In the limit where the spin exchanges are *infinite-ranged*, the Kawasaki model has *global conservation* rather than *local conservation*. In this case, the Kawasaki model is essentially equivalent to the Glauber model [28,29].

It is also of great interest to consider the binary mixture with *antiferromagnetic* interactions, $J < 0$. In this case, there is a phase transition from a high-temperature disordered phase to a low-temperature ordered phase, where the A- and B-atoms order on alternate sub lattices. The appropriate order parameter is now the *staggered magnetization*, which is the difference between the two sub lattice *magnetizations*. This quantity is not conserved by Kawasaki kinetics, though the overall concentration is conserved. For the AB alloy with equal fractions of A and B, the antiferromagnetic case with Kawasaki kinetics is equivalent to the ferromagnetic Ising model with Glauber kinetics [30]. For asymmetric compositions, novel features arise due to the conserved concentration variable.

1.3.3.1 Mean-Field Approximation

As in the Glauber case, Equation 1.51 is the first of a hierarchy of equations involving higher-order correlations of the spin variable. This hierarchy can be truncated by invoking the MF approximation, that is, by replacing the expectation value of a function of spin variables by the function of the expectation values of the spin variables. The resultant MF dynamical model is

$$
2\lambda^{-1}\frac{d}{dt}\langle S_n\rangle = -q\langle S_n\rangle + \sum_{L_n}\langle S_{L_n}\rangle
$$

$$
+ \sum_{k\in L_n}(1 - \langle S_n\rangle\,\langle S_k\rangle)\tanh\left[\beta J\left(\sum_{L_n}\langle S_{L_n}\rangle - \sum_{L_k}\langle S_{L_k}\rangle\right)\right].
$$

(1.52)

Notice that the restrictions on the summations inside the tanh-function have been dropped in the MF approximation. This is necessary for Equation 1.52 to contain the correct MF solution in Equation 1.36 [13]. Recall the MF solution for the $h = 0$ case:

$$
\langle S_k\rangle^{\text{eq}} = \tanh\left(\beta J\sum_{L_k}\langle S_{L_k}\rangle^{\text{eq}}\right).
$$

(1.53)

If we replace this in the RHS of Equation 1.52, we obtain

$$\text{RHS} = -q\langle S_n\rangle^{\text{eq}} + \sum_{L_n}\langle S_{L_n}\rangle^{\text{eq}} + \sum_{k\in L_n}\left(1 - \langle S_n\rangle^{\text{eq}}\langle S_k\rangle^{\text{eq}}\right)$$

$$\times\left[\frac{\tanh\left(\beta J\sum_{L_n}\langle S_{L_n}\rangle^{\text{eq}}\right) - \tanh\left(\beta J\sum_{L_k}\langle S_{L_k}\rangle^{\text{eq}}\right)}{1 - \tanh\left(\beta J\sum_{L_n}\langle S_{L_n}\rangle^{\text{eq}}\right)\tanh\left(\beta J\sum_{L_k}\langle S_{L_k}\rangle^{\text{eq}}\right)}\right]$$

$$= -q\langle S_n\rangle^{\text{eq}} + \sum_{L_n}\langle S_{L_n}\rangle^{\text{eq}} + \sum_{L_n}\left(\langle S_n\rangle^{\text{eq}} - \langle S_{L_n}\rangle^{\text{eq}}\right)$$

$$= 0, \tag{1.54}$$

as expected.

Finally, let us derive a partial differential equation for the order parameter. This is the conserved counterpart of the TDGL equation we derived for the magnetization in Section 1.3.2. We can simplify the RHS of Equation 1.52 by using the identity

$$\tanh(X - Y) = \frac{\tanh X - \tanh Y}{1 - \tanh X\tanh Y}, \quad\text{where}$$

$$X = \beta J\sum_{L_n}\langle S_{L_n}\rangle,$$

$$Y = \beta J\sum_{L_k}\langle S_{L_k}\rangle. \tag{1.55}$$

We are interested in the late-stage dynamics, where the system has equilibrated locally and Equation 1.53 applies. Then, we make the approximation:

$$(1 - \langle S_n\rangle\langle S_k\rangle)\left(\frac{\tanh X - \tanh Y}{1 - \tanh X\tanh Y}\right) \simeq \tanh X - \tanh Y. \tag{1.56}$$

Therefore, we can rewrite Equation 1.52 as

$$2\lambda^{-1}\frac{d}{dt}\langle S_n\rangle \simeq \sum_{L_n}\left(\langle S_{L_n}\rangle - \langle S_n\rangle\right)$$

$$+ \sum_{k\in L_n}\left[\tanh\left(\beta J\sum_{L_n}\langle S_{L_n}\rangle\right) - \tanh\left(\beta J\sum_{L_k}\langle S_{L_k}\rangle\right)\right]$$

$$= \Delta_D\left[\langle S_n\rangle - \tanh\left(\beta J\sum_{L_n}\langle S_{L_n}\rangle\right)\right], \tag{1.57}$$

where Δ_D denotes the discrete Laplacian operator. We can use the Taylor expansion in Equation 1.38 to obtain the coarse-grained version of Equation 1.57 as

$$2\lambda^{-1}\frac{\partial}{\partial t}\psi(\vec{r}, t) = -a^2\nabla^2 \left[\left(\frac{T_c}{T} - 1\right)\psi - \frac{1}{3}\left(\frac{T_c}{T}\right)^3 \psi^3 + \frac{T_c}{qT}a^2\nabla^2\psi\right]$$

$$+ \text{ other terms}, \tag{1.58}$$

where a is the lattice spacing.

Equation 1.58 is known as the Cahn–Hilliard (CH) equation and is the standard model for phase separation driven by diffusion. In Section 1.5.1, we will derive the CH equation using phenomenological arguments.

1.4 DOMAIN GROWTH IN SYSTEMS WITH NONCONSERVED KINETICS

1.4.1 CASE WITH SCALAR ORDER PARAMETER

In Figure 1.2, we had shown the phase diagram for a ferromagnet. The corresponding ordering problem considers a paramagnetic system at $T > T_c, h = 0$ for time $t < 0$. At $t = 0$, the system is rapidly quenched to $T < T_c$, where the preferred equilibrium state is spontaneously magnetized. The *far-from-equilibrium* disordered system evolves toward its new equilibrium state by separating into domains that are rich in either up or down spins (see Figure 1.4). These domains coarsen with time and are characterized by a growing length scale $L(t)$. A finite system becomes ordered in either of the two equivalent states (up or down) as $t \to \infty$.

At the microscopic level, this evolution can be described by an Ising model with Glauber spin-flip kinetics, as discussed in Section 1.3.2. At the coarse-grained level, the appropriate order parameter to describe the system is the local magnetization $\psi(\vec{r}, t)$. In Section 1.3.2, we had used the Glauber–Ising model to derive the TDGL equation 1.39, which governs the evolution of the order parameter. More generally, the TDGL equation models the dissipative (over-damped) relaxation of a ferromagnetic system to its free-energy minimum:

$$\frac{\partial}{\partial t}\psi(\vec{r}, t) = -\Gamma\frac{\delta G[\psi]}{\delta\psi} + \theta(\vec{r}, t). \tag{1.59}$$

In Equation 1.59, Γ denotes the inverse damping coefficient; and $\delta G/\delta\psi$ is the functional derivative of the free-energy functional:

$$G[\psi] = \int d\vec{r}\left[g(\psi) + \frac{1}{2}K(\vec{\nabla}\psi)^2\right]. \tag{1.60}$$

Typical forms of the local free energy $g(\psi)$ are given in Equations 1.8 and 1.9. The second term on the RHS of Equation 1.60 accounts for surface tension due to inhomogeneities in the order parameter. The parameter K (>0) measures the strength of the surface tension.

FIGURE 1.4 Evolution of a disordered ferromagnet, which is quenched to $T < T_c$ at time $t = 0$. These pictures were obtained from a Euler-discretized version of the dimensionless TDGL equation 1.66 with $h = 0$ and no thermal fluctuations ($\epsilon = 0$). The discretization mesh sizes were $\Delta t = 0.1$ and $\Delta x = 1$ in time and space, respectively. The initial condition $\psi(\vec{r}, 0)$ consisted of small-amplitude fluctuations about $\psi = 0$. The lattice size was 256^2, and periodic boundary conditions were applied in both directions. Regions with up spins ($\psi > 0$) and down spins ($\psi < 0$) are marked black and white, respectively.

The noise term in Equation 1.59 is also space- and time-dependent and satisfies the fluctuation-dissipation relation:

$$\overline{\theta(\vec{r}, t)} = 0,$$

$$\overline{\theta(\vec{r}', t')\theta(\vec{r}'', t'')} = 2\Gamma k_B T \delta(\vec{r}' - \vec{r}'')\delta(t' - t''), \tag{1.61}$$

where the bars denote an average over the Gaussian noise ensemble. The presence of the noise term ensures that the system equilibrates to the correct Boltzmann distribution at temperature T. Equations 1.59 through 1.61 are also referred to as *Model A* of order-parameter kinetics, as discussed by Hohenberg and Halperin [31] in the context of dynamic critical phenomena.

Recall the TDGL equation 1.39, which was derived in Section 1.3.2. We identify it as the deterministic version of the general form in Equation 1.59. Further, the damping coefficient $\Gamma = \beta\lambda$, where λ is the inverse timescale of Glauber spin-flips. Finally, the form of the free-energy functional that gives rise to Equation 1.39 is

$$\beta G[\psi] = \int d\vec{r} \left[-\frac{1}{2}\left(\frac{T_c}{T} - 1\right)\psi^2 + \frac{1}{12}\left(\frac{T_c}{T}\right)^3 \psi^4 + \frac{T_c}{2qT}a^2(\vec{\nabla}\psi)^2 \right]. \quad (1.62)$$

For our subsequent discussion, it is convenient to use the general form of the ψ^4-free energy:

$$G[\psi] = \int d\vec{r} \left[-\frac{a(T_c - T)}{2}\psi^2 + \frac{b}{4}\psi^4 - h\psi + \frac{K}{2}(\vec{\nabla}\psi)^2 \right], \quad (1.63)$$

where we have introduced the parameters $a, b > 0$ and a term proportional to the magnetic field; and neglected terms of $O(\psi^6)$ and higher. The parameters a, b can be identified by a comparison with the explicit form of the free energy in (say) Equation 1.62. However, it is more appropriate to think of them as phenomenological parameters, without any reference to an underlying microscopic model.

For the ψ^4-free energy in Equation 1.63, the TDGL equation 1.59 has the form:

$$\frac{\partial}{\partial t}\psi(\vec{r}, t) = \Gamma\left[a(T_c - T)\psi - b\psi^3 + h + K\nabla^2\psi\right] + \theta(\vec{r}, t). \quad (1.64)$$

The parameters in Equation 1.64 can be absorbed into the definitions of space and time by introducing the rescaled variables (for $T < T_c$)

$$\psi' = \frac{\psi}{\psi_0}, \quad \psi_0 = \sqrt{\frac{a(T_c - T)}{b}},$$

$$t' = a(T_c - T)\Gamma t,$$

$$\vec{r}' = \sqrt{\frac{a(T_c - T)}{K}}\,\vec{r}, \quad \xi_b = \sqrt{\frac{2K}{a(T_c - T)}}, \quad (1.65)$$

$$h' = \frac{h}{a(T_c - T)\psi_0},$$

$$\theta' = \frac{\theta}{a(T_c - T)\Gamma\psi_0}.$$

Dropping primes, we obtain the dimensionless TDGL equation:

$$\frac{\partial}{\partial t}\psi(\vec{r}, t) = \psi - \psi^3 + h + \nabla^2\psi + \theta(\vec{r}, t), \quad (1.66)$$

where

$$\overline{\theta(\vec{r}, t)} = 0,$$

$$\overline{\theta(\vec{r}', t')\theta(\vec{r}'', t'')} = 2\epsilon\delta(\vec{r}' - \vec{r}'')\delta(t' - t''),$$

$$\epsilon = \frac{k_B T b \left[a(T_c - T)\right]^{(d-4)/2}}{K^{d/2}}. \tag{1.67}$$

We will focus on the case with $h = 0$ (shown in Figure 1.4), where the system evolves into two competing states. There is a *domain boundary* or *interface* that separates regions enriched in the two states. Our analytical understanding of domain growth problems is based on the dynamics of these interfaces.

1.4.1.1 Static Interfaces or Kinks

Consider the deterministic version of the TDGL equation with $h = 0$:

$$\frac{\partial}{\partial t}\psi(\vec{r}, t) = \psi - \psi^3 + \nabla^2\psi, \tag{1.68}$$

where we have set $\epsilon = 0$ in Equation 1.66. The static solution of this equation corresponds to a uniform state with $\psi_0 = +1$ or $\psi_0 = -1$. Another static solution (with higher energy than that of the uniform state) is the *interface* or *kink*, which is obtained as the solution of

$$\frac{d^2\psi_s}{dz^2} + \psi_s - \psi_s^3 = 0. \tag{1.69}$$

The kink solution is

$$\psi_s(z) = \tanh\left[\pm\frac{(z - z_0)}{\sqrt{2}}\right], \tag{1.70}$$

where z_0 (the center of the kink) is arbitrary. The solutions with a positive sign (kink) and negative sign (anti-kink) are shown in Figure 1.5. The kink (anti-kink) goes from $\psi = -1$ ($\psi = +1$) at $z = -\infty$ to $\psi = +1$ ($\psi = -1$) at $z = \infty$. The solution differs from $\psi \simeq \pm 1$ in a small interfacial region only, whose width defines the correlation length $\xi_b = \sqrt{2}$ (in dimensionless units).

The free energy associated with a configuration $\psi(\vec{r})$ is (in dimensionless units)

$$G[\psi] = \int d\vec{r}\left[-\frac{\psi^2}{2} + \frac{\psi^4}{4} + \frac{1}{2}(\vec{\nabla}\psi)^2\right]. \tag{1.71}$$

Therefore, the free-energy difference between the kink solution and the homogeneous solution $\psi = \psi_0$ is

$$\Delta G = A\int_{-\infty}^{\infty} dz\left[-\frac{1}{2}\left(\psi_s^2 - \psi_0^2\right) + \frac{1}{4}\left(\psi_s^4 - \psi_0^4\right) + \frac{1}{2}\left(\frac{d\psi_s}{dz}\right)^2\right], \tag{1.72}$$

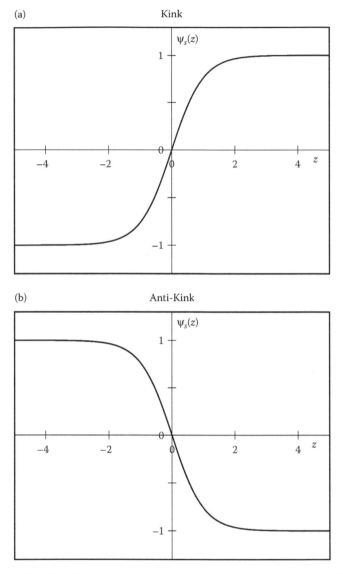

FIGURE 1.5 Variation of the order parameter [$\psi_s(z)$ versus z] for the (a) static kink profile and (b) static anti-kink profile.

where A is the area in the directions perpendicular to the z-axis. This integral is evaluated as follows. Multiply both sides of Equation 1.69 by $d\psi_s/dz$ to obtain

$$\frac{d}{dz}\left[\frac{\psi_s^2}{2} - \frac{\psi_s^4}{4} + \frac{1}{2}\left(\frac{d\psi_s}{dz}\right)^2\right] = 0. \tag{1.73}$$

We integrate this equation to obtain

$$\frac{\psi_s^2}{2} - \frac{\psi_s^4}{4} + \frac{1}{2}\left(\frac{d\psi_s}{dz}\right)^2 = C \text{ (constant)}$$

$$= \frac{\psi_0^2}{2} - \frac{\psi_0^4}{4}, \qquad (1.74)$$

where we have used

$$\lim_{z \to \infty} \psi_s = \pm\psi_0,$$

$$\lim_{z \to \infty} \frac{d\psi_s}{dz} = 0. \qquad (1.75)$$

Replacing Equation 1.74 in Equation 1.72, we obtain the surface tension σ as

$$\sigma = \frac{\Delta G}{A} = \int_{-\infty}^{\infty} dz \left(\frac{d\psi_s}{dz}\right)^2$$

$$= \frac{1}{2}\int_{-\infty}^{\infty} dz \, \text{sech}^4\left(\frac{z}{\sqrt{2}}\right)$$

$$= \frac{2\sqrt{2}}{3}. \qquad (1.76)$$

The above discussion applies to a flat interface. Clearly, the interfaces in Figure 1.4 are not flat. However, in the late stages of evolution, we expect the local order parameter to have equilibrated to a kink profile. We can introduce the local coordinates n (perpendicular to the interface) and $\vec{a}$ (tangential to the interface), as shown in Figure 1.6. The corresponding increase in free energy is

$$\Delta G = \int d\vec{a} \int dn \left(\frac{d\psi_s}{dn}\right)^2$$

$$= \sigma \int d\vec{a}. \qquad (1.77)$$

1.4.1.2 Equation of Motion for Interfaces and Growth Laws

Next, let us derive the Allen–Cahn equation of motion for the interfaces [32]. For this, we compute various terms in the TDGL equation 1.68 in terms of the interfacial coordinates $(n, \vec{a})$. We have

$$\vec{\nabla}\psi = \left.\frac{\partial\psi}{\partial n}\right|_t \hat{n}, \qquad (1.78)$$

where $\hat{n}$ is the unit vector normal to the interface in the direction of increasing ψ. Further

$$\nabla^2\psi = \left.\frac{\partial^2\psi}{\partial n^2}\right|_t \hat{n}\cdot\hat{n} + \left.\frac{\partial\psi}{\partial n}\right|_t \vec{\nabla}\cdot\hat{n}. \qquad (1.79)$$

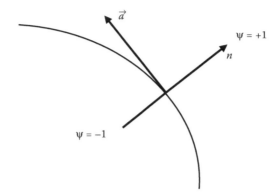

FIGURE 1.6 Curvilinear coordinates with reference to the interface. The normal coordinate is denoted as n and points from $\psi = -1$ to $\psi = +1$. The tangential coordinate points along the interface and is denoted as $\vec{a}$.

Finally, we use the identity

$$\left.\frac{\partial \psi}{\partial t}\right|_n \left.\frac{\partial t}{\partial n}\right|_\psi \left.\frac{\partial n}{\partial \psi}\right|_t = -1 \tag{1.80}$$

to obtain (from the TDGL equation)

$$-\left.\frac{\partial \psi}{\partial n}\right|_t \left.\frac{\partial n}{\partial t}\right|_\psi = \psi - \psi^3 + \left.\frac{\partial^2 \psi}{\partial n^2}\right|_t + \left.\frac{\partial \psi}{\partial n}\right|_t \vec{\nabla} \cdot \hat{n}$$

$$\simeq \left.\frac{\partial \psi}{\partial n}\right|_t \vec{\nabla} \cdot \hat{n}. \tag{1.81}$$

In the above simplification, we have used the fact that the interfaces are locally equilibrated to the static kink profile. We recognize that $\partial n/\partial t|_\psi = v(\vec{a})$, the normal interfacial velocity in the $\hat{n}$-direction. This yields the Allen–Cahn equation:

$$v(\vec{a}) = -\vec{\nabla} \cdot \hat{n} = -K(\vec{a}), \tag{1.82}$$

where K denotes the local curvature of the interface.

It is useful to examine the growth of a droplet of (say) $\psi = -1$ immersed in a matrix of $\psi = +1$. We consider the 3-d case of a spherical droplet with radius R. The normal unit vector at a point (x, y, z) on the surface of the sphere is

$$\hat{n} = \frac{x}{R}\hat{i} + \frac{y}{R}\hat{j} + \frac{z}{R}\hat{k}, \tag{1.83}$$

so that

$$\vec{\nabla} \cdot \hat{n} = \frac{1}{R} - \frac{x^2}{R^3} + \frac{1}{R} - \frac{y^2}{R^3} + \frac{1}{R} - \frac{z^2}{R^3}$$

$$= \frac{2}{R}. \tag{1.84}$$

Further, $v(\vec{a}) = dR/dt$, and Equation 1.82 becomes

$$\frac{dR}{dt} = -\frac{2}{R},$$ (1.85)

with the solution

$$R(0)^2 - R(t)^2 = 4t.$$ (1.86)

Thus, the droplet collapses on a timescale $t_c \sim R(0)^2$. In arbitrary dimensions d, the corresponding form of Equation 1.85 is

$$\frac{dR}{dt} = -\frac{d-1}{R}.$$ (1.87)

More generally, we can use Equation 1.82 to obtain the growth law for the domains in Figure 1.4. For a domain of characteristic size L, we have $v \sim dL/dt$ and curvature $K \sim 1/L$. This yields the diffusive growth law $L(t) \sim t^{1/2}$, which is valid for nonconserved scalar fields.

Before proceeding, it is important to consider the role of thermal fluctuations in the dynamics shown in Figure 1.4. It turns out that thermal noise is asymptotically irrelevant for ordering in systems that are free of disorder. This is because fluctuations only affect the interfacial profile. However, the fixed length scale of the interface becomes irrelevant in comparison with the diverging domain scale [33]. An equivalent argument is due to Bray [34,35], who used a renormalization-group (RG) approach to demonstrate that domain growth is driven by a fixed point at $T = 0$.

1.4.1.3 Correlation Function and Structure Factor

Now, if the system is characterized by a single length scale, the morphology of the domains does not change with time, apart from a scale factor. Therefore, the *order-parameter correlation function* exhibits a dynamical-scaling property [36]:

$$C(\vec{r}, t) \equiv \frac{1}{V} \int d\vec{R} \left[\langle \psi(\vec{R}, t)\psi(\vec{R} + \vec{r}, t) \rangle - \langle \psi(\vec{R}, t) \rangle \langle \psi(\vec{R} + \vec{r}, t) \rangle \right]$$

$$= g\left(\frac{r}{L}\right),$$ (1.88)

where V is the system volume, and the angular brackets denote an averaging over independent initial conditions and thermal fluctuations. This equal-time correlation function is a nonequilibrium quantity as domain growth is a nonequilibrium process. In Equation 1.88, $g(x)$ is a time-independent scaling function.

Actually, most experiments (e.g., neutron or light scattering) probe the time-dependent *structure factor*, which is the Fourier transform of the real-space correlation function:

$$S(\vec{k}, t) = \int d\vec{r} e^{i\vec{k} \cdot \vec{r}} C(\vec{r}, t),$$ (1.89)

where $\vec{k}$ is the wave-vector of the scattered beam. The corresponding dynamical-scaling form for $S(\vec{k}, t)$ is

$$S(\vec{k}, t) = L^d f(kL), \tag{1.90}$$

where $f(p)$ is a scaling function obtained as

$$f(p) = \int d\vec{x} e^{i\vec{p} \cdot \vec{x}} g(x). \tag{1.91}$$

The scaling functions $g(x)$ and $f(p)$ characterize the morphology of the ordering system. In experiments or simulations of domain growth, one usually attempts to obtain the functional forms of $g(x)$ and $f(p)$. Of course, a complete description of the morphology would require knowledge of all higher-order structure factors also, but these have limited experimental relevance.

The case with $d = 1$ does not obey these general arguments, and we discuss it separately here. For domain scales $L(t) \gg \xi_b$, there is only an exponentially decaying interaction of order e^{-L/ξ_b} between domain walls. This results in a logarithmic growth law $L(t) \sim \xi_b \ln t$. The corresponding scaling functions have been explicitly obtained by Nagai and Kawasaki [37].

1.4.1.4 Short-Distance Singularities and Porod's Law

Let us now discuss some general properties of the correlation function and the structure factor. The presence of sharp interfaces (defects) in the phase-ordering system results in a short-distance singularity of the correlation function. This can be obtained as follows. For simplicity, we first consider the 1-d case with a kink defect of size L. We are interested in short distances x such that $L \gg x \gg \xi_b$, where ξ_b is the correlation length. Therefore, we can approximate the kink defect by the sign-function:

$$\psi(x) = \text{sgn}(x), \quad x \in \left[-\frac{L}{2}, \frac{L}{2}\right]. \tag{1.92}$$

The corresponding correlation function is obtained from Equation 1.88 as

$$\begin{aligned}
C(x) &= \frac{1}{L} \int_{-L/2}^{L/2} d\bar{x}\, \text{sgn}(\bar{x})\, \text{sgn}(\bar{x} + x) \\
&= \frac{1}{L} \left[\int_0^{L/2} d\bar{x}\, \text{sgn}(\bar{x} - x) + \int_0^{L/2} d\bar{x}\, \text{sgn}(\bar{x} + x) \right] \\
&= 1 - \frac{2|x|}{L}.
\end{aligned} \tag{1.93}$$

Notice that this function is non-analytic at $x = 0$. This short-distance singularity has important implications for the behavior of the structure factor at large wave-vectors. We will discuss this shortly, but let us first generalize the result in Equation 1.93 to arbitrary d.

We consider a d-dimensional kink with the interface located at $x = 0$. The order-parameter field is perfectly correlated in the $(d - 1)$ dimensions that are perpendicular to x. Therefore, the correlation function at short distances is merely

$$C(\vec{r}) = 1 - \frac{2|x|}{L}, \tag{1.94}$$

where x now denotes the x-component of $\vec{r}$. In the isotropic case, the interface is randomly oriented in d-dimensional space. The corresponding correlation function is

$$C(\vec{r}) = 1 - \frac{2\langle|x|\rangle}{L}, \tag{1.95}$$

where $\langle\cdot\rangle$ denotes an average over the $(d - 2)$ polar angles $\{\theta_1, \theta_2, \ldots, \theta_{d-2}\}$ and one azimuthal angle ϕ. This average is obtained as (using $x = r \sin\theta_1 \sin\theta_2 \ldots \sin\theta_{d-2} \cos\phi$)

$$
\begin{aligned}
\langle|x|\rangle &= \frac{\int_0^\pi d\theta_1 \sin^{d-2}\theta_1 \int_0^\pi d\theta_2 \sin^{d-3}\theta_2 \ldots \int_0^\pi d\theta_{d-2} \sin\theta_{d-2} \int_0^{2\pi} d\phi \, r \, |\sin\theta_1| \ldots |\sin\theta_{d-2}|| \cos\phi|}{\int_0^\pi d\theta_1 \sin^{d-2}\theta_1 \int_0^\pi d\theta_2 \sin^{d-3}\theta_2 \ldots \int_0^\pi d\theta_{d-2} \sin\theta_{d-2} \int_0^{2\pi} d\phi} \\
&= r \, 2^{d-2} B\left(\frac{d}{2}, \frac{d}{2}\right) \cdot \frac{2}{\pi} \\
&= r \frac{1}{\sqrt{\pi}} \frac{\Gamma(d/2)}{\Gamma[(d+1)/2]}. \tag{1.96}
\end{aligned}
$$

In obtaining the above result, we have used the identities [38]

$$
\int_0^{\pi/2} d\theta \sin^{n-1}\theta = 2^{n-2} B\left(\frac{n}{2}, \frac{n}{2}\right)
$$
$$
= \frac{1}{2} B\left(\frac{1}{2}, \frac{n}{2}\right), \tag{1.97}
$$

where $B(x, y)$ is the beta function

$$B(x, y) = \frac{\Gamma(x)\Gamma(y)}{\Gamma(x + y)}. \tag{1.98}$$

Putting the result for $\langle|x|\rangle$ into Equation 1.95, we obtain the short-distance behavior of the correlation function for interface defects in d dimensions:

$$C(r) = 1 - \frac{2}{\sqrt{\pi}} \frac{\Gamma(d/2)}{\Gamma[(d+1)/2]} \frac{r}{L}, \quad \frac{r}{L} \to 0. \tag{1.99}$$

The short-distance singularity in $C(r)$ as $r/L \to 0$ gives rise to a power-law decay of the structure-factor tail. Recall that the structure factor is merely the Fourier transform of the correlation function. From power-counting of the singular term in Equation 1.99, the functional form of the structure-factor tail is

$$S(k) = L^d \frac{A_d}{(kL)^{d+1}}, \quad kL \to \infty. \tag{1.100}$$

This important result is referred to as the Porod law and was first obtained for scattering from two-phase systems [39]. Notice that $S(k)$ satisfies the dynamical-scaling form in Equation 1.90. Here, A_d denotes the amplitude of the Porod tail, which can be extracted as follows. We have the inverse Fourier transform

$$C(r) = \int \frac{d\vec{k}}{(2\pi)^d} e^{-i\vec{k}\cdot\vec{r}} S(k)$$

$$= 1 - \int \frac{d\vec{k}}{(2\pi)^d} \left(1 - e^{-i\vec{k}\cdot\vec{r}}\right) S(k)$$

$$= 1 - I(r), \tag{1.101}$$

where we have used the fact that $C(0) = 1$.
We can decompose the above integral as

$$I(r) = \int_0^K dk\, k^{d-1} \int \frac{d\Omega_k}{(2\pi)^d} \left(1 - e^{-i\vec{k}\cdot\vec{r}}\right) S(k)$$

$$+ \int_K^\infty dk\, k^{d-1} \int \frac{d\Omega_k}{(2\pi)^d} \left(1 - e^{-i\vec{k}\cdot\vec{r}}\right) \frac{A_d}{L k^{d+1}}$$

$$= I_1(r) + I_2(r), \tag{1.102}$$

where K is sufficiently large that the Porod tail applies for $k > K$. Finally, notice that we can extend the lower limit of $I_2(r)$ to $k = 0$. [This does not give rise to a divergence as $k \to 0$ because an extra factor of k^2 arises from the $(1 - e^{-i\vec{k}\cdot\vec{r}})$ term.] Then

$$I(r) = \int_0^K dk\, k^{d-1} \int \frac{d\Omega_k}{(2\pi)^d} \left(1 - e^{-i\vec{k}\cdot\vec{r}}\right) \left[S(k) - \frac{A_d}{L k^{d+1}}\right]$$

$$+ \int_0^\infty dk\, k^{d-1} \int \frac{d\Omega_k}{(2\pi)^d} \left(1 - e^{-i\vec{k}\cdot\vec{r}}\right) \frac{A_d}{L k^{d+1}}$$

$$= I_3(r) + I_4(r). \tag{1.103}$$

Notice that $I_3(r)$ is analytic in r because of the finite upper limit of the k-integral. Clearly, the singular terms in $C(r)$ arise from $I_4(r)$, and we can isolate these as follows:

$$
\begin{aligned}
I_4(r) &= \frac{A_d}{L} \int_0^\infty \frac{d\vec{k}}{(2\pi)^d} \left(1 - e^{-i\vec{k}\cdot\vec{r}}\right) \frac{1}{k^{d+1}} \\
&= \frac{A_d}{L} \int_0^\infty \frac{d\vec{k}}{(2\pi)^d} \left(1 - e^{-i\vec{k}\cdot\vec{r}}\right) \int_0^\infty \frac{du}{\Gamma[(d+1)/2]} u^{((d+1)/2)-1} e^{-uk^2} \\
&= \frac{A_d}{L} \frac{1}{\Gamma[(d+1)/2]} \int_0^\infty du\, u^{(d-1)/2} \int_0^\infty \frac{d\vec{k}}{(2\pi)^d} \left(1 - e^{-i\vec{k}\cdot\vec{r}}\right) e^{-uk^2} \\
&= \frac{A_d}{L} \frac{1}{(4\pi)^{d/2}\Gamma[(d+1)/2]} \int_0^\infty du\, u^{-1/2} \left(1 - e^{-r^2/(4u)}\right).
\end{aligned}
\tag{1.104}
$$

The final step is to evaluate the integral on the RHS. We make the substitution $z = r^2/(4u)$ to obtain

$$
\begin{aligned}
\int_0^\infty du\, u^{-1/2} \left(1 - e^{-r^2/(4u)}\right) &= \frac{r}{2} \int_0^\infty dz\, z^{-3/2} \left(1 - e^{-z}\right) \\
&= -\frac{r}{2} \Gamma\left(-\frac{1}{2}\right) \\
&= \sqrt{\pi}\, r.
\end{aligned}
\tag{1.105}
$$

This yields the following expression for the singular part of $C(r)$:

$$
C_s(r) = -A_d \frac{\sqrt{\pi}}{(4\pi)^{d/2}\Gamma[(d+1)/2]} \frac{r}{L}.
\tag{1.106}
$$

Comparing Equations 1.106 and 1.99, we obtain the exact amplitude of the Porod tail:

$$
A_d = 2^{d+1} \pi^{(d/2)-1} \Gamma\left(\frac{d}{2}\right).
\tag{1.107}
$$

1.4.1.5 Ohta–Jasnow–Kawasaki Theory

Let us next consider the case with multiple defects or interfaces, corresponding to ordering from a random initial condition (see Figure 1.4). In this context, an approximate theory has been developed by Ohta et al. (OJK) [40]. Recall that the ordering system is described by an order parameter field $\psi(\vec{r}, t)$ that obeys the TDGL equation 1.68. In the late stages of domain growth, the system consists of large domains with $\psi = +1$ or $\psi = -1$. The order parameter rapidly changes sign at the domain boundaries. OJK introduced a nonlinear transformation $\psi \equiv \psi(m)$, where $m(\vec{r}, t)$ is an auxiliary field that varies smoothly through the interface. The simplest choice for m is the normal distance from the interface, so that $m(\vec{r}, t) = 0$ on the interface. The corresponding choice for the nonlinear transformation is then $\psi = \text{sgn}(m)$. Let us obtain the equation obeyed by $m(\vec{r}, t)$ as follows. Recall the Allen–Cahn equation $v(\vec{a}) = -\vec{\nabla} \cdot \hat{n}$, where $\hat{n}$ is the unit vector normal to the interface. We use the fact that

$$\hat{n} = \frac{\vec{\nabla} m}{|\vec{\nabla} m|} \tag{1.108}$$

to obtain

$$v = -\vec{\nabla} \cdot \hat{n} = -\frac{\nabla^2 m}{|\vec{\nabla} m|} + \frac{n_i n_j \nabla_i \nabla_j m}{|\vec{\nabla} m|}. \tag{1.109}$$

Further, in a reference frame moving with the interface,

$$\frac{dm}{dt} = \frac{\partial m}{\partial t} + \vec{\nabla} m \cdot \vec{v} = 0. \tag{1.110}$$

We use $\vec{\nabla} m \cdot \vec{v} = |\vec{\nabla} m| v$ to obtain

$$v = -\frac{1}{|\vec{\nabla} m|} \frac{\partial m}{\partial t}. \tag{1.111}$$

Comparing Equations 1.109 and 1.111, we obtain the required equation for $m(\vec{r}, t)$:

$$\frac{\partial}{\partial t} m(\vec{r}, t) = \nabla^2 m - n_i n_j \nabla_i \nabla_j m. \tag{1.112}$$

Equation 1.112 is nonlinear and analytically intractable, as was the case with the original TDGL equation. Then, OJK replaced the term $n_i n_j$ by its spherical average as

$$n_i n_j \simeq \frac{\delta_{ij}}{d}. \tag{1.113}$$

Under this approximation, the auxiliary field $m(\vec{r}, t)$ obeys the diffusion equation:

$$\frac{\partial}{\partial t} m(\vec{r}, t) = \left(\frac{d-1}{d} \right) \nabla^2 m \equiv D \nabla^2 m. \tag{1.114}$$

Now, we are in a position to calculate the correlation function of the ψ-field from the correlation function of the m-field. We model the random initial condition for the field $m(\vec{r}, 0)$ by a Gaussian distribution with zero mean and correlation

$$\langle m(\vec{r}, 0) m(\vec{r}', 0) \rangle = A \delta(\vec{r} - \vec{r}'). \tag{1.115}$$

The appropriate distribution is

$$P[m(\vec{r},0)] = \frac{1}{\sqrt{2\pi\sigma(0)^2}} \exp\left[-\frac{m(\vec{r},0)^2}{2\sigma(0)^2}\right], \quad \sigma(0)^2 = \langle m(\vec{r},0)^2\rangle. \quad (1.116)$$

As the evolution of $m(\vec{r},t)$ is governed by the linear equation 1.114, the distribution of m remains Gaussian for all times:

$$P[m(\vec{r},t)] = \frac{1}{\sqrt{2\pi\sigma(t)^2}} \exp\left[-\frac{m(\vec{r},t)^2}{2\sigma(t)^2}\right], \quad \sigma(t)^2 = \langle m(\vec{r},t)^2\rangle. \quad (1.117)$$

Let us calculate the general correlation function (at unequal space-time points)

$$C(\vec{r}_1,t_1;\vec{r}_2,t_2) = \langle\psi(\vec{r}_1,t_1)\psi(\vec{r}_2,t_2)\rangle$$
$$= \langle \mathrm{sgn}\left[m(\vec{r}_1,t_1)\right]\mathrm{sgn}\left[m(\vec{r}_2,t_2)\right]\rangle. \quad (1.118)$$

To obtain this average, we need the normalized bivariate Gaussian distribution for $m(\vec{r}_1,t_1)$ ($=x$, say) and $m(\vec{r}_2,t_2)$ ($=y$):

$$P(x,y) = \frac{1}{2\pi\sigma(t_1)\sigma(t_2)\sqrt{1-\gamma^2}}$$
$$\times \exp\left[-\frac{1}{2(1-\gamma^2)}\left(\frac{x^2}{\sigma(t_1)^2} + \frac{y^2}{\sigma(t_2)^2} - \frac{2\gamma xy}{\sigma(t_1)\sigma(t_2)}\right)\right], \quad (1.119)$$

where

$$\sigma(t_1)^2 = \langle x^2\rangle = \langle m(\vec{r}_1,t_1)^2\rangle,$$
$$\sigma(t_2)^2 = \langle y^2\rangle = \langle m(\vec{r}_2,t_2)^2\rangle,$$
$$\gamma = \frac{\langle xy\rangle}{\sqrt{\langle x^2\rangle\langle y^2\rangle}} = \frac{\langle m(\vec{r}_1,t_1)m(\vec{r}_2,t_2)\rangle}{\sqrt{\langle m(\vec{r}_1,t_1)^2\rangle\langle m(\vec{r}_2,t_2)^2\rangle}}. \quad (1.120)$$

We will calculate the quantities $\sigma(t_1), \sigma(t_2)$ and γ shortly. Let us first obtain the correlation function

$$C(\vec{r}_1,t_1;\vec{r}_2,t_2) = \frac{1}{2\pi\sigma(t_1)\sigma(t_2)\sqrt{1-\gamma^2}} \int_{-\infty}^{\infty} dx \int_{-\infty}^{\infty} dy \frac{x}{\sqrt{x^2}}\frac{y}{\sqrt{y^2}}$$
$$\times \exp\left[-\frac{1}{2(1-\gamma^2)}\left(\frac{x^2}{\sigma(t_1)^2} + \frac{y^2}{\sigma(t_2)^2} - \frac{2\gamma xy}{\sigma(t_1)\sigma(t_2)}\right)\right],$$

$$(1.121)$$

where we have used $\text{sgn}(z) = z/\sqrt{z^2}$. We rescale as $x' = x/\sigma(t_1)$ and $y' = y/\sigma(t_2)$ to obtain (dropping the primes)

$$C(\vec{r}_1, t_1; \vec{r}_2, t_2) = \frac{1}{2\pi\sqrt{1-\gamma^2}} \int_{-\infty}^{\infty} dx \int_{-\infty}^{\infty} dy \frac{x}{\sqrt{x^2}} \frac{y}{\sqrt{y^2}} \exp\left(-\alpha x^2 - \alpha y^2 + \beta xy\right),$$

(1.122)

where

$$\alpha = \frac{1}{2(1-\gamma^2)},$$

$$\beta = \frac{\gamma}{(1-\gamma^2)}.$$

(1.123)

The integral on the RHS of Equation 1.122 can be simplified by using the identity

$$\frac{1}{\sqrt{z}} = \frac{1}{\sqrt{\pi}} \int_{-\infty}^{\infty} d\theta \exp(-z\theta^2).$$

(1.124)

This yields

$$
\begin{aligned}
C(\vec{r}_1, t_1; \vec{r}_2, t_2) &= \frac{1}{2\pi^2\sqrt{1-\gamma^2}} \int_{-\infty}^{\infty} d\theta \int_{-\infty}^{\infty} d\phi \int_{-\infty}^{\infty} dx \int_{-\infty}^{\infty} dy\, xy \\
&\quad \times \exp\left[-(\alpha+\theta^2)x^2 - (\alpha+\phi^2)y^2 + \beta xy\right] \\
&= \frac{1}{2\pi^2\sqrt{1-\gamma^2}} \frac{\partial}{\partial\beta} \int_{-\infty}^{\infty} d\theta \int_{-\infty}^{\infty} d\phi \int_{-\infty}^{\infty} dx \int_{-\infty}^{\infty} dy \\
&\quad \times \exp\left[-(\alpha+\theta^2)x^2 - (\alpha+\phi^2)y^2 + \beta xy\right].
\end{aligned}
$$

(1.125)

The integrals over x and y are performed using the general expression for the Gaussian integral:

$$\int_{-\infty}^{\infty} dz \exp(-p^2 z^2 \pm qz) = \exp\left(\frac{q^2}{4p^2}\right) \frac{\sqrt{\pi}}{|p|}.$$

(1.126)

Then, the integrand becomes

$$
\begin{aligned}
h(\theta, \phi) &= \int_{-\infty}^{\infty} dx \exp\left[-(\alpha+\theta^2)x^2\right] \int_{-\infty}^{\infty} dy \exp\left[-(\alpha+\phi^2)y^2 + \beta xy\right] \\
&= \int_{-\infty}^{\infty} dx \exp\left[-(\alpha+\theta^2)x^2\right] \exp\left[\frac{\beta^2 x^2}{4(\alpha+\phi^2)}\right] \sqrt{\frac{\pi}{\alpha+\phi^2}} \\
&= \sqrt{\frac{\pi}{\alpha+\phi^2}} \int_{-\infty}^{\infty} dx \exp\left[-\left(\alpha+\theta^2 - \frac{\beta^2}{4(\alpha+\phi^2)}\right)x^2\right]
\end{aligned}
$$

$$= \sqrt{\frac{\pi}{\alpha + \phi^2}} \frac{\sqrt{\pi}}{\left[\alpha + \theta^2 - \dfrac{\beta^2}{4(\alpha + \phi^2)}\right]^{1/2}}. \tag{1.127}$$

Notice that the x-integral is well defined because

$$\alpha + \theta^2 - \frac{\beta^2}{4(\alpha + \phi^2)} > \alpha - \frac{\beta^2}{4\alpha} > 0. \tag{1.128}$$

Finally, we have

$$C(\vec{r}_1, t_1; \vec{r}_2, t_2) = \frac{\beta}{8\pi\sqrt{1 - \gamma^2}} \int_{-\infty}^{\infty} d\theta \int_{-\infty}^{\infty} d\phi \frac{1}{[\alpha + \phi^2]^{3/2}} \frac{1}{\left[\alpha + \theta^2 - \dfrac{\beta^2}{4(\alpha + \phi^2)}\right]^{3/2}}$$

$$= \frac{\beta}{4\pi\sqrt{1 - \gamma^2}} \int_{-\infty}^{\infty} d\phi \frac{1}{(\alpha + \phi^2)^{3/2}} \frac{1}{\left[\alpha - \dfrac{\beta^2}{4(\alpha + \phi^2)}\right]}$$

$$= \frac{\beta}{\pi\sqrt{1 - \gamma^2}} \int_{-\infty}^{\infty} d\phi \frac{1}{(\alpha + \phi^2)^{1/2}} \frac{1}{\left(4\alpha\phi^2 + 4\alpha^2 - \beta^2\right)}. \tag{1.129}$$

Some further algebra yields the final result:

$$C(\vec{r}_1, t_1; \vec{r}_2, t_2) = \frac{2}{\pi} \sin^{-1}(\gamma), \tag{1.130}$$

where γ was defined in Equation 1.120.

Our only remaining task is to obtain the quantity γ, which is the (normalized) correlation function of the auxiliary field $m(\vec{r}, t)$. It is easy to calculate this quantity in momentum space as follows. We note that

$$m(\vec{r}, t) = e^{Dt\nabla^2} m(\vec{r}, 0),$$
$$m(\vec{k}, t) = e^{-Dtk^2} m(\vec{k}, 0). \tag{1.131}$$

Further, we have

$$\left\langle m(\vec{k}, 0) m(\vec{k}', 0) \right\rangle = \int d\vec{r} e^{i\vec{k}\cdot\vec{r}} \int d\vec{r}' e^{i\vec{k}'\cdot\vec{r}'} \left\langle m(\vec{r}, 0) m(\vec{r}', 0) \right\rangle$$

$$= A \int d\vec{r} e^{i(\vec{k} + \vec{k}')\cdot\vec{r}} = A\delta(\vec{k} + \vec{k}'), \tag{1.132}$$

where we have used Equation 1.115. Therefore

$$\left\langle m(\vec{k}_1, t_1) m(\vec{k}_2, t_2) \right\rangle = e^{-Dt_1 k_1^2} e^{-Dt_2 k_2^2} \left\langle m(\vec{k}_1, 0) m(\vec{k}_2, 0) \right\rangle$$

$$= A\delta(\vec{k}_1 + \vec{k}_2) e^{-D(t_1 k_1^2 + t_2 k_2^2)}. \tag{1.133}$$

Finally, we obtain

$$
\langle m(\vec{r}_1, t_1) m(\vec{r}_2, t_2) \rangle = \int \frac{d\vec{k}_1}{(2\pi)^d} e^{-i\vec{k}_1 \cdot \vec{r}_1} \int \frac{d\vec{k}_2}{(2\pi)^d} e^{-i\vec{k}_2 \cdot \vec{r}_2} \left\langle m(\vec{k}_1, t_1) m(\vec{k}_2, t_2) \right\rangle
$$

$$
= A \int \frac{d\vec{k}_1}{(2\pi)^d} e^{-i\vec{k}_1 \cdot (\vec{r}_1 - \vec{r}_2)} e^{-D(t_1+t_2)k_1^2}
$$

$$
= \frac{A}{[4\pi D(t_1 + t_2)]^{d/2}} \exp\left[-\frac{r^2}{4D(t_1 + t_2)} \right], \qquad (1.134)
$$

where $\vec{r} = \vec{r}_1 - \vec{r}_2$. Thus

$$
\gamma = \frac{\langle m(\vec{r}_1, t_1) m(\vec{r}_2, t_2) \rangle}{\sqrt{\langle m(\vec{r}_1, t_1)^2 \rangle \langle m(\vec{r}_2, t_2)^2 \rangle}}
$$

$$
= \left(\frac{2\sqrt{t_1 t_2}}{t_1 + t_2} \right)^{d/2} \exp\left[-\frac{r^2}{4D(t_1 + t_2)} \right]. \qquad (1.135)
$$

Equations 1.130 and 1.135 constitute the OJK result for the correlation function (at unequal space-time points) for the nonconserved ordering problem.

1.4.1.6 Implications of the OJK Function

We are usually interested in the equal-time correlation function ($\vec{r} = \vec{r}_1 - \vec{r}_2$)

$$
C(\vec{r}, t) \equiv C(\vec{r}_1, t; \vec{r}_2, t)
$$

$$
= \frac{2}{\pi} \sin^{-1}\left[\exp\left(-\frac{r^2}{8Dt} \right) \right]. \qquad (1.136)
$$

Notice that the OJK result has the scaling form in Equation 1.88 with the domain scale $L(t) \sim (8Dt)^{1/2}$. This confirms the result we had obtained from dimensional analysis of the Allen–Cahn equation. Further, the OJK function is in excellent agreement with results obtained from experiments and numerical simulations. In Figure 1.7a, we show numerical results for the correlation function of the ordering ferromagnet in Figure 1.4. We plot $C(r, t)$ versus r/L at different times—the resultant data collapse confirms the scaling form in Equation 1.88. The OJK function is also plotted in Figure 1.7a. In Figure 1.7b, we show numerical data for the scaled structure factor, $L^{-d} S(k, t)$ versus kL, at different times. Again, the data collapses onto a scaling function, in accordance with the scaling form in Equation 1.90. The tail of the structure factor decays as a power law, $S(k, t) \sim k^{-(d+1)}$, which is the Porod tail in Equation 1.100. In Figure 1.7b, we also plot the Fourier transform of the OJK function. As expected, this also shows a Porod tail. Our next task is to demonstrate this explicitly. Before

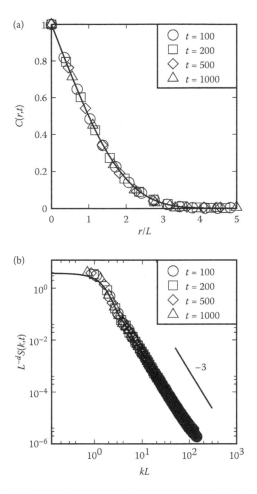

FIGURE 1.7 (a) Scaled correlation function [$C(r, t)$ versus r/L] for the evolution depicted in Figure 1.4. The numerical data is obtained as an average over 5 independent runs for systems of size 1024^2. The length scale $L(t)$ is defined as the distance over which the correlation function falls to half its maximum value. We show data for times $100, 200, 500, 1000$—denoted by the specified symbols. The solid line denotes the OJK function in Equations 1.130 and 1.135, scaled in the same manner as the numerical data. (b) Scaled structure factor [$L^{-d}S(k, t)$ versus kL], corresponding to the data in (a). In this case, we plot the data on a log-log scale. The solid line denotes the Fourier transform of the OJK function. The line with slope -3 corresponds to the Porod tail. (Adapted from Oono, Y. and Puri, S., *Phys. Rev. Lett.*, 58, 836, 1987; *Phys. Rev. A*, 38, 434, 1988; Puri, S. and Oono, Y., *Phys. Rev. A*, 38, 1542, 1988.)

doing that, we present data for the domain growth law [$L(t)$ versus t] in Figure 1.8. We see that the domain-growth process obeys the Allen–Cahn law, $L(t) \sim t^{1/2}$.

In our earlier discussion, we had stressed that the short-distance singularities of the correlation function lead to a power-law decay of the structure factor. In this context, it is useful to undertake a short-distance expansion of the OJK function. We identify

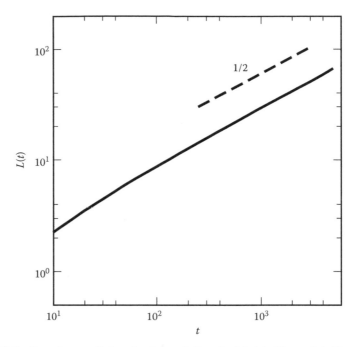

FIGURE 1.8 Domain growth law for the evolution depicted in Figure 1.4. The numerical details are the same as in Figure 1.7. We plot the data on a log-log scale. The dashed line denotes the Allen–Cahn growth law, $L(t) \sim t^{1/2}$.

the scaled variable $x = r/L$, and consider the function

$$g(x) = \frac{2}{\pi} \sin^{-1}\left(e^{-x^2}\right). \tag{1.137}$$

In the limit $x \to 0$, we have $\gamma = e^{-x^2} \to 1$. For a small-x expansion of $g(x)$, it is convenient to use the identity [38]

$$\sin^{-1}\gamma = \gamma F\left(\frac{1}{2}, \frac{1}{2}; \frac{3}{2}; \gamma^2\right), \tag{1.138}$$

where we have introduced the hypergeometric function

$$F(a, b; c; z) = \sum_{n=0}^{\infty} \frac{\Gamma(a+n)\Gamma(b+n)}{\Gamma(c+n)\, n!} \frac{\Gamma(c)}{\Gamma(a)\Gamma(b)} z^n. \tag{1.139}$$

The hypergeometric function satisfies the relation [38]:

$$F(a, b; c; z) = \frac{\Gamma(c)\Gamma(c-a-b)}{\Gamma(c-a)\Gamma(c-b)} F(a, b; a+b-c+1; 1-z) + (1-z)^{c-a-b}$$

$$\times \frac{\Gamma(c)\Gamma(a+b-c)}{\Gamma(a)\Gamma(b)} F(c-a, c-b; c-a-b+1; 1-z). \tag{1.140}$$

Using this relation, we have

$$\sin^{-1} \gamma = \frac{\pi}{2}\gamma F\left(\frac{1}{2},\frac{1}{2};\frac{1}{2};1-\gamma^2\right) - \gamma(1-\gamma^2)^{1/2}F\left(1,1;\frac{3}{2};1-\gamma^2\right) \quad (1.141)$$

and

$$g(x) = e^{-x^2}F\left(\frac{1}{2},\frac{1}{2};\frac{1}{2};1-e^{-2x^2}\right) - \frac{2}{\pi}e^{-x^2}\left(1-e^{-2x^2}\right)^{1/2}F\left(1,1;\frac{3}{2};1-e^{-2x^2}\right)$$

$$\equiv A(x) + B(x). \quad (1.142)$$

The singular terms for $x \to 0$ arise only from the second term (B) on the RHS of Equation 1.142. The first few of these are obtained as

$$B(x) = -\frac{2\sqrt{2}}{\pi}x + \frac{\sqrt{2}}{3\pi}x^3 + O(x^5). \quad (1.143)$$

The first term (A) yields only analytic terms for $x \to 0$. Some calculation shows that the $O(x^2)$ term is missing in the expansion of A, and

$$A(x) = 1 - O(x^4). \quad (1.144)$$

Thus, the overall result for $g(x)$ is

$$g(x) = 1 - \frac{2\sqrt{2}}{\pi}x + \frac{\sqrt{2}}{3\pi}x^3 + O(x^4). \quad (1.145)$$

As before, the linear term in x gives rise to the Porod tail

$$S(k) = L^d \frac{A_{OJK}}{(kL)^{d+1}}, \quad kL \to \infty. \quad (1.146)$$

From Equations 1.106 and 1.143, we observe that

$$A_{OJK}\frac{\sqrt{\pi}}{(4\pi)^{d/2}\Gamma\left(\frac{d+1}{2}\right)} = \frac{2\sqrt{2}}{\pi}, \quad (1.147)$$

so that

$$A_{OJK} = 2^{d+3/2}\pi^{(d-3)/2}\Gamma\left(\frac{d+1}{2}\right). \quad (1.148)$$

The absence of the x^2-term in the expansion of $g(x)$ also has an important consequence for the structure factor. In the limit $x \to 0$, we have

$$\lim_{x \to 0} \nabla^2 g(x) = \lim_{x \to 0}\left(\frac{d^2 g}{dx^2} + \frac{d-1}{x}\frac{dg}{dx}\right)$$

$$= \lim_{x \to 0} -\frac{2\sqrt{2}(d-1)}{\pi x}. \quad (1.149)$$

We can rewrite $\nabla^2 g$ as

$$\nabla^2 g(x) = \nabla^2 \int \frac{d\vec{p}}{(2\pi)^d} e^{-i\vec{p}\cdot\vec{x}} f(p) = -\int \frac{d\vec{p}}{(2\pi)^d} e^{-i\vec{p}\cdot\vec{x}} p^2 f(p), \qquad (1.150)$$

where $\vec{p} = \vec{k}L$ and $f(p)$ is the scaling function for the structure factor—see Equation 1.90.

Further, we have the identity

$$\frac{1}{x} = \int \frac{d\vec{p}}{(2\pi)^d} e^{-i\vec{p}\cdot\vec{x}} \frac{2^{d-1}\pi^{(d-1)/2}\Gamma[(d-1)/2]}{p^{d-1}}. \qquad (1.151)$$

Therefore, Equation 1.150 can be rewritten as

$$\int \frac{d\vec{p}}{(2\pi)^d} \left[p^2 f(p) - \frac{2^{d-1}\pi^{(d-1)/2}\Gamma[(d-1)/2]\,2\sqrt{2}(d-1)}{\pi p^{d-1}} \right] = 0, \qquad (1.152)$$

or

$$\int_0^\infty dp \left[p^{d+1} f(p) - 2^{d+3/2}\pi^{(d-3)/2}\Gamma\left(\frac{d+1}{2}\right) \right] = 0. \qquad (1.153)$$

This reconfirms the result in Equations 1.146 and 1.148, viz.,

$$f(p) = \frac{2^{d+3/2}\pi^{(d-3)/2}\Gamma[(d+1)/2]}{p^{d+1}}, \qquad p \to \infty. \qquad (1.154)$$

Further, Equation 1.153 constitutes a sum rule that must be obeyed by the scaling function. This is referred to as Tomita's sum rule [41,42].

An important extension of the OJK result is due to Oono and Puri (OP) [43], who incorporated the effects of non-zero interfacial thickness into the analytical form for the correlation function. This extension was of considerable experimental and numerical relevance because the non-zero interfacial thickness has a severe impact on the tail of the structure factor. In particular, the power-law decay is replaced by a Gaussian decay for finite times, and the Porod tail is only recovered for $t \to \infty$.

1.4.1.7 Kawasaki–Yalabik–Gunton Theory

An alternative approach to the nonconserved ordering problem is due to Kawasaki et al. (KYG) [44]. The KYG theory yields the same result for the correlation function as does OJK theory. As before, we require an approximate solution to the TDGL Equation 1.68, which we reproduce here for convenience:

$$\frac{\partial}{\partial t}\psi(\vec{r}, t) = \psi - \psi^3 + \nabla^2\psi.$$

Let us consider the Fourier transform of the TDGL equation:

$$\frac{\partial}{\partial t}\psi(\vec{k},t) = (1-k^2)\psi(\vec{k},t) - \int\frac{d\vec{k}_1}{(2\pi)^d}\int\frac{d\vec{k}_2}{(2\pi)^d}\int\frac{d\vec{k}_3}{(2\pi)^d}\delta(\vec{k}-\vec{k}_1-\vec{k}_2-\vec{k}_3)$$

$$\times\,\psi(\vec{k}_1,t)\psi(\vec{k}_2,t)\psi(\vec{k}_3,t). \tag{1.155}$$

The linear part of Equation 1.155 is

$$\frac{\partial}{\partial t}m(\vec{k},t) = (1-k^2)m(\vec{k},t), \tag{1.156}$$

with the solution:

$$m(\vec{k},t) = e^{\gamma(\vec{k})t}m(\vec{k},0), \quad \gamma(\vec{k}) = 1-k^2. \tag{1.157}$$

The formal solution of the nonlinear equation 1.155 is

$$\psi(\vec{k},t) = m(\vec{k},t) - \int_0^t dt'\,e^{\gamma(\vec{k})(t-t')}\int\frac{d\vec{k}_1}{(2\pi)^d}\int\frac{d\vec{k}_2}{(2\pi)^d}\int\frac{d\vec{k}_3}{(2\pi)^d}$$

$$\times\,\delta(\vec{k}-\vec{k}_1-\vec{k}_2-\vec{k}_3)\psi(\vec{k}_1,t')\psi(\vec{k}_2,t')\psi(\vec{k}_3,t'). \tag{1.158}$$

KYG solved this integral equation iteratively, using the linear solution $m(\vec{k},t)$ to obtain an expansion in integrals of products of $m(\vec{k},t)$. Notice that $m(\vec{k},t)$ diverges with time, so each term of this expansion also diverges in time. KYG used a technique developed by Suzuki [45,46] to perform a partial resummation of the terms in this expansion to obtain a well-controlled result:

$$\psi(\vec{r},t) \simeq \frac{m(\vec{r},t)}{\sqrt{1+m(\vec{r},t)^2}}. \tag{1.159}$$

In the large-t limit, $m(\vec{r},t)$ is very large and

$$\psi(\vec{r},t) \simeq \mathrm{sgn}\left[m(\vec{r},t)\right]. \tag{1.160}$$

Notice that this is identical to the nonlinear transformation in OJK theory. In this case also, the auxiliary field $m(\vec{r},t)$ obeys the diffusion equation ($D=1$ here), with the only difference being that there is an additional factor of e^t:

$$m(\vec{r},t) = e^{t(1+\nabla^2)}m(\vec{r},0). \tag{1.161}$$

We assume that $m(\vec{r},0)$ has a Gaussian distribution (cf. Equation 1.116), so that $m(\vec{r},t)$ is also described by a Gaussian distribution. The calculation of the correlation function for the field $\psi(\vec{r},t)$ is identical to that described for OJK theory. The final result is again the OJK function in Equations 1.130 and 1.135 with $D=1$.

Finally, it is useful to examine the KYG theory from an alternative perspective. Let us consider the nonlinear transformation

$$\psi(\vec{r}, t) = \frac{m(\vec{r}, t)}{\sqrt{1 + m(\vec{r}, t)^2}} \tag{1.162}$$

and determine the equation that is satisfied by $m(\vec{r}, t)$. Some simple algebra yields the relevant equation, which is

$$\frac{\partial}{\partial t} m(\vec{r}, t) = m + \nabla^2 m - \frac{3m|\vec{\nabla}m|^2}{1 + m^2}. \tag{1.163}$$

Thus, the KYG approximation is equivalent to discarding the nonlinear term on the RHS of Equation 1.163. There is no justification for such an approximation, so the KYG theory only constitutes an approximate solution to the ordering problem. This can be explicitly demonstrated by comparing the evolution profiles obtained from Equation 1.159 with those obtained numerically from the original TDGL equation. In a sense, it is fortunate that the nonlinear transformation in Equation 1.159 faithfully mimics the actual defect statistics, thereby yielding excellent results for statistical quantities like the correlation function and the structure factor.

In the next subsection, we will see that the KYG "solution" in Equation 1.159 readily generalizes to the problem of domain growth in a system described by a vector order parameter.

1.4.2 CASE WITH VECTOR ORDER PARAMETER

The vector version of the TDGL equation, where ψ is replaced by an n-component vector $\vec{\psi} = (\psi_1, \psi_2, \ldots \psi_n)$, is also of great experimental relevance. In dimensionless units, this has the following form for $T < T_c$ and $\vec{h} = 0$:

$$\frac{\partial}{\partial t} \vec{\psi}(\vec{r}, t) = \vec{\psi} - |\vec{\psi}|^2 \vec{\psi} + \nabla^2 \vec{\psi} + \vec{\theta}(\vec{r}, t), \tag{1.164}$$

where the vector noise $\vec{\theta}(\vec{r}, t)$ satisfies

$$\overline{\vec{\theta}(\vec{r}, t)} = 0,$$
$$\overline{\theta_i(\vec{r}', t')\theta_j(\vec{r}'', t'')} = 2\epsilon\delta_{ij}\delta(\vec{r}' - \vec{r}'')\delta(t' - t''). \tag{1.165}$$

For example, the $n = 2$ case (XY model) is relevant in the ordering of superconductors, superfluids, and liquid crystals. Figure 1.9 shows the evolution of the vector field (ψ_1, ψ_2) for the $d = 2$ XY model from a disordered initial condition. In this case, the topological defects are vortices and anti-vortices, and domain growth is driven by the

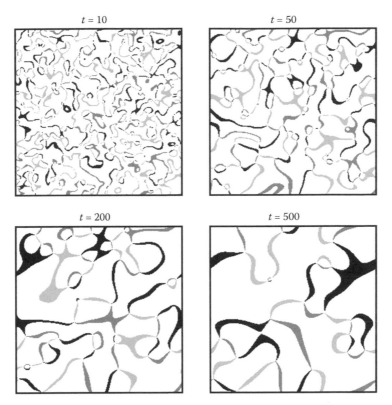

FIGURE 1.9 **(See color insert following page 180.)** Evolution of the dimensionless XY model from a disordered initial condition. The pictures were obtained from a Euler-discretized version of Equation 1.164 with $n = 2$ and $\epsilon = 0$. The numerical details are the same as for Figure 1.4. The snapshots show regions of constant phase $\theta_\psi = \tan^{-1}(\psi_2/\psi_1)$, measured in radians, with the following color coding: $\theta_\psi \in [1.85, 2.15]$ (black); $\theta_\psi \in [3.85, 4.15]$ (red); $\theta_\psi \in [5.85, 6.15]$ (green). Typically, a meeting point of the three colors denotes a vortex defect.

annihilation of these defects. The $n = 3$ case (Heisenberg model) is also of relevance in the ordering of liquid crystals and in the evolution dynamics of the early universe.

1.4.2.1 Generalized Porod's Law

As before, the nature of defects in the ordering system determines general properties of the correlation function and the structure factor. The presence of n-component defects (e.g., vortices for $n = 2, d = 2$; vortex lines for $n = 2, d = 3$; monopoles for $n = 3, d = 3$; etc.) again yields a power-law or *generalized Porod* tail for the scaled structure factor [47,48], that is,

$$f(p) \sim p^{-(d+n)}, \quad p \to \infty. \tag{1.166}$$

The corresponding exact result for the singular behavior of the scaled correlation function of a single defect at short distances is as follows [49]:

$$g(x) = 1 + \pi^{(n-2)/2} \frac{\Gamma(d/2)\,\Gamma[(n+1)/2]^2\,\Gamma(-n/2)}{\Gamma[(d+n)/2)]\,\Gamma(n/2)} x^n + \text{higher-order terms.}$$

$$(1.167)$$

This result is obtained by a generalization (to the n-component case) of the arguments that resulted in Equation 1.99. That result was obtained for interface defects and corresponds to the $n = 1$ case of Equation 1.167. The singular behavior for even values of n has to be carefully extracted from the above expression and involves logarithmic corrections as $1 - g(x) \sim x^n \ln x$. Furthermore, there is no analog of the Tomita sum rule for the vector-ordering problem.

1.4.2.2 Bray–Puri–Toyoki Theory

Bray and Puri (BP) [47] and (independently) Toyoki (T) [48] used a defect-dynamics approach to obtain an approximate solution for the correlation function of the n-component TDGL equation in d-dimensional space. The BPT result is valid for $n \leq d$, corresponding to the case where topological defects are present. The starting point of the BP calculation is a generalization of the KYG solution for the scalar TDGL equation in Equation 1.159 [50]:

$$\vec{\psi}(\vec{r}, t) \simeq \frac{\vec{m}(\vec{r}, t)}{\sqrt{1 + \sum_{i=1}^{n} m_i(\vec{r}, t)^2}}. \qquad (1.168)$$

Here, $\vec{m}(\vec{r}, t)$ is the solution of the linear equation (cf. Equation 1.156)

$$\frac{\partial}{\partial t}\vec{m}(\vec{r}, t) = \vec{m} + \nabla^2 \vec{m}, \qquad (1.169)$$

so that

$$\vec{m}(\vec{r}, t) = e^{t(1+\nabla^2)}\vec{m}(\vec{r}, 0). \qquad (1.170)$$

In the large-t limit, $\vec{m}(\vec{r}, t)$ is very large and

$$\vec{\psi}(\vec{r}, t) \simeq \frac{\vec{m}(\vec{r}, t)}{|\vec{m}(r, t)|}. \qquad (1.171)$$

The subsequent calculation is analogous to that described in Equation 1.116 onwards. BP obtained an explicit scaling form for the correlation function:

$$C(\vec{r}, t) = \langle \vec{\psi}(\vec{r}_1, t) \cdot \vec{\psi}(\vec{r}_2, t) \rangle$$

$$= \frac{n\gamma}{2\pi} \left[B\left(\frac{n+1}{2}, \frac{1}{2}\right) \right]^2 F\left(\frac{1}{2}, \frac{1}{2}; \frac{n+2}{2}; \gamma^2\right), \qquad (1.172)$$

where $\gamma = \exp(-r^2/L^2)$, L being the average defect length-scale. BP demonstrated that the length scale obeys the Allen–Cahn growth law, $L(t) \sim t^{1/2}$. (A more careful calculation demonstrates that there are logarithmic corrections when $n = d = 2$, and $L(t) \sim (t/\ln t)^{1/2}$ [51].) As in the case of the OJK function, Equation 1.172 is valid when the defect core size is identically zero. The nonzero core size in experiments or simulations introduces OP-like corrections in the correlation function and structure factor.

The case with $n > d$ is unusual in that there are no topological defects, and it is not possible to describe the evolution of the system in terms of the annealing of defects. As a matter of fact, systems with $n > d$ need not exhibit dynamical scaling. Some specific results are available for $n = d + 1$, where the system is characterized by the presence of textures [52,53]. In general, systems with textures do not show single-length scaling. Dynamical scaling is restored in the $n \to \infty$ limit, where the correlation function exhibits a Gaussian decay [3]:

$$C(\vec{r}, t) = \exp\left(-\frac{r^2}{L^2}\right), \tag{1.173}$$

where $L(t) \sim t^{1/2}$. There are no general analytical results available for the correlation function of the n-component TDGL equation with $n > d$.

The results of OJK-OP and BPT have been understood to constitute a complete solution of the nonconserved ordering problem. However, the work of Blundell et al. [54] suggests that this is not the case. Generally, one tests for dynamical scaling by plotting $C(\vec{r}, t)$ versus r/L or $S(\vec{k}, t)L^{-d}$ versus kL at different times (see Figure 1.7) [55]. One also compares the analytical results with experimental or numerical results using similar plots. However, there is an arbitrariness in the definition of the characteristic length scale, which is defined as (say) the reciprocal of the first moment of the structure factor, or the distance over which the correlation function falls to half its maximum value. The problem with these definitions, while comparing different results, is that they already build in a high level of agreement on a scaling plot. Blundell et al. [54] have proposed a universal test of dynamical scaling, which does not use an internally defined length scale, but rather uses higher-order structure factors. On such a plot, there is a considerable difference between the OJK/BPT results and the corresponding numerical results. This suggests that our analytical understanding of nonconserved phase ordering may not be so good after all. Subsequently, Mazenko [57–59] showed that the OJK result is exact in the limit $d \to \infty$. There have been attempts to improve on the OJK result for the case of finite d, and these are discussed by Mazenko [60,61].

1.5 DOMAIN GROWTH IN SYSTEMS WITH CONSERVED KINETICS

Let us next consider systems with a conserved order parameter, for example, kinetics of phase separation of a binary (AB) mixture. A typical phase diagram for an AB mixture was shown in Figure 1.3. Though the Hamiltonian of the binary mixture is the same as that for the ferromagnet, recall that the phase diagrams in Figures 1.2 and 1.3 are obtained in different ensembles.

In Figure 1.10, we show the evolution of a binary mixture after a quench from above the coexistence curve (homogeneous phase) to below the coexistence curve (segregated phase). The initially homogeneous system separates into A-rich and B-rich domains, denoted as black and white regions in Figure 1.10. In contrast with the nonconserved case, the evolution in this case must satisfy the constraint that numbers of A and B are constant, that is, the order parameter is conserved. In the corresponding kinetic Ising model in Section 1.3.3, recall that the conservation law was implemented via Kawasaki spin-exchange kinetics.

Experimentalists distinguish between shallow quenches (just below the co-existence curve) and deep quenches (far below the coexistence curve). For shallow quenches, in the region between the coexistence line and the spinodal lines in

FIGURE 1.10 Evolution of a homogeneous binary (AB) mixture, which is quenched below the coexistence curve at time $t = 0$. These pictures were obtained from a numerical solution of the CHC equation 1.182 with $\epsilon = 0$. The discretization mesh sizes were $\Delta t = 0.01$ and $\Delta x = 1$, and the lattice size was 256^2. The random initial condition consisted of small-amplitude fluctuations about $\psi = 0$, corresponding with a mixture with equal amounts of A and B (critical composition). Regions with $\psi > 0$ (A-rich) and $\psi < 0$ (B-rich) are marked in black and white, respectively.

Figure 1.3, the homogeneous system is metastable and decomposes by the *nucleation and growth* of droplets. For deep quenches, into the region below the spinodal lines, the homogeneous system spontaneously decomposes into A-rich and B-rich regions, a process referred to as *spinodal decomposition*. However, there is no sharp physical distinction between the *nucleation and growth* and *spinodal decomposition* regions of the phase diagram [4]. This will be discussed at length in Chapter 2.

1.5.1 SEGREGATION IN BINARY ALLOYS

First, we focus on the kinetics of phase separation in a binary mixture where hydrodynamic effects are not relevant, for example, binary alloys. In this case, the primary mechanism for phase separation is diffusion and aggregation. As in the nonconserved case, we introduce a space-time–dependent order parameter $\psi(\vec{r}, t) = n^A(\vec{r}, t) - n^B(\vec{r}, t)$, where $n^\alpha(\vec{r}, t)$ is the local density of species α. The evolution of ψ is described by the continuity equation:

$$\frac{\partial}{\partial t}\psi(\vec{r}, t) = -\vec{\nabla} \cdot \vec{J}(\vec{r}, t),$$ (1.174)

where $\vec{J}(\vec{r}, t)$ denotes the current. Further, because the current is driven by concentration fluctuations, we expect

$$\vec{J}(\vec{r}, t) = -D\vec{\nabla}\mu(\vec{r}, t),$$ (1.175)

where D is the diffusion coefficient and $\mu(\vec{r}, t)$ is the chemical potential. Finally, the chemical potential is determined as

$$\mu(\vec{r}, t) = \frac{\delta F[\psi]}{\delta \psi},$$ (1.176)

where F refers to the Helmholtz potential, which is the appropriate thermodynamic potential for the binary mixture. This is obtained from Equation 1.60 with $g(\psi)$ replaced by $f(\psi)$.

Combining Equations 1.174 through 1.176, we obtain the CH equation [62,63] for the phase separation of a binary mixture:

$$\frac{\partial}{\partial t}\psi(\vec{r}, t) = D\nabla^2\left(\frac{\delta F[\psi]}{\delta \psi}\right).$$ (1.177)

In Section 1.3.3, we had derived this equation from the spin-exchange kinetic Ising model (cf. Equation 1.58). Notice that Equation 1.177 corresponds to the case of a constant diffusion coefficient. There have also been studies of systems where the diffusion coefficient depends on the local order parameter [64].

The effects of thermal fluctuations can be incorporated in the CH equation by including a noise term in the definition of the current in Equation 1.175 [65]. The resultant model is the Cahn–Hilliard–Cook (CHC) equation:

$$\frac{\partial}{\partial t}\psi(\vec{r}, t) = \vec{\nabla} \cdot \left\{ D\vec{\nabla}\left(\frac{\delta F[\psi]}{\delta \psi}\right) + \vec{\theta}(\vec{r}, t) \right\}.$$ (1.178)

The vector noise satisfies the usual fluctuation–dissipation relation:

$$\overline{\vec{\theta}(\vec{r}, t)} = 0,$$

$$\overline{\theta_i(\vec{r}', t')\theta_j(\vec{r}'', t'')} = 2Dk_BT\delta_{ij}\delta(\vec{r}' - \vec{r}'')\delta(t' - t'').$$

(1.179)

Equations 1.178 and 1.179 are also referred to as *Model B* in the classification scheme of Hohenberg and Halperin [31].

For the ψ^4-form of the free-energy functional (Equation 1.63 without the magnetic field term), the CHC equation has the following form:

$$\frac{\partial}{\partial t}\psi(\vec{r}, t) = \vec{\nabla} \cdot \left\{ D\vec{\nabla}\left[-a(T_c - T)\psi + b\psi^3 - K\nabla^2\psi\right] + \vec{\theta}(\vec{r}, t) \right\}.$$

(1.180)

In analogy with the TDGL equation (cf. Equation 1.64), we rescale variables as follows (for $T < T_c$):

$$\psi' = \frac{\psi}{\psi_0}, \quad \psi_0 = \sqrt{\frac{a(T_c - T)}{b}},$$

$$t' = \frac{Da^2(T_c - T)^2}{K} t,$$

$$\vec{r}' = \sqrt{\frac{a(T_c - T)}{K}} \vec{r}, \quad \xi_b = \sqrt{\frac{2K}{a(T_c - T)}},$$

$$\vec{\theta}' = \frac{\sqrt{bK}}{Da^2(T_c - T)^2} \vec{\theta}.$$

(1.181)

We then drop primes to obtain the dimensionless CHC equation:

$$\frac{\partial}{\partial t}\psi(\vec{r}, t) = \vec{\nabla} \cdot \left[\vec{\nabla}\left(-\psi + \psi^3 - \nabla^2\psi\right) + \vec{\theta}(\vec{r}, t) \right],$$

(1.182)

where

$$\overline{\vec{\theta}(\vec{r}, t)} = 0,$$

$$\overline{\theta_i(\vec{r}', t')\theta_j(\vec{r}'', t'')} = 2\epsilon\delta_{ij}\delta(\vec{r}' - \vec{r}'')\delta(t' - t''),$$

$$\epsilon = \frac{k_BTb\left[a(T_c - T)\right]^{(d-4)/2}}{K^{d/2}}.$$

(1.183)

The evolution depicted in Figure 1.10 is obtained from a numerical solution of the dimensionless CHC equation with a random initial condition, which mimics the disordered state prior to the quench. Regions that are A-rich ($\psi = +1$) are marked black, and regions that are B-rich ($\psi = -1$) are not marked. The mixture has a critical composition with 50% A–50% B, that is, the average value of the order parameter

$\psi_0 = V^{-1} \int d\vec{r} \; \psi(\vec{r},0) = 0$. This composition is maintained during the evolution due to the conservation constraint. As in the case of the TDGL equation, thermal noise is asymptotically irrelevant for the CHC equation—the evolution shown in Figure 1.10 corresponds to the deterministic case with $\epsilon = 0$. In Figure 1.11, we show the evolution for an off-critical composition with 30% A and 70% B, that is, $\psi_0 = -0.4$. This composition still lies inside the spinodal curve in Figure 1.3, so the evolution corresponds to spinodal decomposition. However, the morphology is characterized by the growth of A-rich (minority phase) droplets in a B-rich (majority phase) background.

Before we proceed, it is relevant to discuss the applicability of the CHC model to real binary alloys. Typically, lattice parameter mismatches in alloys can set up large strain fields in the intermediate and late stages of phase separation [66–69] These strain fields drastically modify the results we quote below and must be accounted for in any realistic description of phase separation in alloys. Chapter 8 of this book addresses this problem in considerable detail.

In the absence of strain effects, the phase-separating system is characterized by a unique length scale, $L(t) \sim t^{1/3}$ in $d \geq 2$. This power-law behavior was first derived

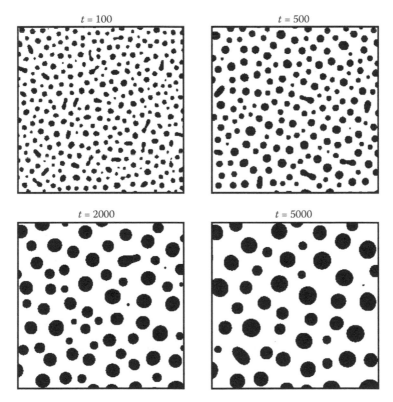

FIGURE 1.11 Analogous to Figure 1.10, but for the case of an off-critical binary mixture. The initial condition consisted of small-amplitude fluctuations about $\psi = -0.4$, corresponding to a mixture with 30% A and 70% B.

by Lifshitz and Slyozov (LS) [70] for extremely off-critical systems, where one of the components is present in a vanishing fraction, and the evolution proceeds via the nucleation and growth of minority droplets. Huse [71] demonstrated that the same law is applicable to spinodal decomposition, where there are approximately equal fractions of the two components, and the coarsening structure is bi-continuous. Typically, the chemical potential on the surface of a domain of size L is $\mu \sim \sigma/L$, where σ is the surface tension. The concentration current is obtained as $D|\vec{\nabla}\mu| \sim D\sigma/L^2$, where D is the diffusion constant. Therefore, the domain size grows as $dL/dt \sim D\sigma/L^2$, or $L(t) \sim (D\sigma t)^{1/3}$. In Figure 1.12, we show the domain growth law for the case with $\psi_0 = 0$ (shown in Figure 1.10)—the asymptotic growth law is seen to be consistent with the LS law.

As in the nonconserved case, the quantities that characterize the evolution morphology in Figures 1.10 and 1.11 are the correlation function and the structure factor. The existence of a characteristic length scale results in the dynamical scaling of these quantities. For the critical quench shown in Figure 1.10, we demonstrate the dynamical scaling of the correlation function and structure factor in Figure 1.13. The presence of interfacial defects results in a Porod tail for the scaled structure factor, $f(p) \sim p^{-(d+1)}$ as $p \to \infty$; and a singular short-distance behavior of the correlation function as in Equation 1.99. The structure factor also obeys the Tomita sum rule [41].

The above properties are common to the nonconserved and conserved cases. There are some additional features due to the conservation law. For example, the

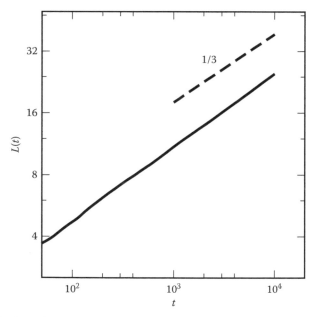

FIGURE 1.12 Domain growth law for the evolution depicted in Figure 1.10. The numerical data was obtained as an average over 10 independent runs for systems of size 512^2. The data is plotted on a log-log scale. The dashed line denotes the Lifshitz–Slyozov growth law, $L(t) \sim t^{1/3}$.

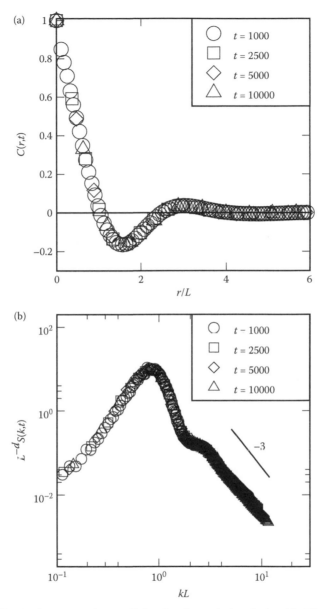

FIGURE 1.13 Analogous to Figure 1.7, but for the evolution depicted in Figure 1.10. The numerical data was obtained as an average over 10 independent runs for systems of size 512^2.

conservation constraint dictates the sum rule:

$$\int d\vec{r}\, C(\vec{r}, t) = 0, \quad \text{or} \quad S(0, t) = 0. \tag{1.184}$$

Furthermore, the conservation law also fixes the $p \to 0$ behavior of the scaled structure factor as $f(p) \sim p^4$ [72,73].

Clearly, we have a good understanding of various general features of the morphology. There have also been extensive simulations of phase-separation kinetics, which have provided detailed results for the late-time behavior. These studies are based on *cell dynamical system* (CDS) models [49,50]; discrete simulations of the CHC equation [74]; and Monte Carlo (MC) simulations of the spin-exchange kinetic Ising model [75,76]. A detailed description of MC simulation techniques for domain growth problems is provided in Chapter 3. However, in spite of many attempts [77–79], there is still no comprehensive theory for the scaling form of the correlation function. A major analytical obstacle in this regard is the strongly correlated motion of interfaces, resulting from the conservation law. In particular, the *Gaussian closure* techniques introduced in Section 1.4.1, which involve linearization of the dynamical equations via a nonlinear transformation, have not worked well in the conserved case [82].

There have also been some studies of the case with conserved vector order parameter, that is, the conserved counterpart of Equation 1.164. In the case where topological defects are present ($n \leq d$), Puri et al. [83,84] demonstrated that Gaussian closure techniques work better than in the case with scalar order parameter. They used this approach to obtain approximate analytical results for the correlation functions of the conserved XY and Heisenberg models. The corresponding growth law is $L(t) \sim t^{1/4}$, which is slower than the LS-growth for the scalar case. In the limit $n \to \infty$, an important result is due to Coniglio and Zannetti [85], who showed that the structure factor exhibits *multiple-length* scaling rather than single-length scaling. However, Bray and Humayun [86] have demonstrated that multi-scaling is a singular property of the $n = \infty$ case.

1.5.2 SEGREGATION IN BINARY FLUID MIXTURES

1.5.2.1 Dimensional Form of Model H

Let us next consider the phase-separation kinetics of immiscible binary fluids [3–6]. In this case, the flow field enables advective transport of the segregating components. The appropriate equation for the order-parameter evolution is [31]

$$\frac{D\psi}{Dt} = \frac{\partial \psi}{\partial t} + \vec{v} \cdot \vec{\nabla} \psi = \vec{\nabla} \cdot \left(D\vec{\nabla}\mu + \vec{\theta} \right), \qquad (1.185)$$

where $\vec{v}(\vec{r}, t)$ denotes the fluid velocity field. We assume that the fluid is incompressible (with constant density ρ). Recall that the density obeys the continuity equation:

$$\frac{\partial \rho}{\partial t} + \vec{\nabla} \cdot (\rho\vec{v}) = 0. \qquad (1.186)$$

The requirement that ρ is constant imposes a constraint on the velocity field as $\vec{\nabla} \cdot \vec{v} = 0$.

The corresponding equation for the velocity field is the Navier–Stokes equation:

$$\rho \frac{D\vec{v}}{Dt} = \rho \left[\frac{\partial \vec{v}}{\partial t} + (\vec{v} \cdot \vec{\nabla})\vec{v} \right] = \eta \nabla^2 \vec{v} - \vec{\nabla} p - \psi \vec{\nabla} \mu + \vec{\zeta}, \tag{1.187}$$

where η is the viscosity, and p is the pressure. The additional term on the RHS, $-\psi \vec{\nabla} \mu$, describes the force exerted on the fluid by the segregating mixture.

The Gaussian white noise $\vec{\theta}$ in Equation 1.185 satisfies the fluctuation-dissipation relation (see Equation 1.179):

$$\overline{\vec{\theta}(\vec{r}, t)} = 0,$$
$$\overline{\theta_i(\vec{r}', t')\theta_j(\vec{r}'', t'')} = 2Dk_B T \delta_{ij} \delta(\vec{r}' - \vec{r}'')\delta(t' - t''). \tag{1.188}$$

Equation 1.187 consists of three equations (for v_x, v_y, v_z) with noises $\zeta_x, \zeta_y, \zeta_z$. Each of these independent noises is obtained as

$$\zeta_i = \vec{\nabla} \cdot \vec{\sigma}, \tag{1.189}$$

where $\vec{\sigma}$ obeys the fluctuation-dissipation relation:

$$\overline{\vec{\sigma}(\vec{r}, t)} = 0,$$
$$\overline{\sigma_i(\vec{r}', t')\sigma_j(\vec{r}'', t'')} = 2\eta k_B T \delta_{ij} \delta(\vec{r}' - \vec{r}'')\delta(t' - t''). \tag{1.190}$$

Our first task is to eliminate the pressure term in Equation 1.187 by using the incompressibility condition $\vec{\nabla} \cdot \vec{v} = 0$. We apply the operator $\vec{\nabla} \cdot$ to both sides of Equation 1.187 to obtain

$$\nabla^2 p = \vec{\nabla} \cdot \vec{F} = \vec{\nabla} \cdot \left[-\psi \vec{\nabla} \mu - \rho(\vec{v} \cdot \vec{\nabla})\vec{v} + \vec{\zeta} \right]. \tag{1.191}$$

In Fourier space, this equation has the solution

$$p(\vec{k}, t) = i \frac{\vec{k} \cdot \vec{F}(\vec{k}, t)}{k^2}. \tag{1.192}$$

We can replace $p(\vec{r}, t)$ [the inverse Fourier transform of $p(\vec{k}, t)$] in Equation 1.187 to obtain

$$\rho \frac{\partial}{\partial t} \vec{v}(\vec{r}, t) = \eta \nabla^2 \vec{v} + \left[-\psi \vec{\nabla} \mu - \rho(\vec{v} \cdot \vec{\nabla})\vec{v} + \vec{\zeta} \right]_\perp. \tag{1.193}$$

In Equation 1.193, we have introduced the following notation: $[\vec{F}(\vec{r}, t)]_\perp$ denotes the transverse part of the vector $\vec{F}(\vec{r}, t)$. In momentum space, this is computed as

$$[\vec{F}(\vec{k}, t)]_\perp = \vec{F}(\vec{k}, t) - \frac{\vec{k} \cdot \vec{F}(\vec{k}, t)}{k^2} \vec{k} = \vec{F}(\vec{k}, t) - \left[\hat{k} \cdot \vec{F}(\vec{k}, t) \right] \hat{k}. \tag{1.194}$$

Equations 1.185 and 1.193 are referred to as *Model H* in the classification scheme of Hohenberg and Halperin [31]. In analogy with Models A and B, we can rescale variables in Model H to obtain a dimensionless version. However, to clarify the domain growth laws in this model, it is convenient to work with the dimensional form.

1.5.2.2   Domain Growth Laws in Binary Fluids

To understand the domain growth laws, we consider the deterministic case with $T = 0$. In the initial stages, the dynamics of the velocity field is much faster than that of the order-parameter field, and we ignore the inertial terms in Equation 1.193. The resultant equation is solved for $\vec{v}$ (in Fourier space) as

$$v_i(\vec{k}, t) = T_{ij}(\vec{k})X_j(\vec{k}, t)$$

$$= \frac{1}{\eta k^2}\left(\delta_{ij} - \frac{k_i k_j}{k^2}\right)X_j(\vec{k}, t),$$

$$\vec{X}(\vec{k}, t) = \int d\vec{r} e^{i\vec{k}\cdot\vec{r}}\left(-\psi\vec{\nabla}\mu\right). \tag{1.195}$$

In Equation 1.195, repeated indices are summed over, and $T_{ij}(\vec{k})$ denotes the Oseen tensor in momentum space. In $d = 3$, the real-space Oseen tensor is

$$T_{ij}(\vec{r}) = \frac{1}{8\pi\eta r}\left(\delta_{ij} + \frac{r_i r_j}{r^2}\right). \tag{1.196}$$

As $\vec{\nabla}\cdot\vec{v} = 0$, we can rewrite Equation 1.185 as

$$\frac{\partial}{\partial t}\psi(\vec{r}, t) = \vec{\nabla}\cdot\left(D\vec{\nabla}\mu - \psi\vec{v}\right), \tag{1.197}$$

where we have set $\vec{\theta} = 0$. Replacing the expression for the velocity from Equation 1.195 in Equation 1.197, we obtain the evolution equation

$$\frac{\partial}{\partial t}\psi(\vec{r}, t) = D\nabla^2\mu + \nabla_i\left[\psi(\vec{r}, t)\int d\vec{r}' T_{ij}(\vec{r} - \vec{r}')\psi(\vec{r}', t)\nabla'_j\mu(\vec{r}', t)\right]. \tag{1.198}$$

By partial integration, we can transform Equation 1.198 into a more symmetric form:

$$\frac{\partial}{\partial t}\psi(\vec{r}, t) = D\nabla^2\mu - \left[\nabla_i\psi(\vec{r}, t)\right]\int d\vec{r}' T_{ij}(\vec{r} - \vec{r}')\left[\nabla'_j\psi(\vec{r}', t)\right]\mu(\vec{r}', t) \tag{1.199}$$

where we have used the property

$$\nabla_i T_{ij}(\vec{r} - \vec{r}') = \nabla'_j T_{ij}(\vec{r} - \vec{r}') = 0. \tag{1.200}$$

Equation 1.198 or 1.199 can be used to understand domain growth laws for coarsening in binary fluids. At early times, growth is driven by diffusion, as in the case of binary alloys. However, there is a crossover to a hydrodynamic growth regime, where material is rapidly transported along domain boundaries by advection [87,88]. Let us estimate the various terms in Equation 1.199. The term on the left-hand side (LHS) is estimated as

$$\frac{\partial \psi}{\partial t} \sim \frac{1}{L}\frac{dL}{dt}, \tag{1.201}$$

where $L(t)$ is the domain length scale. Next, consider the terms on the RHS. The diffusive term is of order $D\sigma/L^3$, where σ is the surface tension. (Recall that $\mu = \sigma/L$ on the surface of a domain with size L.) The advective term is of order $\sigma/(\eta L)$, where we have estimated the Oseen tensor as $T \sim 1/(\eta L)$ in $d = 3$. Thus, the diffusive term is dominant when

$$\frac{D\sigma}{L^3} \gg \frac{\sigma}{\eta L}, \tag{1.202}$$

or $L \ll (D\eta)^{1/2}$. The growth law in the diffusive regime is the usual LS law, $L(t) \sim (D\sigma t)^{1/3}$. At later times, the advective term is dominant, yielding the following growth law for the so-called *viscous hydrodynamic* regime:

$$L(t) \sim \frac{\sigma t}{\eta}. \tag{1.203}$$

At even later times, the approximation of neglecting the inertial terms in Equation 1.193 is no longer valid [88]. To understand this, let us estimate the terms in Equation 1.193, which we repeat here for ease of reference:

$$\rho\frac{\partial \vec{v}}{\partial t} + [\rho(\vec{v} \cdot \vec{\nabla})\vec{v}]_\perp = \eta\nabla^2\vec{v} - \left[\psi\vec{\nabla}\mu\right]_\perp, \tag{1.204}$$

where we have set $\vec{\zeta} = 0$. The fluid velocity scale is estimated as L/t. Then, the first term on the LHS is of order $\rho L/t^2$, and the second (nonlinear) term is also of the same order. The viscous term on the RHS is of order $\eta/(Lt)$, and the force term is of order σ/L^2. The terms on the LHS become comparable with those on the RHS at length scales of order $\eta^2/(\rho\sigma)$. The asymptotic growth law in the *inertial hydrodynamic* regime is then obtained from

$$\frac{\rho L}{t^2} \sim \frac{\sigma}{L^2}, \tag{1.205}$$

as

$$L(t) \sim \left(\frac{\sigma t^2}{\rho}\right)^{1/3}. \tag{1.206}$$

The growth laws for the different regimes are summarized as follows:

$$L(t) \sim (D\sigma t)^{1/3}, \quad L \ll (D\eta)^{1/2}, \quad \text{(diffusive regime)}$$

$$\sim \frac{\sigma t}{\eta}, \quad (D\eta)^{1/2} \ll L \ll \frac{\eta^2}{\rho\sigma}, \quad \text{(viscous hydrodynamic regime)} \quad (1.207)$$

$$\sim \left(\frac{\sigma t^2}{\rho}\right)^{1/3}, \quad \frac{\eta^2}{\rho\sigma} \ll L, \quad \text{(inertial hydrodynamic regime)}$$

Grant and Elder [89] have argued that scaling requires that the asymptotic growth exponent $\phi \leq 1/2$. Otherwise, the system enters a turbulent regime in which it undergoes remixing. Alternatively, they suggest that a faster growth law than $L(t) \sim t^{1/2}$ can be sustained if the system does not exhibit dynamical scaling. However, the Grant–Elder arguments have been criticized by Kendon et al. [90], whose lattice-Boltzmann simulations support the inertial hydrodynamic growth law, $L(t) \sim t^{2/3}$. (A detailed description of the lattice-Boltzmann approach to study the phase separation of fluids is provided in Chapter 4.)

We should also stress that domain connectivity plays a crucial role in enabling hydrodynamic transport. In highly off-critical quenches, the morphology consists of droplets of the minority phase in a matrix of the majority phase (see Figure 1.11). Then, the hydrodynamic mechanism is disabled and domain growth is analogous to that for binary alloys. Furthermore, thermal fluctuations can drive the Brownian motion and coalescence of droplets. This also gives rise to a LS-like growth law, $L(t) \sim (k_B T t / \eta)^{1/3}$ [87], though the physical mechanism is quite different.

Finally, we should remark on the morphology, as characterized by the correlation function or the structure factor. These quantities and the domain growth laws have been investigated in many numerical studies, for example, CDS simulations [91–94]; molecular dynamics (MD) studies [95,96]; lattice-Boltzmann simulations [90,97,98], etc. As regards analytical results, the general morphological features (Porod's law, Tomita sum rule, etc.) discussed in Section 1.4.1 arise in the current context also. However, as in the case of binary alloys, there is no satisfactory theory for the scaling form of the correlation function. This remains one of the major unsolved problems in the field of phase-ordering kinetics.

1.6   SUMMARY AND DISCUSSION

Let us conclude this chapter with a summary and discussion. We have presented a broad overview of the *kinetics of phase transitions*. There are two prototypical problems in this area: (a) the ordering of a ferromagnet, which is an example of the case with nonconserved order parameter; and (b) the phase separation of a binary mixture, which is an example of the case with conserved order parameter. One is interested in the evolution of a thermodynamically unstable disordered or homogeneous state. This evolution is characterized by the emergence and growth of domains that are enriched in different components of the mixture.

We have a good understanding of the morphologies and growth laws that characterize domain-growth kinetics. The domain size $L(t)$ usually grows as a power-law in time, $L(t) \sim t^{\phi}$, where the exponent ϕ depends on (a) the nature of conservation laws, (b) the defects that drive the ordering process, and (c) the relevance of flow fields. The morphology is quantitatively characterized by the correlation function or its Fourier transform, the structure factor. There is a good theoretical understanding of the correlation function for nonconserved dynamics with both scalar and vector order parameters. Unfortunately, our understanding of the conserved problem is not so good. Thus, we know the various limiting behaviors of the correlation function or structure factor. However, the overall functional form is not well understood, and this remains one of the outstanding problems in this area.

We should stress that the concepts and paradigms discussed here have been applied in a broad variety of physical problems. The examples used in this chapter are primarily drawn from metallurgy and materials science. However, the phenomena of phase separation, clustering, and aggregation are ubiquitous in nature.

The subsequent chapters of this book will discuss different aspects of the kinetics of phase transitions. All the chapters are written in a pedagogical manner and will be accessible to an advanced undergraduate student. At this stage, it is useful to provide a brief overview of these chapters. We have mentioned earlier that the distinction between *spinodal decomposition* and *nucleation and growth* is not as sharp as the phase diagram in Figure 1.3 would suggest. This issue is discussed in detail by Binder in Chapter 2. Chapters 3 and 4 are dedicated to a discussion of simulation techniques in this area. In Chapter 3, Barkema describes Monte Carlo simulations of the kinetic Ising models introduced in Section 1.3. In Chapter 4, Gonnella and Yeomans discuss lattice Boltzmann simulations, which have proved very useful in understanding the late stages of phase separation in fluid mixtures (cf. Section 1.5.2). Numerical simulations have played a crucial role in developing our understanding of domain growth problems. The methodology described in Chapters 3 and 4 will prove very useful for a researcher entering this area.

In Chapter 5, Zannetti discusses *slow relaxation* and *aging* in phase-ordering systems. These phenomena are well known in the context of structural glasses and spin glasses. Recent studies indicate that these concepts are also highly relevant in domain growth problems—Zannetti provides an overview of these studies.

Recent interest in this area has focused on incorporating various experimentally relevant features in studies of phase-ordering systems. In this context, Chapter 6 (by Khanna, Agnihotri, and Sharma) discusses the *kinetics of dewetting* of liquid films on surfaces. In Chapter 7, Ohta reviews studies of phase separation in diblock copolymers. In these systems, the polymers A and B (which repel each other) are jointed, so that the system can only undergo segregation on microscales. Finally, in Chapter 8 (written by Onuki, Minami, and Furukawa), there is a detailed discussion of phase separation in solids. In Section 1.5.1, we had emphasized that strain fields play an important role in the segregation kinetics of alloys. Onuki et al. discuss how elastic fields can be incorporated into the Ginzburg–Landau description of solid mixtures.

ACKNOWLEDGMENTS

The author's understanding of the problems discussed here has grown through many fruitful collaborations. He would like to thank his research collaborators for many stimulating interactions and inputs. He is also grateful to Anupam Mukherjee for assistance in preparing the figures.

REFERENCES

1. Stanley, H.E., *Introduction to Phase Transitions and Critical Phenomena*, Oxford University Press, Oxford, 1971.
2. *Phase Transitions and Critical Phenomena*, Domb, C. and Lebowitz, J.L., Eds., Academic Press, London.
3. Bray, A.J., Theory of phase ordering kinetics, *Adv. Phys.*, 43, 357, 1994.
4. Binder, K. and Fratzl, P., Spinodal decomposition, in *Materials Science and Technology* Vol. 5, Kostorz, G., Ed., Wiley-VCH, Weinheim, 2001, chap. 6.
5. Onuki, A., *Phase Transition Dynamics*, Cambridge University Press, Cambridge, 2002.
6. Dattagupta, S. and Puri, S., *Dissipative Phenomena in Condensed Matter Physics*, Springer-Verlag, Heidelberg.
7. Plischke, M. and Bergersen, B., *Equilibrium Statistical Physics*, Second Edition, World Scientific, Singapore, 1994.
8. Glauber, R.J., Time-dependent statistics of the Ising model, *J. Math. Phys.*, 4, 294, 1963.
9. Kawasaki, K., Kinetics of Ising models, in *Phase Transitions and Critical Phenomena*, Vol. 2, Domb, C. and Green, M.S., Eds., Academic Press, London, 1972, 443.
10. Kawasaki, K., Diffusion constants near the critical point for time-dependent Ising models. I, *Phys. Rev.*, 145, 224, 1966.
11. Van Kampen, N.G., *Stochastic Processes in Physics and Chemistry*, North-Holland, Amsterdam, 1981.
12. Suzuki, M. and Kubo, R., Dynamics of the Ising model near the critical point. I, *J. Phys. Soc. Jpn.*, 24, 51, 1968.
13. Binder, K., Kinetic Ising model study of phase separation in binary alloys, *Z. Phys.*, 267, 313, 1974.
14. Binder, K., Theory for the dynamics of clusters. II. Critical diffusion in binary systems and the kinetics of phase separation, *Phys. Rev. B*, 15, 4425, 1977.
15. Gouyet, J.-F., Atomic mobility and spinodal-decomposition dynamics in lattice gases. Simple discrete models, *Europhys. Lett.*, 21, 335, 1993.
16. Gouyet, J.-F., Generalized Allen–Cahn equations to describe far-from-equilibrium order–disorder dynamics, *Phys. Rev. E*, 51, 1695, 1995.
17. Plapp, M. and Gouyet, J.-F., Dendritic growth in a mean-field lattice gas model, *Phys. Rev. E*, 55, 45, 1997.
18. Puri, S., Dynamics of vacancy-mediated phase separation, *Phys. Rev. E*, 55, 1752, 1997.
19. Puri, S. and Sharma, R., Phase ordering dynamics in binary mixtures with annealed vacancies, *Phys. Rev. E*, 57, 1873, 1998.
20. Landau, L.D. and Lifshitz, E.M., *Statistical Physics*, Pergamon Press, Oxford, 1980.
21. Huang, K., *Statistical Mechanics*, Second Edition, Wiley, New York, 1987.
22. Manning, J.R., *Diffusion Kinetics for Atoms in Crystals*, Van Nostrand, Princeton, 1968.
23. Flynn, C.P., *Point Defects and Diffusion*, Clarendon, Oxford, 1972.
24. Yaldram, K. and Binder, K., Spinodal decomposition of a two-dimensional model alloy with mobile vacancies, *Acta Metall.*, 39, 707, 1991.

25. Yaldram, K. and Binder, K., Monte Carlo simulation of phase separation and clustering in the ABV model, *J. Stat. Phys.*, 62, 161, 1991.
26. Tafa, K., Puri, S., and Kumar, D., Kinetics of domain growth in systems with local barriers, *Phys. Rev. E*, 63, 046115, 2001.
27. Tafa, K., Puri, S., and Kumar, D., Kinetics of phase separation in ternary mixtures, *Phys. Rev. E*, 64, 056139, 2001.
28. Bray, A.J., Comment on "Critical dynamics and global conservation laws," *Phys. Rev. Lett.*, 66, 2048, 1991.
29. Annett, J.F. and Banavar, J.R., Critical dynamics, spinodal decomposition, and conservation laws, *Phys. Rev. Lett.*, 68, 2941, 1992.
30. Phani, M.K., Lebowitz, J.L., Kalos, M.H., and Penrose, O., Kinetics of an order-disorder transition, *Phys. Rev. Lett.*, 45, 366, 1980.
31. Hohenberg, P.C. and Halperin, B.I., Theory of dynamic critical phenomena, *Rev. Mod. Phys.*, 49, 435, 1977.
32. Allen, S.M. and Cahn, J.W., A microscopic theory for antiphase boundary motion and its application to antiphase domain coarsening, *Acta Metall.*, 27, 1085, 1979.
33. Puri, S. and Oono, Y., Effect of noise on spinodal decomposition, *J. Phys. A*, 21, L755, 1988.
34. Bray, A.J., Exact renormalization-group results for domain-growth scaling in spinodal decomposition, *Phys. Rev. Lett.*, 62, 2841, 1989.
35. Bray, A.J., Renormalization-group approach to domain-growth scaling, *Phys. Rev. B*, 41, 6724, 1990.
36. Binder, K. and Stauffer, D., Theory for the slowing down of the relaxation and spinodal decomposition of binary mixtures, *Phys. Rev. Lett.*, 33, 1006, 1974.
37. Nagai, T. and Kawasaki, K., Statistical dynamics of interacting kinks II, *Physica A*, 134, 483, 1986.
38. Gradshteyn, I.S. and Ryzhik, I.M., *Table of Integrals, Series, and Products*, Academic Press, London, 1994.
39. Porod, G., in *Small-Angle X-Ray Scattering*, Glatter, O. and Kratky, O., Eds., Academic Press, New York, 1982, 42.
40. Ohta, T., Jasnow, D., and Kawasaki, K., Universal scaling in the motion of random interfaces, *Phys. Rev. Lett.*, 49, 1223, 1982.
41. Tomita, H., Sum rules for small angle scattering by random interface, *Prog. Theor. Phys.*, 72, 656, 1984.
42. Tomita, H., Statistical properties of random interface system, *Prog. Theor. Phys.*, 75, 482, 1986.
43. Oono, Y. and Puri, S., Large wave number features of form factors for phase transition kinetics, *Mod. Phys. Lett. B*, 2, 861, 1988.
44. Kawasaki, K., Yalabik, M.C., and Gunton, J.D., Growth of fluctuations in quenched time-dependent Ginzburg–Landau model systems, *Phys. Rev. A*, 17, 455, 1978.
45. Suzuki, M., Scaling theory of non-equilibrium systems near the instability point. I. General aspects of transient phenomena, *Prog. Theor. Phys.*, 56, 77, 1976.
46. Suzuki, M., Scaling theory of non-equilibrium systems near the instability point. II. Anomalous fluctuation theorems in the extensive region, *Prog. Theor. Phys.*, 56, 477, 1976.
47. Bray, A.J. and Puri, S., Asymptotic structure factor and power-law tails for phase ordering in systems with continuous symmetry, *Phys. Rev. Lett.*, 67, 2670, 1991.
48. Toyoki, H., Structure factors of vector-order-parameter systems containing random topological defects, *Phys. Rev. B*, 45, 1965, 1992.

49. Bray, A.J. and Humayun, K., Universal amplitudes of power-law tails in the asymptotic structure factor of systems with topological defects, *Phys. Rev. E*, 47, R9, 1993.

50. Puri, S. and Roland, C., Approximate solutions of the two-component Ginzburg–Landau equation, *Phys. Lett. A*, 151, 500, 1990.

51. Pargellis, A.N., Finn, P., Goodby, J.W., Panizza, P., Yurke, B., and Cladis, P.E., Defect dynamics and coarsening dynamics in Smectic-C films, *Phys. Rev. A*, 46, 7765, 1992.

52. Newman, T.J., Bray, A.J., and Moore, M.A., Growth of order in vector spin systems and self-organized criticality, *Phys. Rev. B*, 42, 4514, 1990.

53. Zapotocky, M. and Zakrzewski, W., Kinetics of phase ordering with topological textures, *Phys. Rev. E*, 51, R5189, 1995.

54. Blundell, R.E., Bray, A.J., and Sattler, S., Absolute test for theories of phase-ordering dynamics, *Phys. Rev. E*, 48, 2476, 1993.

55. Oono, Y. and Puri, S., Computationally efficient modeling of ordering of quenched phases, *Phys. Rev. Lett.*, 58, 836, 1987.

56. Puri, S. and Oono, Y., Study of phase-separation dynamics by use of cell dynamical systems. II. Two-dimensional results, *Phys. Rev. A*, 38, 1542, 1988.

57. Mazenko, G.F., Theory of unstable thermodynamic systems, *Phys. Rev. Lett.*, 63, 1605, 1989.

58. Mazenko, G.F., Theory of unstable growth, *Phys. Rev. B*, 42, 4487, 1990.

59. Mazenko, G.F., Theory of unstable growth. II. Conserved order parameter, *Phys. Rev. B*, 43, 5747, 1991.

60. Mazenko, G.F., Perturbation expansion in phase-ordering kinetics: I. Scalar order parameter, *Phys. Rev. E*, 58, 1543, 1998.

61. Mazenko, G.F., Perturbation expansion in phase-ordering kinetics. II. n-vector model, *Phys. Rev. E*, 61, 1088, 2000.

62. Cahn, J.W. and Hilliard, J.E., Free energy of a nonuniform system. I. Interfacial free energy, *J. Chem. Phys.*, 28, 258, 1958.

63. Cahn, J.W. and Hilliard, J.E., Free energy of a nonuniform system. III. Nucleation in a two-component incompressible fluid, *J. Chem. Phys.*, 31, 688, 1959.

64. Puri, S., Bray, A.J., and Lebowitz, J.L., Segregation dynamics in systems with order-parameter dependent mobilities, *Phys. Rev. E*, 56, 758, 1997.

65. Cook, H.E., Brownian motion in spinodal decomposition, *Acta Metall.*, 18, 297, 1970.

66. Nishimori, H. and Onuki, A., Pattern formation in phase-separating alloys with cubic symmetry, *Phys. Rev. B*, 42, 980, 1990.

67. Onuki, A. and Nishimori, H., Anomalously slow domain growth due to a modulus inhomogeneity in phase-separating alloys, *Phys. Rev. B*, 43, 13649, 1991.

68. Langmayr, F., Fratzl, P., Vogl, G., and Miekeley, W., Crossover from omega-phase to alpha-phase precipitation in bcc Ti-Mo, *Phys. Rev. B*, 49, 11759, 1994.

69. Fratzl, P., Penrose, O., and Lebowitz, J.L., Modeling of phase separation in alloys with coherent elastic misfit, *J. Stat. Phys.*, 95, 1429, 1999.

70. Lifshitz, I.M. and Slyozov, V.V., The kinetics of precipitation from supersaturated solid solutions, *J. Phys. Chem. Solids*, 19, 35, 1961.

71. Huse, D.A., Corrections to late-stage behavior in spinodal decomposition: Lifshitz–Slyozov scaling and Monte Carlo simulations, *Phys. Rev. B*, 34, 7845, 1986.

72. Yeung, C., Scaling and the small-wave-vector limit of the form factor in phase-ordering dynamics, *Phys. Rev. Lett.*, 61, 1135, 1988.

73. Furukawa, H., Numerical study of multitime scaling in a solid system undergoing phase separation, *Phys. Rev. B*, 40, 2341, 1989.

74. Rogers, T.M., Elder, K.R., and Desai, R.C., Numerical study of the late stages of spinodal decomposition, *Phys. Rev. B*, 37, 9638, 1988.

75. Amar, J., Sullivan, F., and Mountain, R.D., Monte Carlo study of growth in the two-dimensional spin-exchange kinetic Ising model, *Phys. Rev. B*, 37, 196, 1988.

76. Marko, J.F. and Barkema, G.T., Phase ordering in the Ising model with conserved spin, *Phys. Rev. E*, 52, 2522, 1995.

77. Ohta, T. and Nozaki, H., Scaled correlation function in the late stage of spinodal decomposition, in *Space-Time Organization in Macromolecular Fluids*, Tanaka, F., Doi, M., and Ohta, T., Eds., Springer-Verlag, Berlin, 1989, 51.

78. Tomita, H., A new phenomenological equation of the interface dynamics, *Prog. Theor. Phys.*, 90, 521, 1993.

79. Mazenko, G.F., Growth kinetics for a system with a conserved order parameter, *Phys. Rev. E*, 50, 3485, 1994.

80. Mazenko, G.F. and Wickham, R., Growth kinetics for a system with a conserved order parameter: Off-critical quenches, *Phys. Rev. E*, 51, 2886, 1995.

81. Mazenko, G.F., Spinodal decomposition and the Tomita sum rule, *Phys. Rev. E*, 62, 5967, 2000.

82. Yeung, C., Oono, Y., and Shinozaki, A., Possibilities and limitations of Gaussian-closure approximations for phase-ordering dynamics, *Phys. Rev. E*, 49, 2693, 1994.

83. Puri, S., Bray, A.J., and Rojas, F., Ordering kinetics of conserved XY models, *Phys. Rev. E*, 52, 4699, 1995.

84. Rojas, F., Puri, S., and Bray, A.J., Kinetics of phase ordering in the $O(n)$ model with a conserved order parameter, *J. Phys. A*, 34, 3985, 2001.

85. Coniglio, A. and Zannetti, M., Multiscaling in growth kinetics, *Europhys. Lett.*, 10, 575, 1989.

86. Bray, A.J. and Humayun, K., Scaling and multiscaling in the ordering kinetics of a conserved order parameter, *Phys. Rev. Lett.*, 68, 1559, 1992.

87. Siggia, E.D., Late stages of spinodal decomposition in binary mixtures, *Phys. Rev. A*, 20, 595, 1979.

88. Furukawa, H., Effect of inertia on droplet growth in a fluid, *Phys. Rev. A*, 31, 1103, 1985.

89. Grant, M. and Elder, K.R., Spinodal decomposition in fluids, *Phys. Rev. Lett.*, 82, 14, 1999.

90. Kendon, V.M., Desplat, J. C., Bladon, P., and Cates, M.E., 3D spinodal decomposition in the inertial regime, *Phys. Rev. Lett.*, 83, 576, 1999.

91. Koga, T. and Kawasaki, K., Spinodal decomposition in binary fluids: Effects of hydrodynamic interactions, *Phys. Rev. A*, 44, R817, 1991.

92. Puri, S. and Dünweg, B., Temporally linear domain growth in the segregation of binary fluids, *Phys. Rev. A*, 45, R6977, 1992.

93. Shinozaki, A. and Oono, Y., Asymptotic form factor for spinodal decomposition in three-space, *Phys. Rev. Lett.*, 66, 173, 1991.

94. Shinozaki, A. and Oono, Y., Spinodal decomposition in 3-space, *Phys. Rev. E*, 48, 2622, 1993.

95. Laradji, M., Toxvaerd, S., and Mouritsen, O.G., Molecular dynamics simulation of spinodal decomposition in three-dimensional binary fluids, *Phys. Rev. Lett.*, 77, 2253, 1996.

96. Bastea, S. and Lebowitz, J.L., Spinodal decomposition in binary gases, *Phys. Rev. Lett.*, 78, 3499, 1997.

97. Higuera, F.J., Succi, S., and Benzi, R., Lattice gas dynamics with enhanced collisions, *Europhys. Lett.*, 9, 345, 1989.

98. Swift, M.R., Orlandini, E., Osborn, W.R., and Yeomans, J., Lattice Boltzmann simulations of liquid-gas and binary fluid systems, *Phys. Rev. E*, 54, 5041, 1996.

2 Spinodal Decomposition versus Nucleation and Growth

Kurt Binder

CONTENTS

2.1   INTRODUCTION

In this chapter, we will consider the early stages of the kinetic processes by which the phase transitions caused by sudden changes of external control parameters are effected. As in the first chapter of this book, the generic example is phase separation of a binary (AB) mixture, which is suddenly cooled down from high temperatures (where the mixture is thermodynamically stable at arbitrary compositions) to a temperature beneath the miscibility gap. In the final thermodynamic equilibrium state, one has coexistence between macroscopically large domains of A-rich and B-rich phases, with compositions given exactly according to the two branches of the coexistence curve (see Figure 1.3 in Chapter 1), the relative amounts of the two coexisting phases being given by the lever rule.

Now it is common practice to draw inside of such a miscibility gap an additional dividing line, the spinodal curve (denoted by the dashed curve in Figure 1.3). This spinodal curve touches the coexistence curve at the critical point and is thought to provide a distinction between the kinetic mechanisms that start the phase transformation. Inside the spinodal curve, the initially homogeneous system is thought to be unstable against long wavelength concentration fluctuations. The phase transition is initiated by the spontaneous growth of such fluctuations, and (following Cahn [1]) this mechanism is termed *spinodal decomposition*. In between the spinodal and the coexistence curve, however, a free-energy barrier must be overcome, and this requires formation of a "nucleus" of a new phase; this mechanism is termed *nucleation*.

Analogous phenomena and corresponding *spinodals* (distinguishing between different mechanisms for the transition kinetics) exist for many systems. Classic examples of such transitions are [2] the condensation of a vapor or the magnetization reversal of an anisotropic magnet driven by the magnetic field. The treatment of such phase transitions in the framework of statistical thermodynamics is difficult. Drastic simplifications are usually made, typically invoking mean-field theory, which yields unreliable results. The van der Waals description of the vapor–liquid transition is a typical example. The problem that is emphasized here is not the fact that the singularities of the equation of state near the critical point (critical temperature T_c, pressure p_c, volume V_c) are not correctly predicted. When we consider temperatures $T < T_c$ and follow an isotherm in the (pressure p, volume V) plane, we find the well-known loop that traditionally is eliminated by the Maxwell construction. This elimination is necessary because the part of the isotherm in between the minimum at $V_{sp}^{(1)}(T)$ and the maximum at $V_{sp}^{(2)}(T)$ is characterized by $(\partial V/\partial p)_T > 0$, that is the compressibility of the system is negative and the system is mechanically unstable. Such "unstable states" have no physical meaning in the framework of statistical thermodynamics. However, there is a kind of folklore giving a physical interpretation to the branch of the isotherm in between the branches of the coexistence curve $V_{coex}^{(1)}(T)$, $V_{coex}^{(2)}(T)$, and the spinodal curve defined by the locations of the extrema $V_{sp}^{(1)}(T)$, $V_{sp}^{(2)}(T)$ in the (T, V) plane. The homogeneous "states" of the isotherm with $V_{coex}^{(1)}(T) < V < V_{sp}^{(1)}(T)$ are interpreted as metastable undersaturated liquid. Similarly, the homogeneous states in between $V_{sp}^{(2)}(T) < V < V_{coex}^{(2)}(T)$ are interpreted as metastable supersaturated vapor. The true stable equilibrium state requires a two-phase coexistence, of course. Consequently, the spinodal curve [the locus of the points $V_{sp}^{(1)}(T)$, $V_{sp}^{(2)}(T)$ in the (V, T) plane] gets the meaning of a *limit of metastability*.

This "folklore," however, lacks a firm theoretical basis, and it must be emphasized that this mean-field spinodal curve is NOT the actual limit of metastability. Within mean-field-type theories (such as density-functional theories), the spinodal curve also has the meaning that the free-energy barrier against nucleation vanishes there, $\Delta F^* = 0$. Now it is clear, however, that states near the spinodal that then have a rather low free-energy barrier cannot be reached, because they decay too fast by nucleation and growth. Let us rather take as an extreme limit of metastability a state where the free-energy barrier is $\Delta F^*/k_B T = 10$ [i.e., $\exp(-\Delta F^*/k_B T) \simeq 0.0000454$]. For experiments on nucleation of liquid droplets in a supersaturated vapor, it is in fact more appropriate to take $\Delta F^*/k_B T = 30$, but lower barriers are meaningful for computer simulations, where instantaneous quenches from one state to another are possible, and one can take averages of system properties on the nanosecond timescale. From this discussion, it is evident that the notion of a limit of metastability is somewhat ill-defined, at least beyond mean field: it clearly depends on the observation time, and if the latter tends to infinity, there are no metastable states (and equilibrium statistical mechanics, as a matter of principle, cannot tell anything about metastable states, see [2] and references therein).

It needs also to be emphasized that the contour in the (T, V) plane defined by $\Delta F^*/k_B T = 10$ normally is not close to the mean-field spinodal curve, but lies rather close to the coexistence curve. This is demonstrated here for a very simple example, the lattice gas model on the simple cubic lattice with attractive interactions ε between nearest neighbors. The plot shown in Figure 2.1 presents a temperature-density diagram ($\rho = N/V$ where V is the volume of the lattice, N the particle number, i.e., the number of occupied lattice sites).

Because this system possesses an exact symmetry around the line $\rho = 1/2$, which is the critical density, only the region from $0 \le \rho \le 1/2$ is shown. Note that lengths are measured here in units of the lattice spacing, so ρ is dimensionless. Now molecular field theory for this system yields a wrong coexistence curve. (The critical exponent β describing the shape of the coexistence curve is $\beta = 1/2$ rather than the value appropriate for the universality class [3] for one-component order parameters, $\beta \simeq 0.325$ [4,5]. Also, the critical temperature is overestimated, $k_B T_c^{MF}/\varepsilon = 1.5$ in molecular field approximation rather than [6] $k_B T_c/\varepsilon \simeq 1.126$.) Therefore, we have

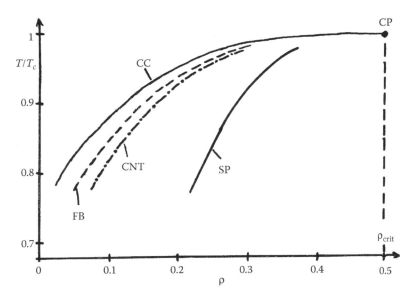

FIGURE 2.1 Part of the coexistence curve (CC) and associated mean-field type spinodal (SP) (cf. text) of the three-dimensional nearest-neighbor Ising-lattice gas model on the simple cubic lattice. In the plane of variables, temperature T (in units of the critical temperature T_c) and density ρ, as well as contours $\Delta F/k_B T = 10$ are drawn, as estimated in Ref. 7 from Monte Carlo simulation (FB, broken curve) and from the classic nucleation theory (CNT, dash-dotted curve), respectively. Because the phase diagram is symmetric around the critical density $\rho_{crit} = 0.5$, only the region to the left of the critical point (CP) is drawn.

not included the equation of the spinodal curve predicted by the molecular field theory

$$\rho_{sp} - \frac{1}{2} = \pm \frac{1}{2} \sqrt{1 - \frac{T}{T_c^{MF}}}. \tag{2.1}$$

Rather, we have used another result of mean-field theory, $\rho_{sp} - 1/2 = (1/\sqrt{3})(\rho_{coex} - 1/2)$, to construct the spinodal curve $\rho_{sp}(T)$ from the coexistence curve $\rho_{coex}(T)$. This follows from the spontaneous magnetization of the Ising model [6] (remember that the lattice gas model is isomorphic to an Ising ferromagnet, if one associates an empty lattice site with an up spin and an occupied lattice site with a down spin, respectively).

Apart from the mean-field spinodal constructed in this way, Figure 2.1 includes two estimates of the contour $\Delta F/k_B T = 10$: one from classic nucleation theory, and the other from a direct Monte Carlo estimation of this free-energy barrier [7]. Both of these estimates are rather close to the coexistence curve and are much more relevant estimates for the regime over which metastable states may occur. In fact, if one imagines a quenching experiment that brings the system instantaneously from a state in the one-phase region (above the coexistence curve) to a state in the two-phase region but somewhat to the left of the spinodal, mean-field theory would predict the kinetics of phase separation proceeds via nucleation and growth, while actually it will proceed via (nonlinear) spinodal decomposition [8]. We emphasize already at this point that for systems with short-range forces, such as a nearest neighbor Ising lattice gas, it is not meaningful to draw a spinodal curve in the phase diagram at all. Later on, it will be discussed to what extent a spinodal is meaningful for systems with a large but finite range of the forces.

2.2 HOMOGENEOUS VERSUS HETEROGENEOUS NUCLEATION

By homogeneous nucleation, one means the process that a droplet of the new (stable) phase is formed from the old (metastable) phase by spontaneous thermal fluctuations [2,9–11]. Such events of homogeneous nucleation often are rather rare, because a high free-energy barrier needs to be overcome. Because the radius R^* of a critical droplet (which has a size that just corresponds with the free-energy barrier) has a nanoscopic linear dimension, direct observation of homogeneous nucleation is very difficult, if at all possible [12,13].

In heterogeneous nucleation, often the free-energy barriers are lower. Consider, for instance, condensation of a liquid from a supersaturated vapor at the walls of a container. The free-energy barrier against droplet formation at the wall surface may be lower than that in the bulk, due to favorable interactions between the fluid atoms in the droplet and the wall [14,15].

However, many basic aspects of homogeneous and heterogeneous nucleation are similar. Hence one cannot understand heterogeneous nucleation if one does not understand homogeneous nucleation for the system considered. In addition, heterogeneous nucleation involves many more parameters: whereas in the simplest case, the nucleation of a liquid droplet from a supersaturated vapor, we know that the average droplet shape is that of a sphere, a liquid droplet at a wall is much more complicated. Assuming that there occurs incomplete wetting [16] of the gas at the wall, a large droplet coexisting with saturated vapor approximately has sphere cap shape. However, even for large droplets we need to know three interfacial energies: $\gamma_{\ell g}$ (the interfacial tension between liquid and gas), $\gamma_{\ell w}$ (the interfacial tension between the liquid and the wall), and γ_{gw} (the interfacial tension between the gas and the wall). Only then do we know the contact angle θ, given by the Young equation [16]

$$\gamma_{gw} - \gamma_{w\ell} = \gamma_{\ell g} \cos \theta. \tag{2.2}$$

This equation is more than 200 years old, but the problem is to predict the relevant interfacial energies reliably. A more fundamental problem also is that results like the Young equation are only true asymptotically, for very large droplets. For not-so-large droplets, corrections come into play due to additional parameters, such as the line tension, an extra free energy contribution due to the three-phase contact line gas–liquid–wall [17]. And the most severe problem, of course, is that all these quasi-macroscopic considerations break down on the nanoscale: it then becomes questionable to separate the free energy of such a nanodroplet into bulk and interfacial contributions. Remember, at the nanoscale interfaces are no longer sharp, but also are spread out over some width [17]. It is not so clear where the interfacial region stops and where the bulk begins. This problem is clearly evident in atomistic simulations, for example, of point-like atoms interacting with Lennard–Jones forces [18]. Irregularities in the shape of such nanodroplets occur, and the surface of the droplet undergoes large statistical fluctuations [18].

When atoms in the gas are close to the droplet surface, there is also ambiguity to decide whether or not such atoms still belong to the gas, or rather to the wings of the liquid–vapor interfacial profile. Although such statistical fluctuations may be less pronounced for other nucleation problems, such as the nucleation of nanocrystals, then there is the additional difficulty that for crystals the surface free energy depends on the crystal surface orientation, and highly nontrivial crystal shapes may result. This is already true for homogeneous nucleation in the bulk, but even more so for heterogeneous nucleation at surfaces. In fact, a classic 2000-page handbook of metallurgy [19] contains no more than 5 pages on heterogeneous nucleation, and some of the key statements are "Heterogeneous nucleation is intellectually attractive, but most of the results do not seem to follow the predictions ... In most cases the validity of the experiments can be questioned ... Thus, at present the first steps in the comprehension of the heterogeneous nucleation phenomena have been taken, ..., but the understanding of the process is still poor."

In view of these problems, the discussion of nucleation in the current chapter will concentrate almost exclusively on homogeneous nucleation. However, here also one has to be very careful: the quoted reference [19] correctly states on homogeneous nucleation that "experiments on nucleation from the melt present as a principal problem the difficulty of removing unknown impurity particles which act as heterogeneous nuclei." In view of the latter problem, the chapter will emphasize tests of the theory by computer simulations, because there we can be absolutely sure that no unknown impurity particles are present. In addition, we will focus on liquid–gas nucleation (or the equivalent problem of nucleation from binary liquid mixtures with a miscibility gap) and even use the simplistic lattice gas model again, because the latter really can provide stringent tests.

2.3 CLASSIC THEORY OF HOMOGENEOUS NUCLEATION: STATIC DESCRIPTION

We wish to achieve an estimation of the free-energy barrier ΔF^* of a critical droplet, treating in parallel the nucleation of a liquid droplet from the gas, and the nucleation of a B-rich droplet from an A-rich background in the case of a binary (AB) liquid mixture, as these systems possess a qualitatively similar phase diagram [2].

Nucleation occurs when we enter the metastable region of the phase diagram, performing at time $t = 0$ a sudden cooling from an initial state in the one-phase region to the state X in the metastable region of the phase diagram (at a temperature distance δT from the coexistence curve, while the distance of the temperature of the considered state temperature T from the critical temperature T_c is denoted as ΔT). We assume that $\Delta T/T_c \ll 1$, as then droplets are sufficiently large to justify a macroscopic description, and also $\delta T/\Delta T \ll 1$ to ensure that we study nucleation rather than spinodal decomposition.

Experimental phase diagrams of such systems would typically include the spinodal curves; but, as has already been emphasized above, spinodal curves as well as the hypothetical free energy F' of one-phase states in the two-phase region are an ill-defined mean-field concept. This concept is not needed at this point and hence avoided in our discussion. We need to come back to this problem, when we discuss the idea due to Cahn and Hilliard [20] that the spinodal curve is useful for distinguishing two mechanisms for the initial stages of phase separation, namely nucleation and growth on the one side of the spinodal, and spinodal decomposition (spontaneous amplification of long-wavelength fluctuations of density or concentration, respectively) on the other [8]. However, when one treats homogeneous nucleation near the coexistence curve, no knowledge that a spinodal curve might exist (or not exist) is invoked.

The conventional ("classic") nucleation theory [2,9–11] is based on the following drastic approximations: It is assumed that nucleation proceeds via the formation of spherical droplets, and these droplets are treated as macroscopic objects, splitting $\Delta F(R)$ into bulk and surface terms:

$$\Delta F(R) = \Delta g \left(\frac{4}{3} \pi R^3 \right) + \gamma_{g\ell} (4\pi R^2). \tag{2.3}$$

For the interfacial free energy $\gamma_{g\ell}$ per unit area, one takes the same result as for a flat planar interface (*capillarity approximation* [2,9–11]). For liquid–vapor nucleation, the difference in thermodynamic potential Δg between the liquid drop and surrounding supersaturated vapor is nothing but the pressure difference Δp. For a state near the coexistence curve we have

$$p_{gas} = p_{coex} + \left(\frac{\partial p}{\partial \mu}\right)_T (\mu - \mu_{coex})$$

$$= p_{coex} + \rho_{gas}(\mu - \mu_{coex}), \tag{2.4}$$

where p_{coex} is the pressure at the coexistence curve, μ the chemical potential, and μ_{coex} its value at the coexistence curve. Similarly,

$$p_{liquid} = p_{coex} + \rho_{liquid}(\mu - \mu_{coex}), \tag{2.5}$$

and hence

$$\Delta p = -(\rho_{liquid} - \rho_{gas})\delta\mu, \quad \delta\mu \equiv \mu - \mu_{coex}. \tag{2.6}$$

Analogously, for a liquid binary mixture we have

$$\Delta g = -\left(c_{coex}^{II} - c_{coex}^{I}\right)\delta\tilde{\mu}, \quad \delta\tilde{\mu} = \Delta\mu - \Delta\mu_{coex}, \tag{2.7}$$

where $\Delta\mu$ is the chemical potential difference between the two species A, B, and c_{coex}^{I}, c_{coex}^{II} denote the concentrations of species B in the binary mixture AB at the coexistence curve.

The resulting free energy $\Delta F(R)$ has a maximum at $R = R^*$,

$$R^* = \frac{2\gamma_{g\ell}}{[(\rho_\ell - \rho_g)\delta\mu]}, \tag{2.8}$$

and the height of this maximum diverges when the coexistence curve is approached,

$$\Delta F^* = \frac{16\pi}{3} \frac{\gamma_{g\ell}^3}{[(\rho_\ell - \rho_g)\delta\mu]^2} \propto (\rho_1 - \rho_g)^{-2}, \tag{2.9}$$

because for a density ρ_1 of the supersaturated gas near the coexistence density ρ_g one has $\rho_1 - \rho_g \propto \delta\mu$.

According to this formula, ΔF^* decreases monotonically with increasing density ρ_1, but there is no supersaturation at which ΔF^* would vanish strictly. No assumption on the existence of a spinodal curve was built into classic nucleation theory, which should be used very close to the coexistence curve, however. After all, we have considered a linear expansion of the thermodynamic potential around phase coexistence only.

One then assumes an Arrhenius formula to describe the nucleation rate, the number of nuclei formed per unit volume and unit time,

$$J = v^* \exp(-\Delta F^*/k_B T), \tag{2.10}$$

v^* being some attempt frequency (all the information on the actual kinetics of the process is "hidden" here).

When one tries to estimate v^* for gas–liquid nucleation, one finds that cases of practical interest (say, one nucleus being formed per cm^3-second) correspond with about $\Delta F^* \simeq$ 30–40 $k_B T$. This large number also means that one needs to know the quantities that go into the calculation of ΔF^* rather precisely. For example, a 10% error of $\gamma_{g\ell}$ implies a 30% error of ΔF^*. If we get 30 or 50 instead of 40 for $\Delta F^*/(k_B T)$, this implies a difference of $\exp(\pm 10)$ for the prediction of the nucleation rate, that is, many orders of magnitude.

However, because at the temperatures of interest, not close to the gas–liquid critical point, the "critical droplet" (which has the radius R^*) typically contains of the order 100 atoms or molecules, a 10% deviation of the surface tension of such a small droplet from its value for a flat interface between coexisting bulk phases cannot be a surprise at all. For instance, a useful phenomenological expression for the interfacial free energy $\gamma_{g\ell}(R)$ of a small droplet of radius R is as follows:

$$\gamma_{g\ell}(R) = \frac{\gamma_{g\ell}(\infty)}{(1 + 2\delta/R)},$$

δ being the so-called Tolman length [21]. However, there is not really a sound basis in statistical mechanics for this formula, and no way to predict the value of this length from first principles.

Of course, there are also great uncertainties to accurately estimate the pre-exponential factor v^*. And these difficulties proliferate when one considers more complicated cases, as occur for heterogeneous nucleation. For instance, the most well-known example is the nucleation of water droplets around charged particles, of charge q and radius r_0 in the atmosphere: this problem was already considered by Thomson (Lord Kelvin) as early as 1870! He suggested that one should add a term to the droplet free energy [22]:

$$\Delta F_{\text{charge}} = \frac{1}{2}q^2 \left(1 - \frac{1}{\varepsilon}\right) \left(\frac{1}{R} - \frac{1}{r_0}\right), \qquad (2.11)$$

where ε is the dielectric constant of the fluid. While for uncharged droplets the relation between $\delta\mu$ and R^* is simply monotonically decreasing, $\delta\mu \propto 1/R^*$, one now obtains a non-monotonic behavior when Equation 2.11 is included. However, despite much experimental effort, the situation regarding the validity of the description based on Equations 2.3, 2.5, and 2.11 remained controversial [22]. And similar difficulties concern the experimental verification of predictions concerning nucleation at dislocation lines in solids, where a strain energy term needs to be included [23]. In the following, we shall not consider these complicated problems of heterogeneous nucleation at various kinds of defects.

2.4 KINETICS OF NUCLEATION AND GROWTH

Already in the early days of nucleation theory [24], a kinetic atomistic interpretation of homogeneous nucleation was introduced in terms of the growth and shrinking of clusters containing ℓ atoms (or molecules, respectively) by condensation or evaporation of single atoms. Denoting the number of clusters of *size* ℓ at time t per unit volume as $\bar{n}_\ell(t)$, in terms of rate constants for condensation (a_ℓ) and evaporation (b_ℓ) one can write down the following set of equations

$$\frac{d}{dt}\bar{n}_\ell(t) = -a_\ell\bar{n}_\ell - b_\ell\bar{n}_\ell + a_{\ell-1}\bar{n}_{\ell-1} + b_{\ell+1}\bar{n}_{\ell+1}, \quad \ell \geq 2, \tag{2.12}$$

$$\frac{d}{dt}\bar{n}_1(t) = -a_1\bar{n}_1 + b_2\bar{n}_2. \tag{2.13}$$

Note that the last two terms on the right-hand side of Equation 2.12 describe the gain resulting at size ℓ due to a condensation at size $\ell - 1$ or an evaporation at size $\ell + 1$, respectively. However, solving this set of equations exactly is a difficult problem [25], and hence we shall describe here only a simple approximate treatment.

A hypothetical steady-state cluster concentration n_ℓ results as the solution of Equation 2.12 when we require that $d\bar{n}_\ell/dt = 0$ for all ℓ. Then we must have as a result $a_{\ell-1}\,n_{\ell-1} = b_\ell n_\ell$, that is, b_ℓ can be eliminated in favor of a_ℓ and n_ℓ. Transforming from finite differences to differentials, $\bar{n}_{\ell+1} = \bar{n}_\ell + (\partial\bar{n}_\ell/\partial\ell) + \cdots$, one can formulate the description of nucleation kinetics as a continuity equation for the *current* J_ℓ in the (one-dimensional) cluster size space $\{\ell\}$:

$$\frac{\partial\bar{n}_\ell}{\partial t} + \frac{\partial J_\ell}{\partial\ell} = 0. \tag{2.14}$$

Here, the current $J_\ell(t)$ contains a diffusive term and a drift term,

$$J_\ell(t) = -a_\ell\left(\frac{\partial\bar{n}_\ell(t)}{\partial\ell}\right) + \left(a_\ell\frac{\partial\ell n(n_\ell)}{\partial\ell}\right)\bar{n}_\ell(t). \tag{2.15}$$

Writing $n_\ell = \exp(-F_\ell/k_BT)$, F_ℓ being essentially the same free energy $F(R)$ as considered before, the critical radius R^* corresponds with the critical cluster size ℓ^*. We recognize (see Figure 2.2) that the drift term changes sign at $\ell = \ell^*$: for $\ell < \ell^*$, the drift acts against the diffusion, therefore subcritical clusters tend to disintegrate again, whereas for $\ell > \ell^*$ the drift acts in the same direction as the diffusion, that is, supercritical clusters will grow and hence create the new phase. Now one neglects the depletion of monomers, because then the problem is simply solved in terms of a steady-state solution $J_\ell(t) = J$, independent of ℓ. J is nothing but the nucleation rate,

$$J = \left[\int_0^\infty (a_\ell n_\ell)^{-1}\,d\ell\right]^{-1}, \quad \frac{\bar{n}_\ell(t)}{n_\ell} = J\int_\ell^\infty (a_\ell n_\ell)^{-1}\,d\ell. \tag{2.16}$$

Figure 2.2a suggests expanding the free energy F_ℓ quadratically at ℓ^* as

$$F_\ell = F_{\ell^*} - \frac{1}{2}g\left(\ell - \ell^*\right)^2 + \cdots, \tag{2.17}$$

and approximating $a_\ell = a_{\ell^*}$, one obtains an explicit formula for the nucleation rate:

$$J \simeq a_{\ell^*}\sqrt{g}\exp\left[-F(\ell^*)/k_BT\right]. \tag{2.18}$$

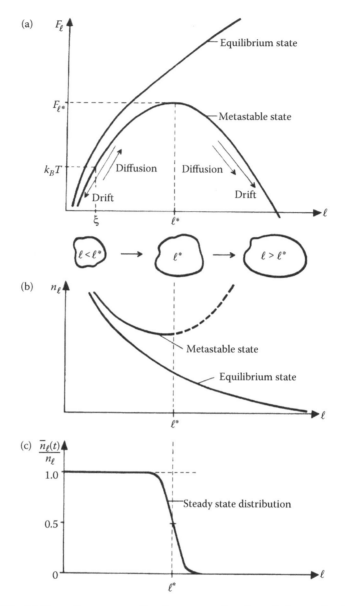

FIGURE 2.2 (a) Formation energy of a cluster as a function of cluster size (ℓ) and the associated mechanism of cluster growth over the critical size ℓ^*. (b) Cluster concentration n_ℓ for metastable and equilibrium states. (c) Nonequilibrium cluster distribution in a steady-state nucleation process (schematic). (From Binder, K., *Fluctuations, Instabilities, and Phase Transitions*, Riste, T., Ed. Plenum Press, New York, 1975, 53. With permission.)

Notice that a_{ℓ^*} is still a kinetic factor (cf. Equation 2.10), describing the probability that a liquid droplet grows from size ℓ^* to size $\ell^* + 1$ by incorporating a gas atom. However, the factor $\sqrt{g}$ is a static factor: it describes the curvature of the potential barrier over which the flow of clusters has to go. Clearly, when the maximum is very broad (i.e., when g is small), it takes more time to diffuse from the left "border" of the maximum to the right "border" (in a regime of order $\delta F = k_B T$ around $F(\ell^*)$, the motion in the cluster size space is essentially diffusive). This factor is called the *Zeldovich factor* [2,9–11].

Another correction, which becomes particularly important near the critical point of the fluid, concerns the kinetic prefactor a_{ℓ^*}: rather than assuming that only single atoms one by one are incorporated into a large cluster, one can consider *cluster reactions* such that a cluster of size ℓ' is incorporated into a cluster of size ℓ to yield a cluster of size $\ell + \ell'$. Instead of the coefficients a_ℓ, one obtains a *cluster reaction matrix* $W(\ell, \ell')$ [27]. However, one can show that for large enough ℓ^*, Equations 2.14 through 2.18 remain valid, if a_ℓ is replaced by the *cluster reaction rate* R_ℓ defined as the second moment of $W(\ell, \ell')$,

$$R_\ell = \sum_{\ell'} W(\ell, \ell')(\ell')^2 / n_\ell. \tag{2.19}$$

This description was tested [27] by Monte Carlo simulations for a two-dimensional lattice gas model, at an inverse reduced temperature $J/k_B T = 0.46$, close to the critical temperature ($J/k_B T_c \simeq 0.44$, where J is the exchange constant of the nearest neighbor Ising square lattice). Figure 2.3 shows that, using incorporation of monomers only, one would underestimate the nucleation rate by a factor of 4; thus the correction due to Equation 2.19 clearly is not very important. On the other hand, Equation 2.3 is rather inaccurate here {remember that the interfacial tension $\gamma_{g\ell}$ vanishes at the critical point, so the result of the classic theory $n_\ell \propto \exp[-(36\pi)^{1/3}\ell^{2/3}\gamma_{g\ell}/k_B T]$ does not make any sense right at T_c, and a model involving a power-law correction due to Fisher [28], $n_\ell \propto \ell^{-\tau}$ where τ is related to critical exponents, works much better}. It turns out that using the reaction rate R_ℓ and cluster size distribution n_ℓ, as observed in the simulations, as an input in the theoretical formulas, Equation 2.16, one can predict the steady-state cluster-size distribution $\bar{n}_\ell(t)/n_\ell$ in good agreement with the direct observation from the simulation [27]. This shows that nucleation theory is rather good, even for relatively small droplets, if the droplet free energy F_ℓ and droplet growth rate R_ℓ are known well enough.

However, there are many cases where one needs to explicitly include more cluster properties than merely their size. For example, one may wish to allow explicitly for different shapes other than spherical clusters. Formally this can be done by generalizing the scalar variable ℓ to a whole vector $\vec{\ell}$. Then the problem of nucleation no longer amounts to a flow over a simple maximum in a one-dimensional cluster size space, but rather to a flow over a saddle point in a multi-dimensional space. Mathematically the problem resembles Maxwell's equations of electrodynamics [10]:

$$\vec{\nabla} \cdot \vec{j}(\vec{\ell}) = 0,$$

$$\vec{E}(\vec{\ell}) = -\vec{\nabla}\phi,$$

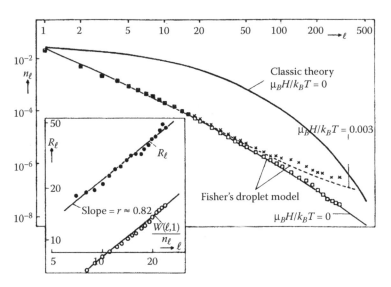

FIGURE 2.3 Cluster concentration n_ℓ versus ℓ for the nearest-neighbor Ising ferromagnet on the square lattice at $J/k_BT = 0.46$ (points) at two fields, compared with Fisher's droplet model [28] and classic nucleation theory (using the exactly known interfacial tension of the Ising model). The inset shows the cluster reaction rate R_ℓ and the reaction rate due to monomers only [denoted as $W(\ell, 1)/n_\ell$]. (From Binder, K. and Müller-Krumbhaar, H., *Phys. Rev. B, 9*, 2194, 1974. With permission.)

$$\phi = -\bar{n}\left(\vec{\ell}\right)/n\left(\vec{\ell}\right), \tag{2.20}$$

$$\vec{\nabla} \times \vec{E}\left(\vec{\ell}\right) = 0,$$

$$n\left(\vec{\ell}\right) = \exp\left[-F\left(\vec{\ell}\right)/k_BT\right]. \tag{2.21}$$

The nucleation rate is then the total cluster current at large distances from the origin,

$$J = \oint \vec{j}\left(\vec{\ell}\right) \cdot d\vec{f}. \tag{2.22}$$

Expanding then again quadratically around the saddle point $\vec{\ell}^*$ one finds, in addition to the analog of the Zeldovich factor, a correction proportional to the effective cross-sectional area of the saddle point region [10,29] (Figure 2.4). Such a more-dimensional nucleation theory in fact is indispensable, when one considers nucleation in systems with several chemical components, for example, gas–liquid nucleation in binary mixtures [30]. A well-known example is also the nucleation of voids in metals under radioactive irradiation: the voids are actually filled by helium gas, and both the volume of a void and the amount of helium gas filling it (as a consequence of the radiation damage) are required for a minimal description of the problem. This is also an example, where the large disparity between kinetic prefactors may cause the

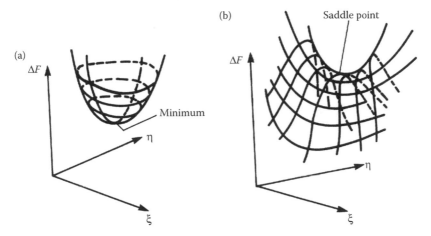

FIGURE 2.4 Free-energy functional ΔF schematically displayed as a function of two phase space coordinates, η and ξ; (a) shows a minimum (corresponding with a stable or metastable homogeneous phase), and (b) exhibits a saddle point configuration, representing a metastable phase plus one droplet (here ξ and η may actually represent coordinates of the droplet). (From Binder, K., *Phys. Rev. A*, 29, 341, 1984. With permission.)

dominant part of the nucleation current not to pass over near the saddle points but rather in a region higher up at the ridge [30,31].

Finally, we emphasize that most experiments on nucleation are unable to directly measure the nucleation rate itself, because the nuclei formed are too small and too rare to be detected at sizes close to the critical sizes. One can infer the nucleation rate only indirectly from observations at much later time, when the nuclei have triggered the transformation of the metastable phase to the stable one to an appreciable extent. This consideration prompts us to consider the combined effect of nucleation and growth, and we ask what is the *completion time* of a phase transformation [32]. In an experiment where from a supersaturated vapor some liquid condenses, we ask the question what is the time τ_{cond} after which an appreciable fraction of the metastable supersaturated phase has condensed? A simple argument to answer this question introduces the radius $R(t - t')$ of a droplet at time t that was nucleated at time t'. Then the fraction of liquid that has condensed is

$$\frac{\delta\rho(t)}{(\rho_{liquid} - \rho_{gas})} = \int_0^t dt' J(t') \frac{4}{3}\pi[R(t - t')]^3. \tag{2.23}$$

Neglecting both the *time lag* [10] necessary to reach steady-state nucleation and the decrease of supersaturation during the phase transformation, one has $J(t) \simeq J =$ const. and

$$R^2(t - t') = R^{*2} + \text{const.} \left(\frac{\delta T}{\Delta T}\right) D(t - t'), \tag{2.24}$$

where D is the diffusion constant, which scales as $D \propto (\Delta T)^{\nu}$ near T_c, ν being the critical exponent of the correlation length [3,33]. One thus finds

$$\frac{\delta\rho(t)}{\left(\rho_{\text{liquid}} - \rho_{\text{gas}}\right)} \propto J \left(\frac{\delta T}{\Delta T} D\right)^{3/2} t^{5/2} = \left(\frac{t}{\tau_{\text{cond}}}\right)^{5/2}. \qquad (2.25)$$

The condensation time is thus defined by the condition that in this law, the transformed fraction is of order unity. This yields a result compatible with dynamic scaling [10,33]

$$\tau_{\text{cond}} = (\Delta T)^{3\nu} \left[\left(\frac{\delta T}{\Delta T}\right)^{3/2} \tilde{J}\left(\frac{\delta T}{\Delta T}\right)\right]^{2/5}, \quad J = (\Delta T)^{2+3\nu} \tilde{J}\left(\frac{\delta T}{\Delta T}\right), \qquad (2.26)$$

where we have made use of the idea [10] that the nucleation rate near the critical point is a scaled function of the reduced undercooling. Using, for a rough estimation, classic nucleation theory to calculate $\tilde{J}$, and asking what is the reduced undercooling such that $\tau_{\text{cond}} = 1$ sec, one finds a critical enhancement of undercooling (see Figure 2.5), due to the critical slowing down [the factor $(\Delta T)^{3\nu}$ in Equation 2.26]. This rough theory explains qualitatively the critical enhancement of metastability seen in the experiments, which for a long time has been a great puzzle. For a more accurate

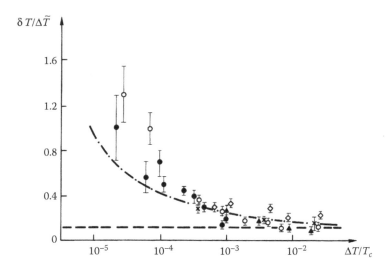

FIGURE 2.5 Reduced supercooling $\delta T/\Delta\tilde{T} = \delta T/|T_c - T - \delta T|$ plotted against the relative distance from the critical point, $\Delta T/T_c = |T_c - T|/T_c$, for binary mixtures, C_7F_{14}-C_7H_{14} (open diamonds); 2,6 lutidine-water (full and open circles); and simple fluids, ^{3}He (crosses) and CO_2 (triangles). The horizontal broken line is the result of the classic nucleation theory, with parameters relevant to CO_2. The dash-dotted curve is the theoretical result of Binder and Stauffer [10], that is, Equation 2.26 calculated for a completion time $\tau_{\text{cond}} = 1$ s. The experimental data in this picture were taken from Goldburg [34]. (From Binder, K., *Rep. Progr. Phys.*, 50, 783, 1987. With permission.)

description, the decrease of supersaturation during the growth of the nucleated droplets needs to be taken into account, requiring extensive numerical calculations [35]. A similar calculation is also very well able to account for the lifetime of metastable states in the two-dimensional Ising model [27].

We have emphasized the case of nucleation near a bulk critical point, because this is the rare case in which one has a known nontrivial control parameter, namely the distance ΔT from T_c. In addition, near the critical point the characteristic correlation length of density fluctuations diverges $\left[\xi \propto (\Delta T)^{-\nu}\right]$, also the droplets get very large and then a universal quasi-macroscopic rather than a system-specific atomistic description is better justified.

Nucleation theory has also been extensively studied for another limiting case, namely nucleation of small clusters of overturned spins in an Ising ferromagnet at very low temperatures [36–38]. In this limit, where the critical cluster size is very small (monomers in the extreme case, or dimers, trimers, quadrumers, etc.), special techniques become available [37], and the discrete spectrum of cluster sizes and shapes leads to intricate phenomena [38], which are, however, outside the scope of this chapter. We also shall not discuss the many attempts (e.g., [39,40]) to make the phenomenological theory of nucleation more rigorous, nor the many studies of nucleation barriers of diverse systems with designated computer simulation techniques (see e.g., [41] and references therein).

2.5 THE LINEARIZED THEORY OF SPINODAL DECOMPOSITION

We now consider a quenching experiment (or simulation) that either brings a simple fluid or a binary mixture rapidly (instantaneously in the ideal case) from a state in the one-phase region to a state underneath the spinodal curve in the phase diagram describing the coexistence between gas and liquid (for the simple fluid) or between unmixed A-rich and B-rich phases (for the binary mixture), respectively (see Figure 1.3). We have mentioned that the existence of a spinodal curve in the phase diagram is doubtful, as this is a somewhat ill-defined mean-field concept. Nevertheless the related mean-field concept describing the decay of the *unstable states* that occur within mean-field theory underneath the spinodal is useful as a starting point for more refined descriptions, as will be seen below.

In the fluid, we are interested in describing the time evolution of the local density $\rho(\vec{r}, t)$, $\vec{r}$ denoting the spatial coordinate, and t the time after the quench. In the binary mixture, the corresponding quantity is the local (relative) concentration $c(\vec{r}, t)$ of one species, say A. In order to treat both systems on the same footing, we describe them by the order parameter field $\phi(\vec{r}, t)$, which is taken relative to the density ρ_{crit} or concentration c_{crit} at the critical point. Subsequently, we derive the Cahn–Hilliard (CH) equation obeyed by $\phi(\vec{r}, t)$, in parallel with the discussion in Section 1.5.1.

In such systems, the average order parameter $\bar{\phi}$ obtained by integration over the total volume V is a conserved quantity, independent of time

$$\bar{\phi} = V^{-1} \int d\vec{r}\phi(\vec{r}, t). \tag{2.27}$$

A consequence of this conservation law is the existence of a continuity equation (cf. Equation 1.174), linking the change of $\phi(\vec{r}, t)$ with time to a current density $\vec{J}$,

$$\frac{\partial}{\partial t}\phi(\vec{r}, t) + \vec{\nabla} \cdot \vec{J}(\vec{r}, t) = 0. \tag{2.28}$$

The basic assumption that we now make is that $\phi(\vec{r}, t)$ is the only slowly varying field that matters for our problem: this assumption is well justified for a solid binary mixture, or a lattice gas fluid that physically could be realized in an adsorbed layer on a surface where particles can "sit" only on the preferred sites of a lattice created by the corrugation potential of the crystal surface. It is not justified for fluid systems, of course, where the velocity field $\vec{v}(\vec{r}, t)$ also needs to be taken into account. This complication was discussed in Section 1.5.2.

The current density $\vec{J}$ is proportional to the gradient of the local chemical potential (or chemical potential difference, respectively), as described by Equation 1.175. This equation is not even exact for solid mixtures, as it is a deterministic equation, neglecting statistical fluctuations. Further, it is supposed to hold only for very small deviations from equilibrium.

The chemical potential is obtained as a functional derivative (cf. Equation 1.176) of the free-energy functional $F\left[\phi(\vec{r})\right]$:

$$F\left[\phi(\vec{r})\right] = \int d\vec{r} \left[f(\phi) + \frac{1}{2d}\left(r\vec{\nabla}\phi\right)^2\right], \tag{2.29}$$

where $f(\phi)$ is the (coarse-grained) free-energy density, and d is the dimensionality of the system. Here the coefficient r in the gradient term can be interpreted [2,42–44] as the range of (pairwise) interactions (between molecules, when we describe the vapor to liquid transition of a fluid; between spins on a lattice, when we describe an Ising ferromagnet; etc.). The $[r(\vec{\nabla}\phi)]^2$ term represents the free-energy cost due to order parameter inhomogeneities, within mean-field theory [2,42–44]. Typically, the shape of $f(\phi)$ versus ϕ for temperatures below T_c has the familiar double-well form, where the two minima describe the two coexisting phases, and the inflection points of the $f(\phi)$ versus ϕ curve describe the (mean-field) spinodals—we will come back to a discussion of $f(\phi)$ in Section 2.9. Of course, Equation 2.29 should be viewed as an expansion in powers of $\vec{\nabla}\phi$, and hence requires a slow variation of $\phi(\vec{r})$ with $\vec{r}$, because it keeps only the first term in this expansion.

Using Equation 1.176, we obtain the chemical potential as

$$\mu(\vec{r}, t) = \left(\frac{\partial f(\phi)}{\partial \phi}\right)_T - \frac{1}{d}r^2\nabla^2\phi(\vec{r}, t). \tag{2.30}$$

This yields the CH equation (cf. Equation 1.177) [1,42]

$$\frac{\partial}{\partial t}\phi(\vec{r}, t) = D\nabla^2\left[\left(\frac{\partial f(\phi)}{\partial \phi}\right)_T - \frac{1}{d}r^2\nabla^2\phi(\vec{r}, t)\right]. \tag{2.31}$$

Because $f'(\phi)$ is highly nonlinear, Equation 2.31 is not analytically soluble. In the initial stages of unmixing, $\phi(\vec{r}, t)$ everywhere in the system is very close to its average value $\bar{\phi}$ (Equation 2.27). Then it makes sense to expand Equation 2.31 linearly in the deviation of the local order parameter from its average, $\delta\phi(\vec{r}, t) = \phi(\vec{r}, t) - \bar{\phi}$,

$$\frac{\partial}{\partial t}\delta\phi(\vec{r}, t) = D\nabla^2\left[\left(\frac{\partial^2 f(\phi)}{\partial\phi^2}\right)_T\Bigg|_{\phi=\bar{\phi}} - \frac{1}{d}r^2\nabla^2\right]\delta\phi(\vec{r}, t). \qquad (2.32)$$

The linearity of Equation 2.32 allows solution by Fourier transformation. Defining

$$\delta\phi_k(t) = \int d\vec{r}\, e^{i\vec{k}\cdot\vec{r}}\,\delta\phi(\vec{r}, t), \qquad (2.33)$$

one readily finds from Equation 2.32

$$\delta\phi_k(t) = \delta\phi_k(0)\exp[\omega(k)t], \qquad (2.34)$$

where the amplification factor $\omega(k)$ is

$$\omega(k) = -Dk^2\left[\left(\frac{\partial^2 f(\phi)}{\partial\phi^2}\right)_T\Bigg|_{\phi=\bar{\phi}} + \frac{r^2k^2}{d}\right]. \qquad (2.35)$$

Note that inside the spinodal, the second derivative of the (coarse-grained) free energy $f(\phi)$ is negative. Hence for small enough k (where the term r^2k^2/d does not yet outweigh the first term in the square bracket in Equation 2.35), $\omega(k)$ is positive: indeed fluctuations $\delta\phi_k(0)$ present in the initial state get amplified, they grow exponentially in time as long as the linearization approximation is still valid.

Such a growth of fluctuations can be detected experimentally by scattering of neutrons, X-rays, or light, which yield the equal-time structure factor $S(k, t)$, defined in Equations 1.88 and 1.89. This structure factor is just the mean square thermal average of these fluctuations,

$$S(k, t) = \left\langle|\delta\phi_k(t)|^2\right\rangle_T = \left\langle|\delta\phi_k(0)|^2\right\rangle_T\exp[2\omega(k)t], \qquad (2.36)$$

where the prefactor

$$\left\langle|\delta\phi_k(0)|^2\right\rangle_T = \langle\phi_k\phi_{-k}\rangle_{T_0} = S_{T_0}(k) \qquad (2.37)$$

is the equilibrium static structure factor in the state at the temperature T_0 in the one-phase region, where the (instantaneous) quench to the state of temperature T inside the coexistence curve was started.

It is clear that $\omega(k)$ is positive only for $0 < k < k_c$, where the critical wave-vector k_c is given by

$$k_c = \frac{2\pi}{\lambda_c} = \left[-\left(\frac{\partial^2 f(\phi)}{\partial\phi^2}\right)_T\Bigg|_{\phi=\bar{\phi}}\frac{d}{r^2}\right]^{1/2}. \qquad (2.38)$$

Thus, fluctuations with wavelengths λ larger than the critical wavelength λ_c defined in Equation 2.38 grow with the time t after the quench, whereas fluctuations with wavelengths $\lambda < \lambda_c$ should decay, during the initial stages. For $k = k_c$, one should have $S(k_c, t) = S(k_c, 0) = \text{const.}$, that is, plotting $S(k, t)$ vs. k at different times, one should observe a time-independent intersection point, from which one could read off the value of k_c. We also note from Equation 2.35 that $\omega(k)$ has a maximum at $k_{\max} = k_c/\sqrt{2}$.

However, such a time-independent intersection point has not been observed in experiments or simulations [8]. The reason for this failure of the simplified Cahn [1] theory of spinodal decomposition is readily recognized, the treatment presented in Equations 2.27 through 2.38 misses the effect of thermal fluctuations in the state to which the quench leads. This defect is remedied by adding a random force term $\vec{\theta}(\vec{r}, t)$ to the current in Equation 1.178, or a noise term $\eta(\vec{r}, t)$ on the right-hand side of Equation 2.31 [45]. Here $\eta(\vec{r}, t)$ is assumed to be delta-correlated Gaussian noise, and the mean-square amplitude obeys the fluctuation-dissipation relation (cf. Equation 1.179)

$$\langle \eta(\vec{r}', t')\eta(\vec{r}'', t'')\rangle_T = 2DT\nabla^2\delta(\vec{r}' - \vec{r}'')\delta(t' - t''). \tag{2.39}$$

One then can still work out the linearization approximation, in analogy with Equations 2.32 through 2.38. It turns out that one can cast the result in a simple linear differential equation for $S(k, t)$, namely [45]

$$\frac{d}{dt}S(k.t) = -2Dk^2\left\{\left[\left(\frac{\partial^2 f(\phi)}{\partial\phi^2}\right)_T\bigg|_{\phi=\bar{\phi}} + \frac{r^2k^2}{d}\right]S(k,t) - 1\right\}. \tag{2.40}$$

For $k > k_c$, the square bracket is always positive, and the $S(k, t)$ relaxes to an Ornstein–Zernike form, as in a quench leading from one equilibrium state to another equilibrium state,

$$S(k, \infty) = \left[\left(\frac{\partial^2 f(\phi)}{\partial\phi^2}\right)_T\bigg|_{\phi=\bar{\phi}} + \frac{r^2k^2}{d}\right]^{-1}. \tag{2.41}$$

For $k < k_c$, however, it is useful to define an effective diffusion constant for *uphill diffusion* as follows [8]:

$$D_{\text{eff}}(k, t) \equiv \frac{1}{2k^2}\frac{d}{dt}\ln S(k, t), \tag{2.42}$$

with

$$D_{\text{eff}}(k, t) = -D_0\left(1 - \frac{k^2}{k_c^2}\right) + \frac{D}{S(k, t)}, \tag{2.43}$$

$$D_0 = -D\left(\frac{\partial^2 f(\phi)}{\partial\phi^2}\right)_T\bigg|_{\phi=\bar{\phi}}. \tag{2.44}$$

If the linearized theory would hold for all times, one would still observe an exponential growth of fluctuations at late times, when the last term on the right side of Equation 2.43 can be neglected, and hence $S(k, t) \simeq S(k, 0) \exp\left[D_0 k^2 \left(1 - k^2/k_c^2\right)t\right]$ for $0 < k < k_c$. However, such an exponential growth of fluctuations is hardly ever observed, as will be discussed below. If one nevertheless analyzes the structure factor $S(k, t)$ via Equation 2.42, one would expect a linear variation of $D_{\text{eff}}(k, t)$ with k^2 (*Cahn plot* [8]). This linear variation of the Cahn plot, which would yield information on k_c and hence $\left(\partial^2 f/\partial \phi^2\right)_T$, is also hardly ever observed.

The reason why these predictions almost always fail is the neglect of the nonlinear terms in $\delta\phi$ that were neglected when one approximates Equation 2.31 by Equation 2.32. As was stressed in Chapter 1, derivation of an analytic theory that describes the time evolution of $S(k, t)$ from the initial stages immediately after the quench until the late stages of coarsening is still an unsolved challenge. A very interesting attempt in this direction has been made by Langer et al. [46], and whereas this theory describes the effect of the nonlinearity rather well during the initial stages, it can describe the coarsening stage only qualitatively [8,47,48]. The same caveat applies to extensions dealing with phase separation in polymer mixtures [49] and fluid mixtures considering the effect of hydrodynamics [50]. We shall not discuss this work here further, but rather refer the interested reader to some useful reviews [8,48].

2.6　DENSITY-FUNCTIONAL THEORIES OF NUCLEATION

The simplest version of density-functional theory (DFT) is based on the Ginzburg–Landau theory, where the phase transition is described by an order parameter field $\phi(\vec{r})$ using the free-energy functional given in Equation 2.29.

The purpose of DFT is that one can calculate the nucleation free-energy barrier ΔF^*, and the associated density profile of a critical droplet, in the same way as one computes the interfacial free energy and the order parameter profile of a flat interface between infinitely extended bulk coexisting phases [17,42]. Therefore DFT is suited, within a mean-field framework to compute also critical droplets and their barriers near the spinodal, where these free-energy barriers are very small.

How is such a calculation done? Here one invokes the variational principle of statistical mechanics, one requires that $F\left[\phi(\vec{r})\right]$ becomes a minimum. For a homogeneous state (where $\vec{\nabla}\phi \equiv 0$), this is the ordinary mean-field theory, which yields just the coexisting phases. In the inhomogeneous case, this yields the well-known Ginzburg–Landau equation. The character of such inhomogeneous solutions is determined by the boundary conditions for this differential equation. For instance, when we wish to describe the flat interface (see Figure 1.5 or Figure 2.6a), we consider an inhomogeneity in one direction (the z-direction) across the interface only. Normalizing the order parameter such that the two coexisting phases are described by an order parameter $\pm\phi_{\text{coex}}$, we need boundary conditions $\phi(z \to +\infty) \to \phi_{\text{coex}}$, $\phi(z \to -\infty) \to -\phi_{\text{coex}}$. The resulting order parameter profile across such a flat interface is smooth and has a thickness $2\xi_{\text{coex}}$, where ξ_{coex} is the correlation length at the coexistence curve, which scales as $\xi_{\text{coex}} \propto r(1 - T/T_c)^{-\nu_{\text{MF}}}$ within this mean-field theory. From the free energy excess associated with such an order parameter profile relative to a system where $\phi(z) = +\phi_{coex}$ or $-\phi_{coex}$ uniformly, one readily obtains the interfacial free energy (e.g., $\gamma_{g\ell}$) between the coexisting phases [17,42].

We can also look for solutions that have spherical symmetry (Figure 2.6b,c), such that for large distances ρ from the origin, we end up in a metastable state, $\phi(\rho \to \infty) = \phi_{ms}$ [20,51]. Thus, one sees that, by the density-functional theory, one can account for the fact that on the nanoscale, the interfaces between the phases are

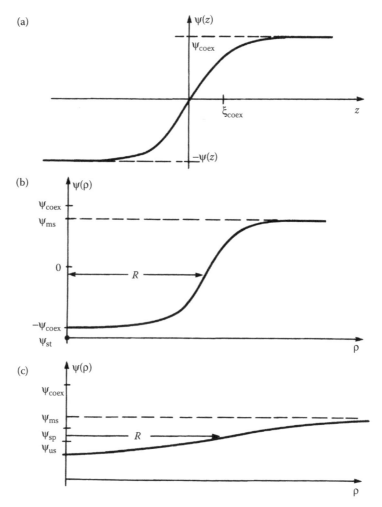

FIGURE 2.6 (a) Order parameter profile $\phi(z)$ across an interface between two coexisting bulk phases, the interface being oriented perpendicular to the z-direction; and the radial order parameter profile for a marginally stable critical droplet (b) near the coexistence curve, and (c) near the spinodal curve. In (a) and (b), the "intrinsic" thickness of the interface is of the order the correlation length ξ_{coex} at coexistence, and in (c) it is of the same order as the critical radius R^*. (From Binder, K., *Phys. Rev. A*, 29, 341, 1984. With permission.)

smeared out and not sharp. And if a Tolman-type correction for the interfacial free energy is present, one can find it from this approach (but it is possible that fluctuation corrections neglected by mean-field theory also make important contributions) [52].

Such density-functional theories of nucleation based on Equation 2.29 or various generalizations thereof are also referred to as *diffuse interface theories* [53,54] or *phase field theories* [55]. These have also been applied to problems of nucleation in systems with elastic forces [56], liquid-crystalline systems, and so on. These extensions are beyond the scope of our current discussion. We also do not discuss density-functional theories that go beyond the gradient square approximation (Equation 2.29), discussed, for example, in [57].

2.7 A GINZBURG CRITERION FOR NUCLEATION NEAR A SPINODAL

A particularly interesting phenomenon occurs for metastable states near the spinodal: note that the theory still invokes existence of a spinodal, it is the inflection point of the double-minimum potential. However, near the spinodal, the droplet profile gets very diffuse and spreads out. The reason for this is that in metastable states near the spinodal, the correlation length ξ would be very large [and ultimately divergent, $\xi(\phi \to \phi_{sp}) \to \infty$]. Both the interfacial thickness and the droplet radius R^* diverge when one approaches the spinodal. But the order parameter variation across the droplet gets smaller and smaller—the physical interpretation is that droplets near the spinodal are no longer compact objects, but rather ramified clusters, and so what one calculates here is just the radial density profile around the center of mass of such a ramified cluster [29,51,58].

From this theory, one can now calculate how the droplet free energy vanishes near the spinodal. However, one needs to be careful: density-functional theories are mean-field theories, and break down, when singularities occur, such as at a spinodal. One can judge the validity of such mean-field theories from a self-consistency condition [29], the so-called *Ginzburg criterion* [59]. Here, we compare the typical amplitude of spontaneous order parameter fluctuations with the scale of the order parameter variation across this profile itself: only when the fluctuation amplitudes are much smaller is the theory justified. This argument shows that, for systems with short-range interactions, the spinodal singularities are wiped out completely. There is a very broad region around the spinodal where a gradual transition from nucleation to spinodal decomposition occurs (Figure 2.7b) [29]. Only if the range r of interactions is very large, the scale for the free-energy barrier $\Delta F^*/k_B T_c$ is magnified by a large factor, the interaction volume r_d (in d dimensions) times the factor $(1 - T/T_c)^{2-d/2}$ [in $d = 3$, this scale factor is $r^3\sqrt{(1 - T/T_c)}$]. It turns out that the mean-field theory breaks down (according to the Ginzburg criterion) when the scale of the free-energy barrier $\Delta F^*/k_B T_c$ becomes of order unity: for short-range interactions, this happens already far from the spinodal (Figure 2.7b), whereas for larger r this happens close to the spinodal, due to this scale factor (Figure 2.7a). From these considerations, one can show that the range over which the spinodal curve is rounded

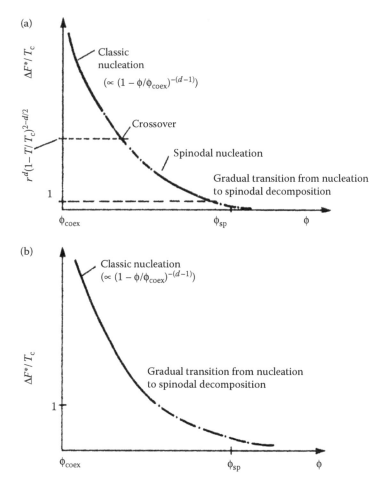

FIGURE 2.7 Schematic plot, of the free-energy barrier for (a) the mean-field critical region, that is, $r^d (1 - T/T_c)^{(4-d)/2} \gg 1$, and (b) the non–mean-field critical region, that is, $r^d (1 - T/T_c)^{(4-d)/2} < 1$. In the mean-field region, the gradual transition from nucleation to spinodal decomposition occurs in a rather narrow region close to the spinodal (which occurs at an order parameter $\phi = \phi_{sp}$), whereas in the non–mean-field region this gradual transition is spread out over a rather wide region. (From Binder, K., *Phys. Rev. A*, 29, 341, 1984. With permission.)

off is given by

$$\frac{\delta\phi_{sp}}{\phi_{sp}} \propto \left[r^d (1 - T/T_c)^{(4-d)/2} \right]^{-2/(6-d)}$$

$$= \left[r^3 (1 - T/T_c)^{1/2} \right]^{-1/3}. \tag{2.45}$$

It has been shown [29] that, instead of systems with long-range interactions, one can also consider symmetrical mixtures of very long flexible polymers: then in the above formula (in $d = 3$ dimensions), the interaction volume r^3 has to be replaced by the square root of the chain length N [29]. This happens because flexible polymers have a loose random walk-like coil structure, all the chains interpenetrate each other, and each chain interacts with $\sqrt{N}$ *neighbors*. In the phase diagram (Figure 2.8), one then has many regions: very close to the critical point, fluctuations always take over, and mean-field theory breaks down (this happens at the horizontal broken line in Figure 2.8). In this critical region, the spinodal curve has no significance whatsoever. For systems with interactions of long but finite range, or for symmetrical polymer mixtures, there is a regime where Equation 2.29 applies and mean-field type critical behavior is found. There one finds two rather distinct regimes of nucleation: classic nucleation based on compact clusters near the coexistence curve, and *spinodal nucleation* (based on ramified clusters) near the spinodal curve. In the "no man's land" in between (the

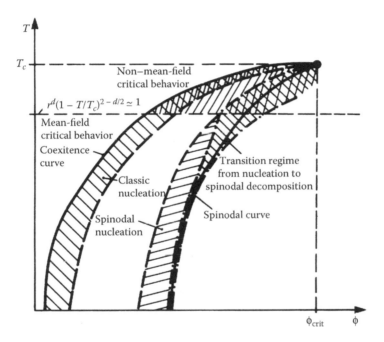

FIGURE 2.8 Various regions in the temperature-order parameter plane (T, ϕ) near the critical temperature T_c. Because of the symmetry around the critical value of the order parameter ϕ_{crit}, only the left half of the phase diagram is shown. Full curves are the coexistence and spinodal curves. The regime inside the two dash-dotted curves around the spinodal curve is the regime where a gradual transition from nucleation to spinodal decomposition occurs. The shaded regime between the coexistence curve and the left of the two broken curves is described by classic nucleation theory. In this regime, a further smooth crossover occurs at $r^d(1 - T/T_c)^{2-d/2} \simeq 1$ from mean-field to non–mean-field critical behavior. The regime between the right broken curve and the left dash-dotted curve (shaded) is the regime of *spinodal nucleation* via ramified clusters. It exists only in the regime of mean-field critical behavior. For clarity, any shift in the location of the spinodal curve relative to its location predicted by mean-field theory has been disregarded. (From Binder, K., *Phys. Rev. A*, 29, 341, 1984. With permission.)

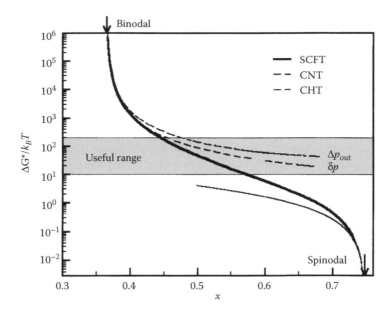

FIGURE 2.9 Plot of the nucleation barrier $\Delta G^*/k_B T$ against homogeneous nucleation versus molar fraction x of carbon dioxide, for a model of a mixture of hexadecane and carbon dioxide, at a reduced pressure $p\sigma^3/k_B T = 0.1$ and a reduced temperature $k_B T/\varepsilon = 0.75$. The full curve is a self-consistent field theory (SCFT) calculation, CNT denotes classic nucleation theory, CHT denotes Cahn–Hilliard theory [20]. (From Müller, M., et al., *J. Chem. Phys.*, 117, 5480, 2002. With permission.)

white region in between the two shaded regions of Figure 2.8), one has to work out the DFT numerically.

Of course, one can ignore these considerations based on the Ginzburg criterion [29] and simply work out numerically the DFT of nucleation over the whole regime, from the coexistence (binodal) all the way to the spinodal. A typical example of such an approach for a model of a mixture of hexadecane and carbon dioxide {the latter molecule is described as a point particle with a Lennard–Jones interaction $U_{LJ}(r) = 4\varepsilon\left[(\sigma/r)^{12} - (\sigma/r)^{6}\right]$, r being the distance between two particles} is shown in Figure 2.9. Note the logarithmic scale of the ordinate: the useful range for nucleation theory, namely $10 \leq \Delta G^*/k_B T \leq 200$, has been shaded. One sees that neither classic nucleation theory (two versions being shown as broken curves) nor the CH theory are accurate in this useful regime. One also sees that the volume fraction where $\Delta G^*/k_B T = 10$ is very far from the spinodal, but this value would not be predicted correctly by the classic nucleation theory here [60]. In the (p, x) diagram, even the contour $\Delta G^*/k_B T = 1$ is not close to the spinodal; the latter curve clearly has no physical significance [60].

As a final point of this section, we comment on the fact that many phase diagrams of real materials contain spinodal curves, claimed to be experimentally determined. It turns out that such spinodals are all *pseudo-spinodals*, found from a suitable extrapolation of data taken in the stable or metastable region of the phase diagram, respectively.

Because the spinodal curve in the phase diagram is like a line of critical points, there is always a response function that would diverge if one could observe metastable states right up to the spinodal. So it is quite common to plot the inverse of such a response function and try to extrapolate it to the point where this inverse response function would vanish. Such a procedure can also be mimicked by computer simulation, of course. As an example, Figure 2.10 shows data [61] for the inverse susceptibility $(k_B T \chi)^{-1}$ of Ising ferromagnets, recorded in metastable states of positive magnetization in a negative magnetic field $H < 0$. Indeed, even for the nearest neighbor case (coordination number $q = 6$), one can see a decrease of $(k_B T \chi)^{-1}$ over some range of fields (Figure 2.10). If we would extrapolate such data, the resulting estimate of the spinodal would be far from the mean-field prediction, however. With increasing q, one observes a gradual convergence toward the mean-field results, as expected from the considerations with the Ginzburg criterion for nucleation [29]. The actual limit

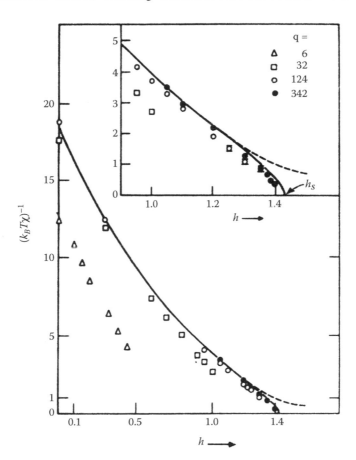

FIGURE 2.10 Inverse susceptibility of Ising ferromagnets plotted against $h = -H/k_B T$ at $T/T_c^{MF} = 4/9$ for various ranges of the exchange interaction: each spin interacts with equal strength with q neighbors. The full curve is the molecular field approximation; the broken curve a fit to a droplet model. (From Heermann, D.W., et al., *Phys. Rev. Lett*, 49, 1262, 1982. With permission.)

of metastability is given by the condition that one stays long enough in a metastable state, before it decays by nucleation and growths, that averages in that metastable state can still be taken with meaningful accuracy [2,26,27].

2.8 THE VALIDITY OF THE LINEAR THEORY OF SPINODAL DECOMPOSITION REVISITED

We have seen, from the point of view of nucleation, that the singular behavior at the spinodal line (i.e., divergence of the critical radius of the droplet, while the free-energy barrier vanishes) is not physically meaningful. The same conclusion emerges, of course, when one approaches the spinodal curve from the point of view of spinodal decomposition: whereas the theoretical prediction is that the critical wavelength λ_c of spinodal decomposition diverges, when the spinodal is reached, usually (in experiments and simulations) no trace of such a behavior is ever seen [2,8,48]. In fact, even the estimation of λ_c is ambiguous, as normally the linearized theory of spinodal decompositions is not valid [2,8,29,48]. We now discuss this problem, using a similar reasoning based on a Ginzburg criterion [29] as in the last section.

Using a free-energy functional (Equation 2.29) that is compatible with the schematic coexistence curve of Figure 2.8, the critical wavelength λ_c can be written as [29]

$$\lambda_c \propto r \left(1 - \frac{T}{T_c}\right)^{-1/2} \left(1 - \frac{\phi}{\phi_{sp}}\right)^{-1/2}. \tag{2.46}$$

Thus λ_c diverges when $\phi \to \phi_{sp}$ in a fashion completely analogous to that of the correlation length ξ when one approaches the spinodal from the metastable side. Now, in order to ensure the validity of the linearized theory of spinodal decomposition, we require

$$\left\langle [\delta\phi(\vec{r}, t)]^2 \right\rangle_{T,L} \ll (\phi - \phi_{sp})^2, \tag{2.47}$$

L being the linear dimension of a coarse-graining cell. Recall that coarse-graining is necessary to transform from a discrete atomistic description, for example, a lattice model of a binary alloy, to the continuum field theory in Equation 2.29—see also Section 1.3 on kinetic Ising models. We then estimate the left-hand side of Equation 2.47 as

$$\left\langle [\delta\phi(\vec{r}, t)]^2 \right\rangle_{T,L} \simeq \left\langle [\delta\phi(\vec{r}, 0)]^2 \right\rangle_{T,L} \exp[2\omega(k_{max})t]. \tag{2.48}$$

In the equation

$$\left\langle [\delta\phi(\vec{r}, 0)]^2 \right\rangle_{T,L} = \left\langle [\delta\phi(\vec{r})]^2 \right\rangle_{T_0,L}, \tag{2.49}$$

we now take the maximum permissible value for L, $L = \lambda_c$. This choice transforms Equation 2.47 into

$$r^{-2}\lambda_c^{2-d} \exp[2\omega(k_{max})t] \ll (\phi - \phi_{sp})^2. \tag{2.50}$$

Using Equation 2.46, we then find

$$\exp[2\omega(k_{max})t] \ll r^d \left(1 - \frac{T}{T_c}\right)^{(4-d)/2} \left(1 - \frac{\phi}{\phi_{sp}}\right)^{(6-d)/2}. \qquad (2.51)$$

Thus the time range over which the linearized theory may hold increases only slowly with the interaction range r, namely logarithmically, $t \propto \ln r$. After this range of time has elapsed, nonlinear effects come strongly into play, limiting the exponential growth, and causing coarsening (i.e., the maximum of the structure factor $S(k,t)$ does not occur at a time-independent value k_m, but rather the position of the maximum $k_m(t)$ of $S(k,t)$ decreases, $k_m(t \to \infty) \to 0$.

In order that Equation 2.51 makes sense at all, it is clear that the right-hand side of this inequality must be much larger than unity. However, this is exactly the same condition as found for the validity of the mean-field theory of nucleation close to the spinodal (cf. Equation 2.45).

In the center of the miscibility gap, on the other hand, $\phi - \phi_{sp}$ is of the same order as the order parameter $\phi_{coex} \propto (1 - T/T_c)^{1/2}$, and λ_c is of the same order as the correlation length of order parameter fluctuations at the coexistence curve, $\xi_{coex} \propto r(1 - T/T_c)^{-1/2}$ (taking mean-field critical exponents $\beta = \nu = 1/2$, compatible with Equation 2.29). Then Equation 2.50 yields

$$\exp[2\omega(k_{max})t] \ll r^2 \phi_{coex}^2 \xi_{coex}^{d-2} = r^d (1 - T/T_c)^{(4-d)/2}. \qquad (2.52)$$

On the other hand, for a system with short-range forces, where $r \simeq 1$, $\xi \propto (1 - T/T_c)^{-\nu}$, $\phi_{coex} \propto (1 - T/T_c)^{\beta}$, we would find from $\langle [\delta\phi(\vec{r})] \rangle_{T_0, L=\xi_{coex}} \propto \xi_{coex}^{\gamma/\nu - d}$ and Equation 2.50 that

$$\exp[2\omega(k_{max})t] \ll \xi_{coex}^{d-(\gamma+2\beta)/\nu}. \qquad (2.53)$$

Remembering the hyperscaling relation between critical exponents [3], $d\nu = \gamma + 2\beta$, we note, however, that the right-hand side of the inequality is unity: so Equation 2.53 is never fulfilled in the non–mean-field critical region! This finding justifies the statements made in the explanation of Figure 2.8.

These results have been confirmed by targeted computer simulations for kinetic Ising models, using the Kawasaki [62] spin exchange model with a long but finite range of the interaction r. It was found [63] that for large r (i.e., Ising models where one spin on the simple cubic lattice is coupled to 124 neighbors rather than only 6 neighbors with equal strength), there is indeed a regime of times, where an exponential growth of $S(k,t)$ with the time after the quench is observed, and the amplification rate $\omega(k)$ in this system agrees with the CH prediction, Equation 2.35.

Though no experimental system directly corresponding with this model is known, it was also pointed out that in three-dimensional symmetric mixtures of flexible polymers, there is a correspondence between the theoretical description in terms of Equation 2.29 and that of a medium-range Ising model, if one identifies the interaction volume r^3 with $\sqrt{N}$, N being the chain length of the polymers [29,64]. As a consequence, it was predicted that the linearized Cahn theory [1] of spinodal decomposition should be observable in such symmetrical mixture of long flexible polymer chains. Indeed, scattering experiments on mixtures of deuterated and protonated polybutadiene [65] did identify successfully a time-regime where the scattering intensity grows exponentially with time, and $\omega(k)/k^2$ varies linearly with k^2 almost to k_c^2. Even for polymer mixtures, however, no experimental evidence exists for a critical increase of the critical wavelength λ_c when the spinodal curve is approached. It also needs to be emphasized that a serious problem in the interpretation of experiments is the fact that the temperature quench (or, alternatively in some systems, a pressure jump) can never be carried out infinitely fast, as assumed by theory and simulation, but only with a finite quench rate. Thus, the system changes already during the quench, and often the early stages considered by the theory are unobservable. Theoretical considerations [66] give evidence that a finite quench rate does affect the behavior of $S(k,t)$ at early times considerably. Further, we also emphasize that an analytic theory describing the gradual crossover from *spinodal nucleation* near the spinodal curve to nonlinear spinodal decomposition inside of the spinodal curve does not exist [8].

2.9 SPINODALS AS A FINITE SIZE EFFECT: THE DROPLET EVAPORATION/CONDENSATION TRANSITION IN A FINITE VOLUME

As is well known, the free-energy functional in Equation 2.29 is a useful starting point to describe the statistical mechanics of phase transitions beyond mean-field theory. What one needs to do is to treat $F[\phi(\vec{r})]$ as an effective Hamiltonian, so the actual free energy becomes $F = -k_B T \ln Z$ with

$$Z = \int D\phi(\vec{r}) \exp\{-F[\phi(\vec{r})]/k_B T\}. \tag{2.54}$$

Then $f(\phi)$ in Equation 2.29 actually obtains the meaning of a coarse-grained free-energy density. The idea is to split the system into cells of linear dimension L: keeping $\phi(\vec{r})$ fixed as a constraint in the cell that has its center of gravity at $\vec{r}$, one averages over all short-range fluctuations of the order parameter within such a cell. On the other hand, the order parameter fluctuations with wavelengths larger than L are considered explicitly in the functional integral, Equation 2.54. In this manner, a discrete lattice model, such as the Ising model of a magnet, may be mapped onto the corresponding continuum field theory.

A crucial point is the appropriate choice of the coarse-graining cell size L. It turns out that, only if L is less than the correlation length ξ of order parameter fluctuations, one can expect that $f(\phi)$ is only weakly L-dependent, and only then it takes the Ginzburg–Landau form,

$$f(\phi) = -\frac{a(T_c - T)}{2}\phi^2 + \frac{b}{4}\phi^4 + \cdots, \tag{2.55}$$

with $a, b > 0$. In contrast, if L exceeds ξ distinctly, the positions $\phi_s^{(1)}(L)$, $\phi_s^{(2)}(L)$ of the inflection points converge toward the coexistence curve. This behavior has been first

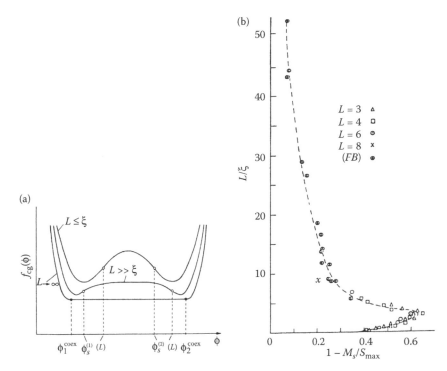

FIGURE 2.11 (a) Schematic plot of the coarse-grained free-energy density $f_{cg}(\phi)$ as a function of the order parameter ϕ. Spinodals $\phi_s^{(1,2)}(L)$ defined from inflection points of $f_{cg}(\phi)$ depend distinctly on the coarse-grained length L. (From Binder, K., *Rep. Progr. Phys.*, 50, 783, 1987. With permission.) (b) Monte Carlo results for the L-dependence of the relative distance of the "spinodal" from the coexistence curve, as deduced from the cell distribution functions of the three-dimensional nearest-neighbor Ising model in the critical region. By scaling L with the correlation length ξ, all temperatures superimpose on one *scaling function*. (From Kaski, K., et al., *Phys. Rev. B*, 29, 3996, 1984. With permission.)

pointed out by Kaski et al. [67] (see Figure 2.11), but the nature of these phenomena have been clarified in detail only recently [18,68,69].

The reason for the behavior seen in Figure 2.11b, namely a drastic change in the size-dependence of the location of the spinodal when L/ξ exceeds 3, is the fact that then the box has become large enough that two-phase coexistence becomes possible within a single box of linear dimension L. For simplicity, we shall discuss this problem no longer for a subsystem of an infinitely large system, but only for the related case of a single system of linear dimension L (with periodic boundary conditions in all directions). That means, we discuss finite size effects on two-phase coexistence.

Using the language appropriate to a liquid–gas transition, the "hump" in the thermodynamic potential in the region from $\rho_{coex}^{(1)}$ to $\rho_{coex}^{(2)}$ (Figure 2.11a) corresponds with five distinct states, separated by transitions that become sharp in the thermodynamic limit, $L \to \infty$: For $\rho_{coex}^{(1)} < \rho < \rho'$, one finds a supersaturated gas. While a supersaturated gas is at best metastable in the thermodynamic limit, for finite L this state is a truly stable phase. For $\rho' < \rho < \rho''$, on the other hand, the equilibrium state is characterized by a liquid droplet surrounded by (weakly supersaturated) gas. As we shall see, the transition at ρ' (where the droplet first appears) is a rounded transition for finite L, but does become a sharp transition in the limit $L \to \infty$ (though $\rho' \to \rho_{\text{coex}}^{(1)}$ in this limit).

For $\rho'' < \rho'''$, the character of phase coexistence changes: rather than having a droplet surrounded by supersaturated gas, we now have a liquid slab surrounded by saturated liquid. For a system with periodic boundary conditions, the slab is connected into itself, and the total area of the two flat vapor–liquid interfaces is (in $d = 3$ dimensions) simply $2L^2$. Consequently, the height of the free-energy hump in this region should be $2L^2 \gamma_{\ell g}/k_B T$, and consequently $f_{\text{cg}}(\rho) = (2/L)\gamma_{\ell g}/k_B T$, independent of ρ in this region. For $\rho''' < \rho''''$, the role of liquid and vapor is simply inverted–we find a gas bubble surrounded by liquid, and for $\rho'''' < \rho < \rho_{\text{coex}}^{(2)}$ we have a liquid (which would be metastable for $L \to \infty$, but in this limit $\rho'''' \to \rho_{\text{coex}}^{(2)}$, and we simply recover the double-tangent construction of the thermodynamic potential).

These different states are by no means hypothetical but can be identified in the two-dimensional Ising model from snapshot pictures of the spin configurations [70]. In fact, the result that the central part of the free-energy hump is independent of ρ and simply related to the interfacial free energy $\gamma_{\ell g}$ has been known for a long time [71] and has become one of the most powerful and standard tools to estimate interfacial free energies from Monte Carlo simulations [72].

Here we rather focus attention to the transition at ρ', which was first mentioned on the basis of phenomenological arguments [73], and qualitative evidence for this transition was given in [7], while a detailed analysis was given only recently [18,69]. For a Lennard–Jones fluid (truncated at $r_c = 2\sigma$ and shifted to zero, so $T_c \simeq 1$ if we choose $\varepsilon = 1$, $\sigma = 1$), typical chemical potential and pressure isotherms are shown in Figure 2.12. One sees that for small L, the $\Delta\mu$ versus ρ curve resembles a van der Waals loop, but there is a distinct size dependence, and for larger L the "loop" has a rather anomalous shape near the maximum and the minimum, respectively.

In order to clarify the behavior near the maximum in the $\Delta\mu$ versus ρ curve in more detail, several states of the system with $T = 0.68, L = 22.5$ were analyzed in detail (Figures 2.13, 2.14). One sees that in the transition region, that is, for values of the particle number N where $(\partial\mu(\rho)/\partial\rho)_T$ has its extremum, both the distribution of the size of the largest droplet N_c^{max} in the system and of the chemical potential are bimodal. Also, snapshots of the configurations reveal clearly that the system sometimes contains a large cluster, sometimes not. Also, the probability distribution of the energy reveals the same bimodal character, typical for a first-order phase transition, rounded by finite size [18].

Comparing the chemical potential versus density loops for different values L (Figure 2.15), one clearly sees that for $L \to \infty$, a sharp spike develops, but the position of the spike tends to the coexistence curve (the origin of the diagram), and the height tends to zero. An analytical model for this transition yields $\left(\rho_c \equiv \rho_{\text{coex}}^{(1)}\right)$ [18,69]

$$\rho_t - \rho_c \propto L^{-3/4}, \quad \Delta\mu_t = \mu_t^u - \mu_t^\ell \propto L^{-3/4}. \tag{2.56}$$

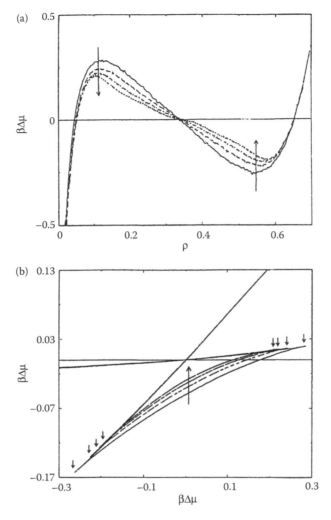

FIGURE 2.12 (a) Chemical potential difference as function of total density at $T = 0.85437$, and (b) pressure difference Δp plotted versus normalized chemical potential difference $\beta\Delta\mu = \Delta\mu/k_B T$, for linear dimensions $L = 6.74, 9.0, 11.3$ and 13.5 [arrows in (a) mark an increase in system size]. The long arrow in (b) marks an increase in system size, and the small arrows indicate the effective spinodal points. Note how these turning points shift toward the thermodynamic limit coexistence point as the system size increases. (From MacDowell, L.G., et al., *J. Chem. Phys.*, 120, 5293, 2004. With permission.)

Here ρ_t is the density of the position at this transition and μ_t^u, μ_t^ℓ the chemical potential values of the two coexisting phases (μ_t^u corresponds with the *metastable* state with no droplet, μ_t^ℓ with the state containing the droplet). Unfortunately, the approach toward the appropriate laws is somewhat slow [18]. But theoretical arguments have been presented to show that the width $\Delta\rho$, over which the transition of Equation 2.56 is rounded, decreases with a much larger power of L than 3/4, so that the transition

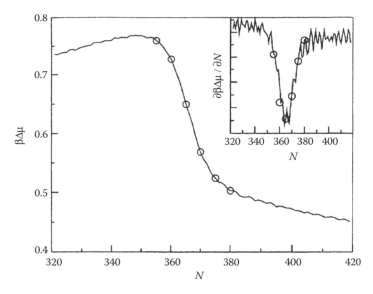

FIGURE 2.13 Chemical potential difference $\beta\Delta\mu$ plotted versus the total particle number N in a box with $L = 22.5$ at a temperature $T = 0.68$. The circles represent those values N for which distributions are shown in Figure 2.14. The inset shows the density derivative of $\mu(\rho)$. (From MacDowell, L.G., et al., *J. Chem. Phys.*, 120, 5293, 2004. With permission.)

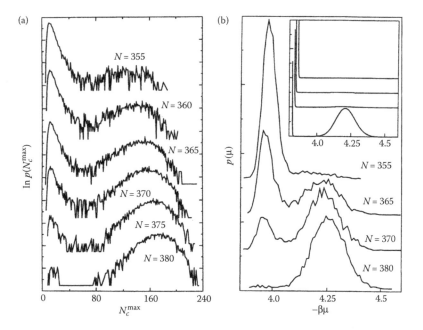

FIGURE 2.14 (a) Probability distribution of finding a state whose largest cluster has the size N_c^{max}, for the states highlighted in Figure 2.13. (b) Probability distribution to find a configuration whose average chemical potential is μ. Inset in (b) shows a corresponding analytic approximation. (From MacDowell, L.G., et al., *J. Chem. Phys.*, 120, 5293, 2004. With permission.)

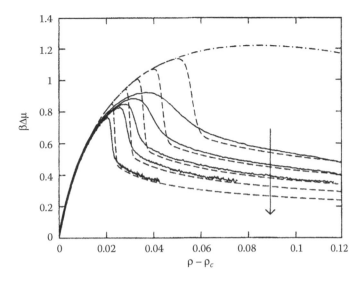

FIGURE 2.15 Chemical potential versus density loops for $T = 0.68$ and several system sizes ($L = 11.3, 13.5, 15.8, 18.3,$ and 22.5, from above to below, full curves). Dashed curves are corresponding predictions from an approximate theory. The dash-dotted line represents the mean-field equation of state as given by the mean spherical approximation. (From MacDowell, L.G., et al., *J. Chem. Phys.*, 120, 5293, 2004. With permission.)

gets sharper and sharper (before it ultimately disappears at phase coexistence $\rho_t = \rho_c$, $\Delta\mu_t = 0$ for $L \to \infty$).

Although no trace of this transition studied in Figures 2.12 through 2.15 is left in the thermodynamic limit, this rounded first-order transition characterized in Equation 2.56 may be of considerable interest when phase coexistence in volumes of nanoscopic size is considered. Figure 2.15 shows that then states are stable (in between ρ_c and ρ_{max} defined by the maximum of the $\Delta\mu(\rho)$ curve) that are only metastable in the thermodynamic limit.

2.10   DISCUSSION

In this chapter, it has been shown that the theory of spinodal decomposition and nucleation phenomena is still rather incomplete. There is no doubt that classic nucleation theory becomes asymptotically correct when the coexistence curve is approached, but for many cases of practical interest, the corresponding free-energy barriers are of order $100\ k_BT$ or higher in the regime where classic nucleation theory is accurate. Although density-functional theories of nucleation can be worked out for arbitrary free-energy barriers, because of their mean-field character their quantitative accuracy is uncertain. Whereas the mean-field spinodal is meaningful in the limit where the range r of interactions becomes infinitely large, for finite r the mean-field spinodal has little significance: pseudo-spinodals found from extrapolation of response functions in (meta-)stable states or spinodals defined from the criterion $\Delta F^*/k_BT = O(1)$ lie much closer to the coexistence curve than to the mean-field spinodal. A true spinodal can be identified for finite systems, but this signifies the onset of two-phase coexistence via the droplet evaporation/condensation transition at the density ρ_t (Equation 2.56). Also, recent work on nucleation kinetics [74] confirms the conclusion that no singularities associated with the mean-field spinodal can be detected in the nucleation process.

Further, with respect to the theory of spinodal decomposition, which addresses the early stages of phase transition kinetics for systems such as fluids or binary mixtures quenched to a state underneath the spinodal curve, the situation is also rather unsatisfactory [8]. The simplest approach (the Cahn linearized theory [1]) yields an exponential growth of the structure factor with time, and singular behavior at the spinodal. From a Ginzburg criterion, one can understand that these predictions become accurate only in the mean-field limit (infinite interaction range, or infinite chain length in a symmetric binary polymer mixture, etc.). For most cases of practical interest, these predictions of the Cahn theory are rather irrelevant, and nonlinear effects (as well as thermal fluctuations) need to be taken into account in order to obtain a quantitatively reliable theory for the initial stages of phase separation kinetics. Unfortunately, a successful theory achieving this task, including a description of the smooth crossover between nucleation and nonlinear spinodal decomposition, is still an unsolved challenge. In this respect, the situation is more difficult than in the late stages of coarsening, considered in other chapters of this book, where the length scales of interest are well-separated from the scale of spontaneous thermal fluctuations (described by the correlation length), and hence thermal fluctuations are negligible, at least asymptotically.

ACKNOWLEDGMENTS

The research reviewed here has profited very much from a fruitful collaboration with H. Furukawa, J.D. Gunton, D.W. Heermann, K. Kaski, L.G. MacDowell, M. Müller, H. Müller-Krumbhaar, D. Stauffer, and P. Virnau. Stimulating interactions with M.H. Kalos, D. Kashchiev, W. Klein, J.S. Langer, J.L. Lebowitz, and H. Reiss are also acknowledged.

REFERENCES

1. Cahn, J.W., On spinodal decomposition, *Acta Metall.*, 9, 795, 1961.
2. Binder, K., Theory of first-order phase transitions, *Rep. Progr. Phys.*, 50, 783, 1987.
3. Fisher, M.E., The renormalization group in the theory of critical behavior, *Rev. Mod. Phys.*, 46, 597, 1974.
4. Binder, K. and Luijten, E., Monte Carlo tests of renormalization-group predictions for critical phenomena in Ising models, *Phys. Rep.*, 344, 179, 2001.
5. Zinn-Justin, J., Precise determination of critical exponents and equations of state by field theory methods, *Phys. Rep.*, 344, 159, 2001.
6. Ferrenberg, A. and Landau, D.P., Critical behavior of the three-dimensional Ising model: A high-resolution Monte Carlo study, *Phys. Rev. B*, 44, 5081, 1991.
7. Furukawa, H. and Binder, K., Two-phase equilibria and nucleation barriers near a critical point, *Phys. Rev. A*, 26, 556, 1982.

8. Binder, K. and Fratzl, P., Spinodal decomposition, in *Phase Transformations in Materials*, Kostorz, G., Ed., Wiley-VCH, Weinheim and New York, 2001, chap. 6.
9. Abraham, F.F., *Homogeneous Nucleation Theory*, Academic Press, New York, 1974.
10. Binder, K. and Stauffer, D., Statistical theory of nucleation, condensation, and coagulation, *Adv. Phys.*, 25, 343, 1976.
11. Kashchiev, D., *Nucleation: Basic Theory with Applications*, Butterworth-Heinemann, Oxford, 2000.
12. Strey, R., Wagner, P. E., and Viisanen, Y., The problem of measuring homogeneous nucleation rates and the molecular contents of nuclei: Progress in the form of nucleation pulse measurements, *J. Phys. Chem.*, 98, 7748, 1994.
13. Wölk, J. and Strey, R., Homogeneous nucleation of H_2O and D_2O in comparison: The isotope effect, *J. Phys. Chem. B*, 105, 11683, 2001.
14. Turnbull, D., Formation of crystal nuclei in liquid metals, *J. Appl. Phys.*, 21, 1022, 1950.
15. Cantor, B. and Doherty, R.D., Heterogeneous nucleation in solidifying alloys, *Acta Metall.*, 27, 33, 1979.
16. Dietrich, S., Wetting phenomena, in *Phase Transitions and Critical Phenomena*, Vol. 12, Domb, C. and Lebowitz, J. L., Eds. Academic Press, New York, 1988, chap. 1.
17. Rowlinson, J.S. and Widom, B., *Molecular Theory of Capillarity*, Oxford University Press, Oxford, 1982.
18. MacDowell, L.G., Virnau, P., Müller, M., and Binder, K., The evaporation/condensation transition of liquid droplets, *J. Chem. Phys.*, 120, 5293, 2004.
19. Cahn, R.W. and Haasen, P., *Physical Metallurgy*, North-Holland, Amsterdam, 1983.
20. Cahn, J.W. and Hilliard, J.E., Free energy of a nonuniform system III. Nucleation in a two-component incompressible fluid, *J. Chem. Phys.*, 31, 688, 1959.
21. Tolman, R.C., The effect of droplet size on surface tension, *J. Chem. Phys.*, 17, 333, 1949.
22. Castleman, A.W., Nucleation and molecular clustering about ions, *Adv. Colloid Interface Sci.*, 10, 13, 1979.
23. Cahn, J.W., Nucleation on dislocations, *Acta Metall.*, 5, 160, 1957.
24. Becker, R. and Döring, W., The kinetic treatment of nucleus formation in supersaturated vapors, *Ann. Physik*, 24, 719, 1935.
25. Kreer, M., Classical Becker–Döring cluster equations: Rigorous results on metastability and long-time behavior, *Ann. Physik*, 2, 398, 1983.
26. Binder, K., Dynamics of first order phase transitions, in *Fluctuations, Instabilities, and Phase Transitions*, Riste, T., Ed. Plenum Press, New York, 1975, 53.
27. Binder, K. and Müller-Krumbhaar, H., Investigation of metastable states and nucleation in the kinetic Ising model, *Phys. Rev. B*, 9, 2194, 1974.
28. Fisher, M.E., Theory of condensation and critical point, *Physics*, 3, 255, 1967.
29. Binder, K., Nucleation barriers, spinodals, and the Ginzburg criterion, *Phys. Rev. A*, 29, 341, 1984.
30. Stauffer, D., Kinetic theory of two-component nucleation and condensation, *J. Aerosol. Sci.*, 7, 319, 1976.
31. Trinkaus, H., Theory of nucleation of multicomponent precipitates, *Phys. Rev. B*, 27, 7372, 1983.
32. Avrami, A., Kinetics of phase change. I. General theory, *J. Chem. Phys.*, 7, 1103, 1939.
33. Hohenberg, P.C. and Halperin, B.I., Theory of dynamic critical phenomena, *Rev. Mod. Phys.*, 49, 435, 1977.
34. Goldburg, W.J., Light scattering studies of phase separation kinetics, in *Scattering Techniques Applied to Supramolecular and Nonequilibrium Systems*, Chen, S.H., Chu, B., and Nossal, R., Eds. Plenum Press, New York, 1981, 383.

35. Langer, J.S. and Schwartz, A.J., Kinetics of nucleation in near-critical fluids, *Phys. Rev. A*, 21, 948, 1980.

36. Shneidman, V.A., Jackson, K.A., and Beatty, K.M., On the applicability of the classical nucleation theory in an Ising system, *J. Chem. Phys.*, 111, 6932, 1989.

37. Novotny, M.A., Low-temperature metastable life times of the square-lattice Ising ferromagnets, in *Computer Simulation Studies in Condensed Matter Physics IX*, Landau, D.P., Mon, K.K., and Schüttler, H.B., Eds., Springer, Berlin, 1997, 182.

38. Shneidman, V.A. and Nita, G.M., Modulation of the nucleation rate pre-exponential in a low-temperature Ising system, *Phys. Rev. Lett.*, 89, 02501, 2002.

39. Ellerby, H.M., Weakliem, C.L., and Reiss, H., Toward a molecular theory of vapor-phase nucleation. I. Identification of the average embryo, *J. Chem. Phys.*, 95, 9209, 1992.

40. Ellerby, H.M. and Reiss, H., Toward a molecular theory of vapor-phase nucleation. II. Fundamental treatment of the cluster distribution, *J.Chem. Phys.*, 97, 5766, 1992.

41. Cacciuto, A., Auer, S., and Frenkel, D., Breakdown of classical nucleation theory near iso-structural phase transitions, *Phys. Rev. Lett.*, 93, 166105, 2004.

42. Cahn, J.W. and Hilliard, J.E., Free energy of a nonuniform system. I. Interfacial free energy, *J. Chem. Phys.*, 28, 258, 1958.

43. Binder, K., Kinetic Ising model study of phase separation in binary alloys, *Z. Physik*, 267, 313, 1974.

44. Binder, K. and Frisch, H.L., Dynamics of surface enrichment. A theory based on the Kawasaki spin exchange model in the presence of a wall, *Z. Physik B*, 84, 402, 1991.

45. Cook, H., Brownian motion in spinodal decomposition, *Acta Metall.*, 18, 297, 1970.

46. Langer, J.S., Bar-on, M., and Miller, H.D., New computational method in the theory of spinodal decomposition, *Phys. Rev. A*, 11, 1417, 1975.

47. Binder, K., Billotet, C., and Mirold, P., On the theory of spinodal decomposition in solid and liquid binary mixtures, *Z. Phys. B*, 90, 183, 1978.

48. Gunton, J.D., San Miguel, M., and Sahni, P.S., The dynamics of first-order phase transitions, in *Phase Transitions and Critical Phenomena*, Vol. 8, Domb, C., and Lebowitz, J.L., Eds., Academic Press, London, 1983, 267.

49. Akcasu, A.Z. and Klein, R., A nonlinear theory of transients following step temperature changes in polymer blends, *Macromolecules*, 26, 1429,1993.

50. Kawasaki, K. and Ohta, T., Theory of early stage spinodal decomposition in fluids near the critical point, *Progr. Theor. Phys.*, 59, 362, 1978.

51. Unger, C. and Klein, W., Nucleation theory near the classical spinodal, *Phys. Rev. B*, 29, 2698, 1984.

52. Granasy, L., Semi-empirical van der Waals/Cahn–Hilliard theory: Size dependence of the Tolman length, *J. Chem. Phys.*, 109, 9660, 1998.

53. Granasy, L., Diffuse interface theory for homogeneous vapor condensation, *J. Chem. Phys.*, 104, 5188, 1996.

54. Granasy, L., Fundamentals of the diffuse interface theory of nucleation, *J. Phys. Chem.*, 100, 10768, 1996.

55. Granasy, L., Börzsönyi, T., and Pusztai, T., Nucleation and bulk crystallization in binary phase field theory, *Phys. Rev. Lett.*, 88, 206105, 2002.

56. Klein, W., Lookman, T., Saxena, A., and Hatch, D.M., Nucleation in systems with elastic forces, *Phys. Rev. Lett.*, 88, 085701, 2002.

57. Granasy, L., Jurek, Z., and Oxtoby, D.W., Analytical density functional theory of homogeneous vapor condensation, *Phys. Rev. E*, 62, 7486, 2000.

58. Heermann, D.W., Coniglio, A., Klein, W., and Stauffer, D., Nucleation and metastability in three-dimensional Ising models, *J. Stat. Phys.*, 36, 447, 1984.

59. Ginzburg, V.L., Some remarks on second order phase transitions and the microscopic theory of ferroelectrics, *Soviet Phys. Solid State*, 2, 1824, 1960.

60. Müller, M., MacDowell, L.G., Virnau, P., and Binder, K., Interface properties and bubble nucleation in compressible mixtures containing polymers, *J. Chem. Phys.*, 117, 5480, 2002.

61. Heermann, D.W., Klein, W., and Stauffer, D., Spinodals in a long-range interaction system, *Phys. Rev. Lett*, 49, 1262, 1982.

62. Kawasaki, K., Kinetics of Ising models, in *Phase Transitions and Critical Phenomena*, Vol. 2, Domb, C. and Green, M. S., Eds., Academic Press, London, 1972, 443.

63. Heermann, D.W., Test of the validity of the classical theory of spinodal decomposition, *Phys. Rev. Lett.*, 52, 1126, 1984.

64. Binder, K., Collective diffusion, nucleation, and spinodal decomposition in polymer mixtures, *J. Chem. Phys.*, 79, 6387, 1983.

65. Bates, F.S. and Wiltzius, P., Spinodal decomposition of a symmetrical critical mixture of deuterated and protonated polymer, *J. Chem. Phys.*, 91, 3258, 1989.

66. Carmesin, H.-O., Heermann, D.W., and Binder, K., Influence of a continuous quenching procedure on the initial stages of spinodal decomposition, *Z. Phys. B*, 65, 89, 1986.

67. Kaski, K., Binder, K., and Gunton, J.D., Study of the cell distribution functions of the three-dimensional Ising model, *Phys. Rev. B*, 29, 3996, 1984.

68. Biskup, M., Chayes, L., and Kotecky, R., Critical region for droplet in the two-dimensional Ising model, *Europhys. Lett.*, 60, 21, 2002.

69. Binder, K., Theory of the evaporation/condensation transition of equilibrium droplets in finite volumes, *Physica A*, 319, 99, 2003.

70. Hunter, J.E. and Reinhardt, W.P., Finite-size scaling behavior of the free-energy barrier between coexisting phases: Determination of the critical temperature and interfacial tension of the Lennard–Jones fluid, *J. Chem. Phys.*, 103, 6627, 1995.

71. Binder, K., Monte Carlo calculation of the surface tension for two- and three-dimensional lattice gas models, *Phys. Rev. A*, 25, 1699, 1982.

72. Landau, D.P. and Binder, K., *A Guide to Monte Carlo Simulations in Statistical Physics*, Second Edition, Cambridge University Press, Cambridge, 2005.

73. Binder, K. and Kalos, M.H., "Critical clusters" in a supersaturated vapor: Theory and Monte Carlo simulation, *J. Stat. Phys.*, 22, 363, 1980.

74. ter Horst, J.H. and Kashchiev, D., Determining the nucleation rate from the dimer-growth probability, *J. Chem. Phys.*, 123, 114507, 2005.

3 Monte Carlo Simulations of Domain Growth

Gerard T. Barkema

CONTENTS

3.1 PHASE SEPARATION IN THE ISING MODEL

In Section 1.3, two important examples of *kinetic Ising models* were introduced, that is, the Ising model with Glauber spin-flip kinetics, and the Ising model with Kawasaki spin-exchange kinetics. In this chapter, we will discuss efficient Monte Carlo techniques for simulating these and related models. As stated earlier, simulations of domain growth are often performed with either the lattice gas or the Ising model. In the lattice gas, a large number of particles (representing atoms or molecules) are placed on the sites of a lattice and interact with each other. Because these simulations are geared toward a qualitative understanding of domain growth rather than a quantitative match with any specific materials, the interactions between the particles are kept as simple as possible, resulting in the following rules:

- **Hard-core repulsion:** No lattice site can be occupied by more than one particle.
- **Short-range attraction:** If two particles occupy nearest-neighbor sites on the lattice, they feel an attractive interaction with a strength $-\epsilon$ ($\epsilon > 0$).
- **Chemical potential:** The introduction of a particle into the system raises the energy by μ.

The state of the system can then be specified by introducing the occupation-number variables n_i for sites $i = 1, \ldots, N$. This variable takes the value $n_i = 1$ when the site i

is occupied by a particle, or $n_i = 0$ if it is empty. With this notation, the Hamiltonian of the system becomes (cf. Equation 1.2)

$$H = -\epsilon \sum_{\langle ij \rangle} n_i n_j + \mu \sum_{i=1}^{N} n_i. \tag{3.1}$$

Here, the notation $\langle ij \rangle$ indicates that the first summation runs over all pairs of nearest-neighbor sites.

Now let us define a new set of spin variables:

$$S_i = 2n_i - 1. \tag{3.2}$$

These new variables take the values $S_i = \pm 1$, just as spins in the familiar Ising model. Inverting Equation 3.2 to get $n_i = (S_i + 1)/2$, and substituting this into Equation 3.1, yields

$$H = -\frac{1}{4}\epsilon \sum_{\langle ij \rangle} (S_i + 1)(S_j + 1) + \frac{1}{2}\mu \sum_{i=1}^{N} (S_i + 1)$$

$$= -\frac{1}{4}\epsilon \sum_{\langle ij \rangle} S_i S_j + \left(\frac{\mu}{2} - \frac{q\epsilon}{4}\right) \sum_{i=1}^{N} S_i + \frac{1}{2}N\mu - \frac{Nq\epsilon}{8}, \tag{3.3}$$

where q is the lattice coordination number (equal to the number of nearest neighbors of each site). If we now define

$$J = \frac{\epsilon}{4},$$

$$h = \frac{q\epsilon}{4} - \frac{\mu}{2}, \tag{3.4}$$

the Hamiltonian becomes

$$H = -J \sum_{\langle ij \rangle} S_i S_j - h \sum_{i=1}^{N} S_i + \text{constant}, \tag{3.5}$$

which is the usual Hamiltonian of the Ising model with coupling constant J and external magnetic field h. Note that the additive constant in the Hamiltonian makes no difference to any physical observable.

Because the Hamiltonians of the lattice gas model and the Ising model are mathematically equivalent, so is their behavior, and any simulation result obtained for one model can be directly translated into the language of the other model. For instance, the average magnetization per site $m = \sum_i \langle S_i \rangle / N$ of the two-dimensional ferromagnetic Ising model without an external field ($h = 0$) is known analytically. Below the critical temperature T_c, the absolute value of the magnetization per site equals [1]

$$|m| = [1 - \sinh^{-4}(2\beta J)]^{1/8}, \tag{3.6}$$

and the magnetization is equally likely to be positive $(m = |m|)$ as negative $(m = -|m|)$. Here, $\beta = (k_B T)^{-1}$ is the inverse temperature with Boltzmann constant k_B. Above the critical temperature, corresponding with $\beta_c J = \log(1 + \sqrt{2})/2$, the magnetization per site is zero.

Consequently, the particle density $\rho = \sum_i \langle n_i \rangle / N$ of the two-dimensional lattice gas at a chemical potential of $\mu = q\epsilon/2$ for $T > T_c$ is 50%. For $T < T_c$, there are two preferred densities for the model:

$$\rho_+ = \frac{1}{2}(1 + |m|),$$

$$\rho_- = \frac{1}{2}(1 - |m|).$$

(3.7)

If the system is forced into a state with an average density ρ between ρ_- and ρ_+ at $T < T_c$, it can rearrange itself by phase separation into domains of these densities (see Section 1.2.2). In this way, the system lowers its free energy with a term proportional to the domain volume, at the expense of a free-energy cost proportional to the total length of the domain boundaries.

As discussed in Sections 1.4 and 1.5, in simulations of phase separation, the system is initially prepared in a state in which domains are either absent or very small. A standard approach to do this is to take a system that is in equilibrium, but at a temperature T_0 much above T_c, possibly even infinitely high. When the simulation is started at time $t = 0$, immediately the temperature is quenched to some value $T < T_c$. Domains will form and gradually grow in size, as that lowers the total length of the domain boundaries and thus the total free energy—see Figures 1.4, 1.10, and 1.11. The manner in which domain growth occurs depends on many properties, for example, the dynamics of the particles/spins, details of the initial condition, and possible variations in temperature or chemical potential. The purpose of Monte Carlo simulations of domain growth is to clarify the domain growth laws and evolution morphologies, as measured by the *correlation function* or the *structure factor*, which were defined in Section 1.4.1.

3.2 DOMAIN GROWTH WITHOUT CONSERVATION LAWS: MODEL A

As we established above, the equilibrium properties of the lattice gas model can be easily mapped onto those of the Ising model. For the determination of equilibrium properties of the Ising model, many different algorithms have been developed [2,3]. One of the simplest approaches to perform Monte Carlo simulations of the Ising model is based on a so-called *Markov process*. It proceeds as follows:

1. The system starts in some spin configuration $C_0 \equiv \{S_i\}$ at $t = 0$.
2. A small change is proposed in configuration C_i, resulting in a trial configuration C'_{i+1}.

3. This small change is accepted with some probability $W(C_i|C'_{i+1})$, in which case the next configuration C_{i+1} is the trial configuration C'_{i+1}. Otherwise, it is rejected, and the next configuration C_{i+1} is simply the old configuration C_i.
4. Steps 2 and 3 are repeated many (billions of) times.

A commonly used *small change* or *move* is Glauber kinetics [4], that is, the reversal (flipping) of a randomly selected spin—see Section 1.3.2. To ensure that the above procedure is correct, and the sampling of the spin configuration occurs in agreement with the Boltzmann distribution, the acceptance probabilities $W(C_i|C'_{i+1})$ have to fulfill the condition of detailed balance. For any two states μ and v with equilibrium probabilities $P_{eq}(\mu)$ and $P_{eq}(v)$, the acceptance probabilities have to satisfy (cf. Equation 1.26)

$$W(\mu|v)P_{eq}(\mu) = W(v|\mu)P_{eq}(v). \tag{3.8}$$

Thus, the ratio of the acceptance probabilities of each move and that of its reverse is fixed by the ratio of the equilibrium probabilities. Another natural constraint is that the acceptance probability can never fall below 0 or exceed 1. In the Metropolis algorithm, one maximizes the acceptance probabilities under these constraints, yielding

$$W(\mu|v) = \min\{1, \exp[-\beta(E_v - E_\mu)]\}. \tag{3.9}$$

Thus, if we select a new state that has an energy lower than or equal to the current one, we should always accept the transition to that state. If the new state has a higher energy, then the probability that we accept it equals the ratio of the Boltzmann weights of states v and μ.

For the simulation of dynamics, we have to assign a timescale to the sequence of states generated by the algorithm above. In the absence of long-range interactions, the rate at which a spin is proposed to be flipped should be independent of whatever happens far away from it. In particular, this rate should not depend on the system size. If the total system contains N lattice sites, the probability that a given spin is proposed to be flipped in a single step of the Markov process equals $1/N$. A constant rate of proposed spin flips is then obtained by assigning a time increment $\Delta t = 1/N$ to each step of the Markov process. The time unit is then called *Monte Carlo step per site,* or MCS, and (statistically) each spin is proposed to be flipped once per time unit. This completes the description of the most commonly used dynamics in Monte Carlo simulations, known as *single-spin-flip dynamics with Metropolis acceptance rates.*

A slightly different (but closely related) approach is referred to as *single-spin-flip dynamics with heat-bath acceptance rates.* One way of presenting this algorithm is that a site is selected randomly, its spin is removed, and a new spin replaces it. Irrespective of the old spin value, the probability that the new spin has value S_i is equal to the Boltzmann weight of the state with this added spin, divided by the sum of the Boltzmann weights of all states considered. Thus, the new spin is drawn from a heat bath, hence the name of this algorithm. For the Ising model, the algorithm is identical to single-spin-flip dynamics with Metropolis acceptance rates, except that

the acceptance probability now becomes

$$W(\mu|\nu) = \frac{\exp(-\beta E_\nu)}{\exp(-\beta E_\mu) + \exp(-\beta E_\nu)}. \tag{3.10}$$

The biggest difference between the acceptance probabilities in Equations 3.9 and 3.10 is a factor of 2, which occurs if the spin flip would not change the energy: the Metropolis acceptance probability equals 1, and the heat-bath acceptance probability is only 0.5. Computationally, the heat-bath algorithm makes more sense if more than two states are part of the heat bath; this would, for instance, occur in Potts models with a large number of spin states.

A typical simulation of domain growth starts with a lattice on which spins are placed randomly at 50% density. (This can also be described as an equilibrium configuration at infinite temperature, because then all configurations are equally likely.) Then, the configuration evolves in time, for instance with spin-flip dynamics with Metropolis acceptance rates. The configuration is stored for later analysis at regular time intervals, often logarithmically spaced. Figure 3.1 shows evolution snapshots obtained from a

FIGURE 3.1 Snapshots of domain growth in the two-dimensional Ising model with spin-flip dynamics and Metropolis acceptance rates. These snapshots are obtained from a simulation of a 5000×5000 system with periodic boundary conditions at a temperature $\beta J = 0.6$: the times are $t = 100 \cdot 10^{i/2}$ MCS with $i = 1, \ldots, 4$. This figure is the microscopic counterpart of the order-parameter evolution shown in Figure 1.4.

simulation of a $d = 2$ system at a temperature $\beta J = 0.6$, quenched well below the critical temperature $\beta_c J = 0.44069$. This is the microscopic counterpart of the order-parameter evolution shown in Figure 1.4. The formation of domains, as well as their growth, is clearly visible.

As discussed in Section 1.4.1, a standard method to quantify the domain growth process is to calculate the correlation function or the structure factor. These quantities are defined as the discrete analogs of Equations 1.88 through 1.89:

$$C(\vec{r}, t) = \frac{1}{N} \sum_{\vec{R}} \left[\langle S_{\vec{R}}(t) S_{\vec{R}+\vec{r}}(t) \rangle - \langle S_{\vec{R}}(t) \rangle \langle S_{\vec{R}+\vec{r}}(t) \rangle \right], \tag{3.11}$$

and

$$S(\vec{k}, t) = \sum_{\vec{r}} e^{i\vec{k}\cdot\vec{r}} C(\vec{r}, t), \tag{3.12}$$

where the angular brackets denote an averaging over independent initial conditions and thermal fluctuations.

Let us first study the structure factor. The lattice causes some angular dependence in the structure factor, particularly at low temperatures. Nevertheless, it is common practice to disregard the angular information in the structure factor and to concentrate on its radial average. To obtain this numerically, the distance in reciprocal space is divided into bins with some size Δk. Next, for each vector $\vec{k}$, the bin number $b = k/\Delta k$ is identified. The structure factor values $S(\vec{k}, t)$ belonging to bin b are then averaged to obtain $S(k, t)$. The results of this procedure, applied to the snapshots shown in Figure 3.1, are plotted in Figure 3.2. The correlation function can also be radially averaged in the same way as the structure factor. The corresponding results for $C(r, t)$ versus r are plotted in Figure 3.3.

To identify a typical length scale in such configurations, various procedures are commonly used. The first one is based on the property that the structure factor crosses over from being flat to Porod's law, $S(k, t) \sim k^{-(d+1)}$ as $k \to \infty$ [5]. Thus, the function $k^{(d+1)/2} S(k, t)$ will have a maximum at a well-defined wavelength k_{max}, which is the location of the crossover. One characterization of the typical length scale would then be inversely proportional to this:

$$L_1(t) = \frac{\overline{N}}{2\pi k_{\text{max}}}. \tag{3.13}$$

Here, $\overline{N}$ is the linear dimension of the system, so that $\overline{N}^d = N$, the total number of lattice sites. This procedure is numerically not very convenient, because the discretization in k-values leads to large unphysical jumps in L_1, in particular for small k-values or large domain sizes.

A procedure that is similar to the above, but suffers less from discretization effects, is to measure the first moment of the rescaled, radially averaged structure factor:

$$\langle k \rangle = \frac{\int dk \, k S(k, t)}{\int dk \, S(k, t)}, \tag{3.14}$$

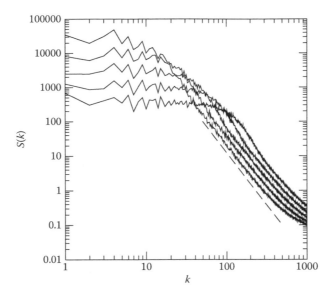

FIGURE 3.2 Radially averaged structure factor, $S(k,t)$ versus k, for the snapshots shown in Figure 3.1, plus one data set at time $t = 10^2$ MCS. These curves are obtained from a simulation of a 5000×5000 system at a temperature of $\beta J = 0.6$. Measurements are taken every half-decade in time, starting at $t = 10^2$ MCS until $t = 10^4$ MCS. As a guide to the eye, a dashed line corresponding with Porod's law, $S(k,t) \sim k^{-(d+1)} = k^{-3}$ in $d = 2$, is added. The curve closest to this line corresponds with the longest simulation time.

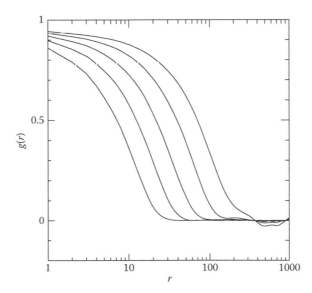

FIGURE 3.3 Radially averaged correlation function, $C(r,t)$ versus r, for the nonconserved evolution in Figure 3.1. The data sets are shown at the same times as in Figure 3.2. The curve at the extreme right corresponds with the longest simulation time.

from which a typical domain size can be obtained as

$$L_2(t) = \frac{\overline{N}}{2\pi\langle k \rangle}.\qquad(3.15)$$

Also, various length scales can be identified from the correlation function. For instance, one can measure the distance L_3 at which the radially averaged correlation function has decayed to some value a for the first time:

$$C(L_3, t) = aC(0, t),\qquad(3.16)$$

with (say) $a = 1/2$. We should stress that all these definitions are equivalent in the scaling regime, where L_1, L_2, L_3, and so on, are proportional to each other.

However, it is not necessary to measure a typical domain size and then plot it as a function of time, as we did in Figures 1.8 and 1.12. The structure is expected to become self-similar at sufficiently long times, with a typical length-scale that grows in time as $L \sim t^\phi$; the two-point correlation functions $C(r, t)$, measured at various times, should then all *collapse* if they are plotted as a function of rescaled distance $t^{-\phi}r$ (cf. Figures 1.7a and 1.13a). This is precisely what is done in Figure 3.4, with the value $\phi = 1/2$. Indeed, all curves collapse very nicely. Alternatively, one could collapse the structure factors in a similar way, by plotting $k^d S(k, t)$ versus $kt^{-\phi}$ with $\phi = 1/2$, as suggested by Equation 1.90.

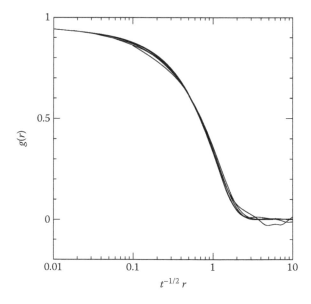

FIGURE 3.4 Radially averaged correlation function, $C(r, t)$ as a function of the rescaled distance $t^{-1/2}r$. The data sets are the same as those in Figure 3.3.

3.3 DOMAIN GROWTH WITH LOCAL CONSERVATION: MODEL B

Flipping individual spins might be the natural kind of dynamics in a magnetic system. However, in the context of a lattice gas, a spin flip corresponds with a highly unnatural phenomenon: the sudden appearance or disappearance of a particle. A more natural move in the lattice gas consists of a particle hopping from its site to a nearest-neighbor site. In the language of the Ising model, this corresponds with the exchange of spins at adjacent sites. As discussed in Section 1.3.3, this stochastic process is usually termed *Kawasaki dynamics* [6,7]. Strictly speaking, one should reserve the term *Kawasaki dynamics* for spin-exchange dynamics with heat-bath acceptance probabilities, as that is what Kawasaki did in his original article [7]. Currently, most people use Metropolis acceptance probabilities instead of the heat-bath algorithm and still refer to the dynamics as *Kawasaki dynamics*.

Figure 3.5 shows snapshots obtained from a simulation of the two-dimensional Ising model with spin-exchange dynamics at Metropolis acceptance probabilities.

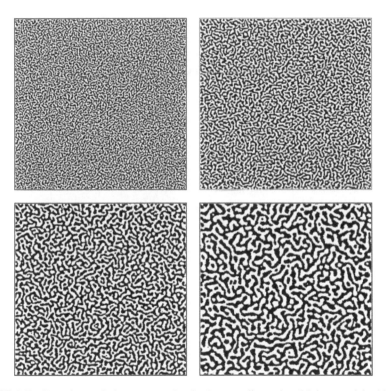

FIGURE 3.5 Snapshots of phase separation in the two-dimensional Ising model with spin-exchange dynamics and Metropolis acceptance rates. The mixture has a critical composition with 50% A (up spins) and 50% B (down spins). These snapshots are obtained from a simulation of a 2000 × 2000 system with periodic boundary conditions at a temperature $\beta J = 0.6$: the times are $t = 10^4 \cdot 10^{i/2}$ MCS with $i = 1, \ldots, 4$. This figure is the microscopic counterpart of the order-parameter evolution shown in Figure 1.10.

This figure is the microscopic counterpart of Figure 1.10, where the order-parameter evolution for phase separation with a critical composition was shown. Note that the times at which the snapshots are taken are orders of magnitude larger than the snapshots in Figure 3.1, while the domains have still not grown as much. Indeed, the coarsening process proceeds much slower if local conservation laws are enforced, and the particles have to diffuse slowly one lattice spacing at a time. Because of the far longer simulation times, the system size in the simulations of Figure 3.5 has been reduced to 2000 × 2000, in order to keep the computational effort tractable on a simple desktop computer.

Apart from the reduced speed of the coarsening process, the snapshots with local conservation are also distinctly different: the domains have a worm-like appearance with approximately uniform width, and there are fewer spherical islands. To better quantify such differences, we again plot the radially averaged structure factor in Figure 3.6. Comparing the structure factors with those plotted in Figure 3.2, the biggest difference is that (as a consequence of the local spin conservation) the structure factor at small wave-vectors is suppressed, and the peak in the structure factor becomes more pronounced—also compare Figure 1.13b with Figure 1.7b. This is consistent with our earlier remark of a more worm-like appearance of the domains.

Another approach to quantify the consequences of local conservation laws on the domain structures is to look at the correlation functions in Figure 3.7. Compared with the nonconserved case in Figure 3.3, the most striking difference is that the correlation

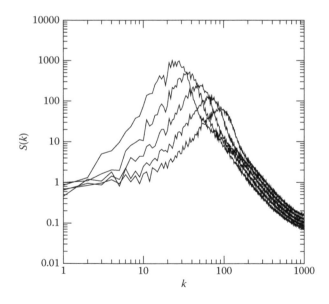

FIGURE 3.6 Radially averaged structure factor, $S(k, t)$ versus k, for the snapshots shown in Figure 3.5, plus one data set at time $t = 10^4$ MCS. These curves are obtained from a simulation of a 2000 × 2000 system at a temperature of $\beta J = 0.6$. Measurements are taken every half-decade in time, starting at $t = 10^4$ MCS until $t = 10^6$ MCS. The curve with the highest peak value corresponds with the longest simulation time.

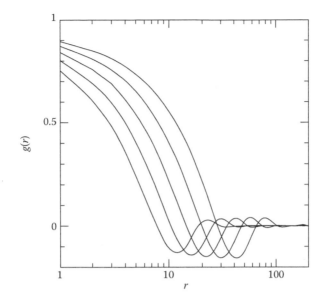

FIGURE 3.7   Radially averaged correlation function, $C(r, t)$ versus r, for the conserved evolution in Figure 3.5. The data sets are shown at the same times as in Figure 3.6. The curve at the extreme right corresponds with the longest simulation time.

function shows an oscillatory behavior with distance as we require

$$\int d\vec{r} \, C(\vec{r}, t) = S(\vec{k} = 0, t) = 0. \tag{3.17}$$

The domain growth law that characterizes diffusion-driven phase separation is the Lifshitz–Slyozov (LS) law, $L(t) \sim t^{\phi}$ with $\phi = 1/3$ [8,9]. The correlation functions $C(r, t)$, measured at various times, should then collapse when plotted as a function of rescaled distance $t^{-\phi}r$. This procedure is followed in Figure 3.8, with the exponent $\phi = 1/3$. The collapse is still not perfect, even though the simulations ran until 10^6 MCS. However, with increasing time, the quality of the collapse improves [10,11]. Alternatively, a similar quality of data collapse can be obtained from the structure factors by plotting $k^d S(k, t)$ versus kt^{ϕ} with $\phi = 1/3$.

3.4   COMPUTATIONAL ADVANCES

3.4.1   MULTISPIN CODING

Simulations of domain growth are often computer-intensive, taking weeks or even months. In a typical program, the state of each lattice site is stored in one byte of data. However, spins can take only two values. This is precisely the amount of information that can be stored in a single bit. Thus, in a standard program, the processor manipulates only single bits, where in fact it could manipulate words of 32 or 64 bits without requiring extra CPU time. There would be a drastic improvement in speed

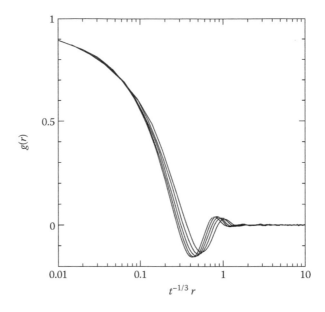

FIGURE 3.8 Radially averaged correlation function, $C(r, t)$, as a function of the rescaled distance $t^{-1/3} r$. The data sets are the same as those in Figure 3.7.

if one could effectively use all bits in a word. An implementation capable of doing precisely this is known as *multispin coding*.

Multispin coding relies heavily on low-level bit manipulation functions. These functions include logical operations like AND ($\wedge$), OR ($\vee$), and exclusive OR ($\oplus$), as well as operations like shifts, additions, and subtractions. Note that these operations are contained in the programming language C as well as in many versions of FORTRAN, and thus there is no need to actually program in assembly language. The low-level C-instructions that we use in this chapter are the logical operations AND, OR, and XOR. Furthermore, we will use the operations LSHIFT and RSHIFT, which shift the bits to the left and right. These operations are listed in Table 3.1.

TABLE 3.1
Logical Operations in Assembly and in the Programming Language C

Logical Operation	Symbol	In Assembly	In C
Bitwise AND	$\wedge$	AND	&
Bitwise OR	$\vee$	OR	|
Exclusive OR	$\oplus$	XOR	$\wedge$
Shift left	$\ll$	LSHIFT	$\ll$
Shift right	$\gg$	RSHIFT	$\gg$

Rather than slowly working toward an actual multispin program, we will postpone the explanation and first present the core of a multispin program. The following program in C simulates the two-dimensional Ising model with spin-flip dynamics and Metropolis acceptance:

```
/* select a random site and its spin                          */
i=N*random();
st=spin[i];
/* determine which neighbors are anti-aligned to it    */
a0=st^spin[nb[i][0]];
a1=st^spin[nb[i][1]];
a2=st^spin[nb[i][2]];
a3=st^spin[nb[i][3]];
/* compute P1 and P2, which tell whether at least 1,
    resp. 2 neighbors are anti-aligned                        */
P2=a1&a2;
P1=a1|a2;
P2 |= P1 & a3;
P1 |= a3;
P2 |= P1 & a4;
P1 |= a4;
/*flip the spin with Metropolis acceptance probability */
bp1=bp[bps++];
bp2=bp[bps++];
spin[i]  ^= P2|(bp1&(P1|bp2));
```

The piece of code presented above presumes that all lattice sites are numbered, such that the spins can be stored in a one-dimensional array called spin. For each site i, the neighbors are stored in an array nb[i]. Moreover, it requires the availability of random bit patterns bp[], which are constructed such that the average density of 1's in this array is $\exp(-4\beta J)$. With these conditions, the values of P1 and P2 indicate whether at least 1, respectively 2, neighboring spins are anti-aligned. It is then relatively straightforward to check whether the spin is flipped with Metropolis acceptance probabilities. If two or more neighbors are anti-aligned, the spin is surely flipped. However, if one neighbor is anti-aligned, the probability of flipping it is $\exp(-4\beta J)$. Finally, if all its neighbors are aligned, the probability of flipping it is $\exp(-8\beta J)$.

This leaves us the explanation of how the random bit patterns bp[] can be generated efficiently. A way to do this is to first generate a few thousand such patterns one bit at a time and then to reuse these patterns by reshuffling the bits in these patterns as well as the order of the patterns.

Note that in the piece of code above, the variables spin[] are not bits, but words of typically 32 or 64 bits. Extracting the k-th bit out of these words yields the configuration of simulation k, for k running up to 32 or 64. Thus, at roughly the computational cost of an ordinary simulation, we have now realized 32 or 64 simulations.*

Occasionally, a single large simulation is preferred to a large number of separate smaller simulations. In these cases, we may wish to use a multispin algorithm in which the different bits within a word represent parts of the same system, so that many parts of one system get updated at each step. The simplest way of doing this is to take as a starting point the approach above, which simulates 32 or 64 systems in parallel. Then, the right-side neighbor of the spins stored in the kth bit of the word corresponding with the right-most lattice sites is identified as the kth bit in the word corresponding with the left-most spins. We then "glue" these systems together in a different way. In the positive (negative) x-direction across the periodic boundary, these words of spins are rotated by $+8$ (-8) bits. On the other hand, in the positive (negative) y-direction across the periodic boundary, these words of spins are rotated by $+1$ (-1) bits modulo 8. Another way of looking at this is that the 32 or 64 separate systems are now connected to each other in a 4×8 or 8×8 superlattice with periodic boundary conditions.

Once the basic idea of multispin coding is clear, extensions of the above code to three dimensions, to other kinds of lattices, or to spin-exchange dynamics are straightforward; we leave that to the reader.

3.4.2 REJECTION-FREE IMPLEMENTATION

Up to now, each spin is proposed to be flipped once per unit of time. However, once the domains are sizable, most spins are surrounded by aligned neighbors, and a spin with four aligned neighbors succeeds in actually flipping only with a probability of $P_{acc} = \exp(-8\beta J)$. In simulations of domain growth, the system's temperature is necessarily below the critical temperature, that is, $\beta J > 0.44069$, and the acceptance probability is thus below 0.03. It is a bit of a waste to do the effort to compute the energy if a spin is flipped, and then not do that 29 out of 30 times.

A completely different approach to improve the computational efficiency of the simulations, while preserving the dynamics, exploits the property that the dynamics is determined by the product of the proposition rate and the acceptance probability, and the computational effort is proportional to the proposition rate. The optimal algorithm would then have the maximal acceptance probability of 1 and a proposition rate proportional to Equation 3.9.

A practical way to implement this is to group spins with the same number of aligned neighbors together in lists. For Ising model simulations on a square lattice with spin-flip dynamics at Metropolis acceptance probabilities, the rate of flipping of a spin with j aligned neighbors is

$$r_j = \min\{1, \exp[(8 - 4j)\beta J]\} . \tag{3.18}$$

---

*Although each of these simulations is correct on its own, the simulations are not completely independent from each other, as the site selection is shared. This does not bias the results, but the possibility of correlations between the results might complicate error estimations.

Denoting the number of spins with j aligned neighbors as N_j, one step of the algorithm then proceeds as follows:

1. Increment the timescale with $\Delta t = \left(\sum_j N_j r_j \right)^{-1}$.
2. Determine from which list one selects the spin to be flipped; the probability that the list with k aligned neighbors is selected is $N_k r_k \Delta t$.
3. Select randomly one of the spins on this list and flip it.

How do we compare the efficiency of this *rejection-free* implementation with a standard implementation? A step of the rejection-free implementation is more complex, but also corresponds with a larger time increment. The latter factor depends on the configuration and temperature; for the snapshots shown in Figure 3.1, the time steps in the rejection-free approach are 9.5, 11.8, 14.6, and 17.7 times bigger than in the standard implementation, respectively. So, if the implementation penalty is a factor of 2, the gain is still close to an order of magnitude. Moreover, because Metropolis dynamics has the maximal acceptance probability and is thus optimized for the standard implementation, the computational gain with any other kind of dynamics (heat-bath dynamics, for instance) is bound to be even higher.

A final remark is that the rejection-free approach cannot be combined with multi-spin coding; which of these two yields the highest computational efficiency depends on temperature and density, and also on the domain size reached. It might be that at early times, right after the quench, multispin coding is faster, but that at later times when the domains are big, a rejection-free approach is preferred.

3.4.3 TUNED DYNAMICS FOR MODEL B

The above subsections started from a specific set of rules determining the dynamics of the system—mostly spin-flip or spin-exchange dynamics with Metropolis acceptance probabilities—and then outlined how simulations based on these rules can be implemented highly efficiently. By doing so, we have been more strict than we need to be. In this subsection, we will explore how the dynamical rules can be altered such that the physics of spinodal decomposition can be studied with greater numerical efficiency.

We will restrict ourselves to variations in the dynamics that leave the equilibrium properties intact. Further, in simulations belonging to the dynamical universality class of Model B [12], it is essential that the magnetization be locally conserved. However, within these restrictions, there is still a lot of freedom to modify the dynamics.

In Section 3.3, we have discussed the simulation of diffusion-driven phase separation. In this case, the asymptotic growth law is the LS law, $L(t) \sim t^{1/3}$. However, even though the domain size has grown to the order of 50 lattice distances in Figure 3.5, the curves in Figure 3.8 are still not completely on top of each other. Why is this?

In the movement of particles or spins in the Ising model with local conservation, we must distinguish between two kinds of processes. The first one is *bulk diffusion*, in which individual particles detach from one domain, then make a random walk inside a domain of opposite spin, until they eventually rejoin a domain. The domain growth proceeds due to a slight bias to leave smaller domains and join bigger domains. A different kind of process is *surface diffusion*, which refers to the sliding of particles along the interface between the two types of domains, without ever detaching from the domain. This will also contribute to domain growth because it makes isolated domains move slowly, until they eventually merge with other domains.

In the early stages of spinodal decomposition, surface diffusion wins over bulk diffusion: the detachment step in the latter costs energy, whereas it costs nothing to slide along an interface. However, in the long run, the fraction of interfaces in the system decreases in time, and with it the role of surface diffusion. Thus, only bulk diffusion contributes to the asymptotic domain growth. The seriousness of the problem becomes clear if we realize that bulk diffusion yields the growth law $L \simeq c_b t^{1/3}$, whereas surface diffusion corresponds with the growth law $L \simeq c_s t^{1/4}$ [13]. Because of the activated nature of bulk diffusion, $c_s > c_b$ and surface diffusion dominates at early times. The crossover between these two regimes occurs at a crossover time t_c, where

$$c_b t_c^{1/3} \simeq c_s t_c^{1/4}, \quad \text{or} \quad t_c \simeq \left(\frac{c_s}{c_b}\right)^{12}. \tag{3.19}$$

Therefore, a relatively modest difference in prefactors c_b and c_s postpones the asymptotic growth regime by orders of magnitude. Moreover, the computational problem is even more severe, because at these large times the domains have grown considerably and will only fit in large system sizes. It thus pays off tremendously if surface diffusion can somehow be reduced.

It turns out that a relatively easy modification in the acceptance probabilities can suppress surface diffusion relative to bulk diffusion. Suppose that we characterize the state of a particle or spin on site i by the number m_i of its neighbors that are aligned—from now on, we will call m_i the coordination number. Nearest-neighbor spin exchanges in the Ising model on a square lattice then correspond with transitions from $\{m_i, m_j\}$ to $\{3 - m_i, 3 - m_j\}$. Because the exchanging spins necessarily have at least one anti-aligned neighbor, the coordination numbers of the exchanging spins are limited to 0, 1, 2, or 3, and thus the spin-exchange dynamics is characterized by 16 numbers. Our dynamics should leave equilibrium properties intact, and hence should obey detailed balance; this fixes the ratio between forward and backward transitions. Combined with up-down symmetry, it reduces the dynamical degrees of freedom from 16 to 6, one of which is a simple overall scaling of time. Especially relevant for surface diffusion are the transitions $\{1,2\} \leftrightarrow \{2,1\}$ and $\{1,1\} \leftrightarrow \{2,2\}$, corresponding with the sliding of a particle along a straight interface. Bulk diffusion, on the other hand, involves predominantly the transitions $\{0,3\} \leftrightarrow \{3,0\}$, corresponding with the diffusion of a free-floating particle; and also $\{0,2\} \leftrightarrow \{3,1\}$, corresponding with detachment and reattachment.

Thus, if we want to suppress surface diffusion, we simply reduce the transitions $\{1,2\} \leftrightarrow \{2,1\}$ and $\{1,1\} \leftrightarrow \{2,2\}$, possibly all the way to strictly forbidding any such exchange. This is what we have done to generate the snapshots in Figure 3.9, with all other parameters equal to those in Figure 3.5. The snapshots in Figures 3.9 and 3.5 look very similar, which confirms that we have not introduced unwanted side-effects. However, an analysis of the configurations in Figure 3.9 reveals that

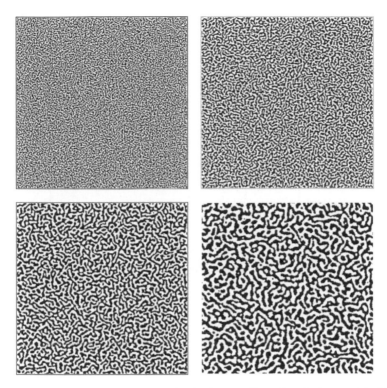

FIGURE 3.9 Snapshots of phase separation in the two-dimensional Ising model with spin-exchange dynamics and Metropolis acceptance probabilities. This simulation is identical to the one that gives Figure 3.5, except that surface diffusion is suppressed (see text). The times at which these snapshots are taken are $t = 10^4 \cdot 10^{i/2}$ MCS with $i = 1, \ldots, 4$.

the domain sizes are slightly smaller than those in Figure 3.5. This is reasonable because the difference between the two simulations lies in the fact that we now forbid certain exchanges; making less moves can hardly be expected to increase the growth rate. More importantly, the numerically determined growth exponent approaches its asymptotic value of 1/3 faster in our new simulation. Let us define the typical domain size as the first zero-crossing of the correlation function. Then, the domain size grows by a factor of 2.04 between $t = 10^4$ and 10^5 MCS in Figure 3.9, instead of a factor of 1.98 in the original simulations. This should be compared with the asymptotic domain growth per decade in time, which gives a factor of $10^{1/3} \simeq 2.15$. The improved convergence to long-time behavior can also be seen if a scaling collapse is made. Figure 3.10 plots the correlation functions $C(r, t)$ as a function of rescaled distance $t^{-1/3}r$. This looks very similar to the earlier scaling collapse in Figure 3.8, except that the locations of zero-crossings and minima are now closer together.

It is also possible to almost completely shut down bulk diffusion by not making any spin exchanges of the type $\{0,3\} \leftrightarrow \{3,0\}$. This model is the microscopic counterpart of a coarse-grained model referred to as *Model S* [13]. The domain growth rate is significantly slower in these simulations and is consistent with the expected

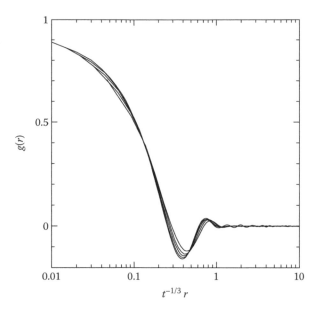

FIGURE 3.10   Analogous to Figure 3.8, but for the evolution shown in Figure 3.9.

behavior of $L \sim t^{1/4}$ for Model S [14]. Interestingly, simulations of this kind lead to domain structures that are morphologically very similar to those with usual Model B dynamics, even though the growth laws are clearly different.

REFERENCES

1. Pathria, R.K., *Statistical Mechanics*, Second Edition, Butterworth-Heinemann, 1996.
2. Binder, K. and Heermann, D.W., *Monte Carlo Simulation in Statistical Physics: An Introduction*, Fourth Edition, Springer-Verlag, Berlin, 2002.
3. Newman, M.E.J. and Barkema, G.T., *Monte Carlo Methods in Statistical Physics*, Oxford University Press, Oxford, 1999.
4. Glauber, R.J., Time-dependent statistics of the Ising model, *J. Math. Phys.*, 4, 294, 1963.
5. Porod, G., General theory, in *Small-Angle X-Ray Scattering*, Glatter, O. and Kratky, O., Eds., Academic Press, New York, 1982, 17–52.
6. Kawasaki, K., Kinetics of Ising models, in *Phase Transitions and Critical Phenomena*, Vol. 2, Domb, C. and Green, M.S., Eds., Academic Press, London, 1972, 443.
7. Kawasaki, K., Diffusion constants near the critical point for time-dependent Ising models. I, *Phys. Rev.*, 145, 224, 1966.
8. Lifshitz, I.M. and Slyozov, V.V., The kinetics of precipitation from supersaturated solid solutions, *J. Phys. Chem. Solids*, 19, 35, 1961.
9. Huse, D.A., Corrections to late-stage behavior in spinodal decomposition: Lifshitz–Slyozov scaling and Monte Carlo simulations, *Phys. Rev. B*, 34, 7845, 1986.
10. Amar, J., Sullivan, F., and Mountain, R.D., Monte Carlo study of growth in the two-dimensional spin-exchange kinetic Ising model, *Phys. Rev. B*, 37, 196, 1988.
11. Marko, J.F. and Barkema, G.T., Phase ordering in the Ising model with conserved spin, *Phys. Rev. E*, 52, 2522, 1995.

12. Hohenberg, P.C. and Halperin, B.I., Theory of dynamic critical phenomena, *Rev. Mod. Phys.*, 49, 435, 1977.
13. Puri, S., Bray, A.J., and Lebowitz, J.L., Segregation dynamics in systems with order-parameter dependent mobilities, *Phys. Rev. E*, 56, 758, 1997.
14. van Gemmert, S., Barkema, G.T., and Puri, S., Phase separation driven by surface diffusion: A Monte Carlo study, *Phys. Rev. E*, 72, 046131, 2005.

4 Using the Lattice Boltzmann Algorithm to Explore Phase Ordering in Fluids

Giuseppe Gonnella and Julia M. Yeomans

CONTENTS

4.1 INTRODUCTION

In Section 1.5.2, arguments were presented for domain growth laws in phase-separating fluids. An important tool for understanding the asymptotic behavior of segregating fluids is the lattice Boltzmann method. In this chapter, we present details of the lattice Boltzmann method to solve hydrodynamic equations of motion. The diversity of the processes and patterns involved is illustrated by Figure 4.1.

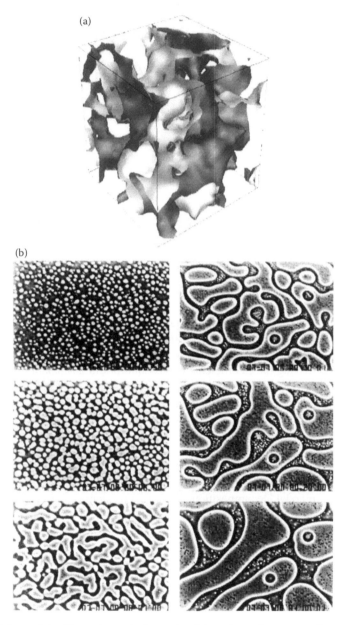

FIGURE 4.1a and b Examples to indicate the rich variety of structures formed as fluids phase-separate. (a) Three-dimensional morphology of a phase-separating dPB (perdeuterated polybutadiene) and PI (polyisoprene) polymer blend. (From Jinnai, H., et al., *Macromolecules*, 30, 130, 1997. With permission.) (b) Time evolution of the phase ordering of a two-dimensional binary mixture of poly(vinyl methyl ether) (PVME) and water. (Reprinted with permission from Tanaka, H., *Phys. Rev. E*, 51, 1313, 1995; © 1995 The American Physical Society, http://link.aps.org/doi/10.1103/PhysRevE.51.1313)

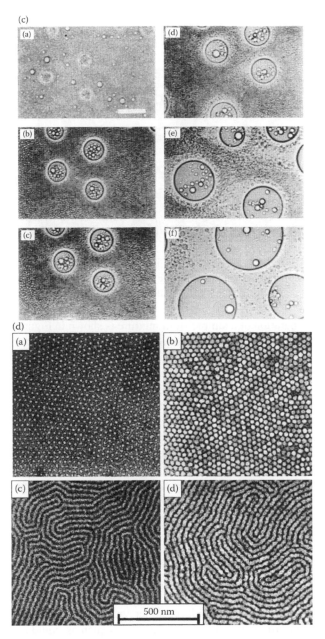

FIGURE 4.1c and d (c) Double phase separation in a two-dimensional binary mixture of ϵ-caprolactone oligomer (OCL) and styrene oligomer (OS). Hydrodynamics is dominant and drives the domains circular long before the order parameter in each domain has reached its equilibrium value. If the order parameter still lies within the spinodal regime, further spinodal decomposition can occur, leading to nested circular domains. (Reprinted with permission from Tanaka, H., *Phys. Rev. E*, 51, 1313, 1995; © 1995 The American Physical Society, http://link.aps.org/doi/10.1103/PhysRevE.51.1313) (d) Patterns formed on a silicon nitride-coated silicon wafer using a microphase-separated diblock copolymer film as a template. (From Park, M., et al., *Science*, 276, 1401, 1997. With permission.)

(e)

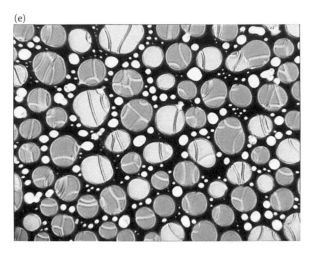

FIGURE 4.1e   (e) Phase transition from the isotropic liquid (black) to the cholesteric phase (bright droplets) of a multi-component commercial liquid-crystal mixture. (From Dierking, I., *Textures of Liquid Crystals*, Wiley-VCH, Weinheim, 2003. With permission.)

The phase separation of a simple fluid mixture is shown in Figure 4.1a and b. The ordering is somewhat analogous to that of a binary alloy described in Section 1.5.1. However, as stressed in Section 1.5.2, because we are considering fluids, there is a new aspect to consider: these materials are liquid, and flow can be important. The simple diffusion equation that describes phase ordering in a binary alloy is replaced by a convection-diffusion equation (cf. Equation 1.185) together with the Navier–Stokes equation of hydrodynamics (cf. Equation 1.187). We have seen earlier that this leads to a richer behavior with several different growth regimes.

Section 4.2 of this chapter discusses the model used to describe the hydrodynamics and phase ordering of binary mixtures of simple fluids. The corresponding equations of motion are complicated, and a numerical approach is called for. One that has proved particularly successful is the lattice Boltzmann algorithm, and we devote the rest of Section 4.2 to explaining the details of this method.

We then describe, in Section 4.3, the current understanding of phase ordering in a binary fluid. Simple scaling arguments are compared with the results from lattice Boltzmann solutions of the equations of motion first in three, and then in two, dimensions. As an example of a driven system, we explore the effect of a shear flow on phase ordering in binary fluids.

Phase ordering in simple liquids is rather well understood, but there is still a lot to learn about ordering in more complex fluids such as liquid crystals, colloids, polymeric fluids, and their mixtures. These materials can have a wide range of different ordered states, for example, the nematic, smectic, cholesteric, and blue phases of liquid crystals. They can also exhibit highly non-Newtonian flow behavior, such as shear-thinning in polymer solutions. Phase ordering in complex fluids is important for material design: the process can be stopped mid-way by an increase in fluid viscosity, a temperature quench, or free-energy barriers to give materials with novel microstructures. For example, Figure 4.1d shows patterns formed on a silicon nitride–coated silicon wafer using a microphase-separated diblock copolymer film as a template [3]. (A detailed discussion of the evolution kinetics of microphase structures is presented in Chapter 7.)

In Section 4.4, we give examples of lattice Boltzmann simulations of phase ordering in complex fluids. We consider lamellar phases, such as those found in amphiphilic fluids or copolymer systems, and ordering in the nematic phase of liquid crystals. We then discuss some examples of very recent numerical and experimental work, thus illustrating a range of questions currently at the forefront of research.

4.2   LATTICE BOLTZMANN FOR BINARY FLUIDS

In this section, we introduce a model for the equilibrium properties and dynamics of a binary fluid mixture. The equilibrium properties will be modeled on a mesoscopic length scale by means of a free-energy functional, which is minimized in equilibrium. As the aim is to describe the universal properties of phase-ordering systems, a mesoscopic approach (which ignores molecular detail) is appropriate. The dynamics is described by the convection-diffusion equation and the Navier–Stokes equation of hydrodynamics, but with the bulk pressure generalized to a pressure tensor. This is needed in a system with concentration gradients.

4.2.1   Thermodynamics and Equations of Motion of a Binary Fluid

Consider a binary fluid, which consists of A and B molecules and phase-separates into an A-rich and a B-rich phase below a critical temperature T_c (see Figure 1.3). Figure 4.2 shows typical evolution snapshots resulting from temperature quenches to different points in the phase-separated region of the phase diagram. The equilibrium properties of the fluid can be described by the free energy [6]

$$F = \int d\vec{r} \left[f(n, \psi, T) + \frac{1}{2} K (\vec{\nabla} \psi)^2 \right]. \tag{4.1}$$

Here, $n = n_A + n_B$ is the total density, where n_A, n_B are the number densities of A and B, respectively. Furthermore, $\psi = n_A - n_B$ is the concentration, which is the order parameter of the binary fluid. T is the temperature, and we take the Boltzmann constant $k_B = 1$. The first term in Equation 4.1 describes the bulk behavior of the fluid. A simple choice is (cf. Equation 1.8)

$$f(n, \psi, T) = \frac{\lambda}{4} n \left(1 - \frac{\psi^2}{n^2} \right) + T \left(\frac{n + \psi}{2} \right) \ln \left(\frac{n + \psi}{2} \right) + T \left(\frac{n - \psi}{2} \right) \ln \left(\frac{n - \psi}{2} \right), \tag{4.2}$$

where λ measures the strength of the interaction between the two components. For T less than a critical temperature $T_c = \lambda/2$ (this value is obtained by setting $a = 0$

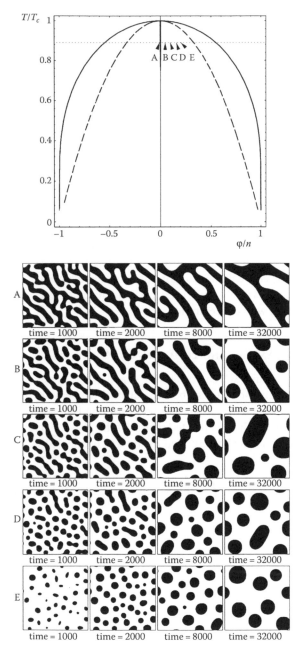

FIGURE 4.2 Phase diagram of a binary fluid and typical phase-separation morphologies. Below a critical temperature T_c, there is phase separation into an A-rich and a B-rich phase. The full curve is the phase boundary, and the dashed curve is the spinodal line within which spinodal decomposition occurs. The frames below show the evolution in time of a two-dimensional binary system quenched to the points A, B . . . E. (After Wagner, A.J., *Theory and Application of the Lattice Boltzmann Method*, D. Phil. Thesis, University of Oxford, 1998.)

in the expansion of $f(n, \psi, T)$ in Equation 4.5), the system phase separates into two phases with concentrations:

$$\psi_\pm = \pm \sqrt{3n^2 \left(\frac{\lambda}{2T} - 1 \right)}. \tag{4.3}$$

In Equation 4.1, the derivative in ψ assigns a free-energy cost to any variation in the order parameter; for example, across an interface between the two phases. Hence, the coefficient K is related to the surface tension [7]

$$\sigma = \int_{-\infty}^{\infty} dz \left[f(n, \psi, T) - f(n, \psi_\pm, T) + \frac{K}{2} \left(\frac{d\psi}{dz} \right)^2 \right]$$

$$= \int_{\psi_-}^{\psi_+} d\psi \sqrt{2K \left[f(n, \psi, T) - f(n, \psi_\pm, T) \right]}. \tag{4.4}$$

As discussed in Section 1.4.1, the first integral is calculated along a direction normal to a flat interface between the two bulk phases $\psi_\pm$. Close to the critical point, $f(n, \psi, T)$ in Equation 4.2 can be approximated as

$$f(n, \psi, T) \simeq \frac{a}{2} \psi^2 + \frac{b}{4} \psi^4,$$

$$a = \frac{T}{n} - \frac{\lambda}{2n}, \quad b = \frac{T}{3n^3}. \tag{4.5}$$

The corresponding concentration profile between two domains is the kink (or anti-kink) solution shown in Figure 1.5:

$$\psi_s(z) = \sqrt{\frac{-a}{b}} \tanh \left(\frac{2z}{\xi_b} \right). \tag{4.6}$$

The interface width is

$$\xi_b = 2 \sqrt{\frac{-2K}{a}}, \tag{4.7}$$

and the surface tension is

$$\sigma = \frac{2}{3} \sqrt{\frac{-2a^3 K}{b^2}}. \tag{4.8}$$

We can use the free energy to obtain other thermodynamic functions by differentiation [7]. We shall need the chemical potential conjugate to the concentration difference and the pressure tensor. The chemical potential is the functional derivative of the free energy with respect to the concentration

$$\mu(\vec{r}) = \frac{\delta F}{\delta \psi}$$

$$= -\frac{\lambda}{2} \frac{\psi}{n} + \frac{T}{2} \ln \left(\frac{n + \psi}{n - \psi} \right) - K \nabla^2 \psi. \tag{4.9}$$

For a homogeneous system, the bulk pressure is the functional derivative of the free energy with respect to the volume, and the condition for thermodynamic equilibrium is that the pressure is constant. However, for a fluid with concentration gradients, we need a more general equilibrium condition

$$\partial_\alpha P_{\alpha\beta} = 0. \tag{4.10}$$

For example, there is a pressure drop across a curved interface that must be balanced by a pressure gradient tangential to the interface. [In Equation 4.10 and in the following, Greek subscripts label Cartesian directions and the usual sum over repeated indices will be assumed.] A suitable choice of pressure tensor, which reduces to the usual bulk pressure if no concentration gradients are present, is [8]

$$P_{\alpha\beta} = \left(n\frac{\delta F}{\delta n} + \psi\frac{\delta F}{\delta \psi} - f \right) \delta_{\alpha\beta} + D_{\alpha\beta}(\psi), \tag{4.11}$$

where f is the free-energy density. Here, a symmetric tensor $D_{\alpha\beta}(\psi)$ is added to ensure that the condition of mechanical equilibrium in Equation 4.10 is satisfied. For the free-energy choice in Equations 4.1 and 4.2,

$$P_{\alpha\beta} = p_B\delta_{\alpha\beta} + P_{\alpha\beta}^{\text{chem}}, \tag{4.12}$$

with

$$p_B = nT. \tag{4.13}$$

Furthermore,

$$P_{\alpha\beta}^{\text{chem}} = -K\left[\psi\partial_\gamma\partial_\gamma\psi + \frac{1}{2}(\partial_\gamma\psi)(\partial_\gamma\psi) \right]\delta_{\alpha\beta} + K(\partial_\alpha\psi)(\partial_\beta\psi), \tag{4.14}$$

where the last term corresponds with the tensor $D_{\alpha\beta}$. In isothermal lattice Boltzmann models for simple fluids T can be identified with $c^2/3$ [9] where c is the lattice velocity defined in the next section. Notice that the pressure tensor satisfies

$$n\partial_\alpha\frac{\delta F}{\delta n} + \psi\partial_\alpha\frac{\delta F}{\delta \psi} = \partial_\beta P_{\alpha\beta}. \tag{4.15}$$

As discussed in Section 1.5.2, the hydrodynamic equations of fluids follow from the conservation laws for mass and momentum. For binary mixtures at constant temperature, the evolution of the density, velocity, and concentration fields is described by the continuity, the Navier–Stokes, and the convection-diffusion equations [10,11]

$$\partial_t n + \partial_\alpha (nu_\alpha) = 0, \tag{4.16}$$

$$\partial_t (nu_\beta) + \partial_\alpha (nu_\alpha u_\beta) = -\partial_\alpha P_{\alpha\beta} + \partial_\alpha\left[\eta\left(\partial_\alpha u_\beta + \partial_\beta u_\alpha - \frac{\delta_{\alpha\beta}}{d}\partial_\gamma u_\gamma \right) + \zeta\delta_{\alpha\beta}\partial_\gamma u_\gamma \right], \tag{4.17}$$

$$\partial_t \psi + \partial_\alpha (\psi u_\alpha) = D\partial_\alpha\partial_\alpha\mu. \tag{4.18}$$

Here, η and ζ are the shear and bulk viscosities, D is the diffusion constant, and d is the dimensionality of the system. Note that the influence of the free energy appears in the equations of motion through the pressure tensor and the chemical potential. In Section 1.5.2, we had imposed the condition that the fluid was incompressible. We will use this condition later on in this chapter.

The model described in this section is a natural one for describing phase ordering because the interfaces between the different phases, which control the ordering process, appear and move naturally as the system evolves toward its free-energy minimum. It should be contrasted with more traditional approaches, where each of the fluid components is described independently by a simple Navier–Stokes equation, and explicit interface tracking is used to follow the dynamics of the interfaces. This makes interface patterns that move and change their topology particularly difficult to handle.

4.2.2 LATTICE BOLTZMANN MODEL FOR A BINARY FLUID

In general, Equations 4.16 through 4.18 must be solved numerically. A widely used approach is the lattice Boltzmann algorithm, and we now describe this in some detail. The remainder of Section 4.2 is somewhat technical and is not needed to understand the rest of the chapter.

The lattice Boltzmann algorithm [9,12–16] is defined in terms of the dynamics of a set of real numbers moving along fixed directions of a lattice in discrete time. To illustrate the method, we shall choose a two-dimensional square lattice with first and second neighbor interactions. This is one of the simplest geometries that can lead to an algorithm that reproduces the Navier–Stokes equations in the continuum limit. Horizontal and vertical links have length Δx and diagonal links $\sqrt{2}\Delta x$. If Δt is the simulation time step and $c = \Delta x / \Delta t$, the lattice vectors (shown in Figure 4.3)

$$\vec{e}_0 = (0,0),$$

$$\vec{e}_1 = (c,0), \quad \vec{e}_2 = (0,c), \quad \vec{e}_3 = (-c,0), \quad \vec{e}_4 = (0,-c),$$

$$\vec{e}_5 = (c,c), \quad \vec{e}_6 = (-c,c), \quad \vec{e}_7 = (-c,-c), \quad \vec{e}_8 = (c,-c) \qquad (4.19)$$

are velocity vectors connecting the first and second neighbors of the lattice, together with the zero velocity vector $\vec{e}_0$.

In lattice Boltzmann models for one-component fluids, a set of distribution functions $\{f_i(\vec{x},t)\}$ is defined on each lattice site $\vec{x}$. Each of these can be interpreted as the density of fluid that will move in direction i at time t. If we want to solve the equations of motion of a binary fluid, we need to consider two sets of distribution functions on each site, $\{f_i\}$ and $\{g_i\}$. These are related to the fluid density n, its momentum $n\vec{u}$, and the concentration field ψ through

$$n(\vec{x},t) = \sum_i f_i(\vec{x},t), \qquad (4.20)$$

$$n(\vec{x},t)u_\alpha(\vec{x},t) = \sum_i f_i(\vec{x},t)e_{i\alpha}, \qquad (4.21)$$

$$\psi(\vec{x},t) = \sum_i g_i(\vec{x},t). \qquad (4.22)$$

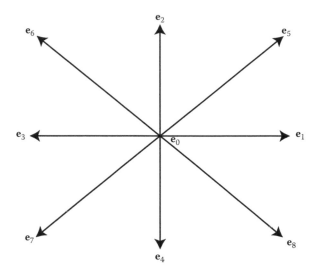

FIGURE 4.3   The velocity vectors used in a two-dimensional lattice Boltzmann algorithm.

In the most usual implementation of the lattice Boltzmann algorithm, the $\{f_i\}$ and the $\{g_i\}$ evolve according to single-relaxation-time Boltzmann equations [17]

$$f_i(\vec{x} + \vec{e}_i \Delta t, t + \Delta t) - f_i(\vec{x}, t) = -\frac{1}{\tau_f}(f_i - f_i^0), \tag{4.23}$$

$$g_i(\vec{x} + \vec{e}_i \Delta t, t + \Delta t) - g_i(\vec{x}, t) = -\frac{1}{\tau_g}(g_i - g_i^0). \tag{4.24}$$

Here, τ_f and τ_g are relaxation times related to the viscosity and diffusion constants, respectively.

The equilibrium distribution functions f_i^0 and g_i^0 must be chosen in such a way that the conservation laws are obeyed. They must also ensure the correct equilibrium properties for the system. It follows immediately from the evolution Equations 4.23 and 4.24 that the total density, momentum, and concentration are conserved in the relaxation step if the zeroth and first moments of the equilibrium distribution function obey

$$\sum_i f_i^0(\vec{x}, t) = n(\vec{x}, t), \tag{4.25}$$

$$\sum_i f_i^0(\vec{x}, t)e_{i\alpha} = n(\vec{x}, t)u_\alpha(\vec{x}, t), \tag{4.26}$$

$$\sum_i g_i^0(\vec{x}, t) = \psi(\vec{x}, t). \tag{4.27}$$

Moreover, it can be shown that the Navier–Stokes and convection-diffusion equations are reproduced in the continuum limit if higher moments of the distribution functions are constrained by

$$\sum_i f_i^0 e_{i\alpha} e_{i\beta} = P_{\alpha\beta} + nu_\alpha u_\beta,$$ (4.28)

$$\sum_i g_i^0 e_{i\alpha} = \psi u_\alpha,$$ (4.29)

$$\sum_i g_i^0 e_{i\alpha} e_{i\beta} = c^2 \Gamma \mu \delta_{\alpha\beta} + \psi u_\alpha u_\beta.$$ (4.30)

Here, $P_{\alpha\beta}$ is the pressure tensor in Equation 4.12, μ is the chemical potential in Equation 4.9, and Γ is a mobility that can be used to tune the diffusion constant. We are considering a fluid where the two components have the same density and viscosity, and Equation 4.29 implies that the two components relax instantaneously to the same velocity.

The constraints in Equations 4.25 through 4.30 can be implemented by considering an expansion of the f_i^0 and g_i^0 to second order in the velocity:

$$f_i^0 = A^\sigma + B^\sigma e_{i\alpha} u_\alpha + C^\sigma u^2 + D^\sigma (e_{i\alpha} u_\alpha)^2 + G_{\alpha\beta}^\sigma e_{i\alpha} e_{i\beta},$$ (4.31)

$$g_i^0 = H^\sigma + J^\sigma e_{i\alpha} u_\alpha + K^\sigma u^2 + Q^\sigma (e_{i\alpha} u_\alpha)^2,$$ (4.32)

where the label σ on the coefficients takes values $0, 1, 2$ corresponding with velocity magnitudes $0, c$, and $\sqrt{2}c$, respectively.

In the next section, we shall show how to find expressions for the coefficients $A^\sigma, B^\sigma, C^\sigma, \ldots$ in the expansion of f_i^0 and g_i^0. We shall then derive the continuum limit of the algorithm and show that it does indeed solve Equations 4.16 through 4.18.

4.2.3 DERIVING THE COEFFICIENTS IN THE EQUILIBRIUM DISTRIBUTION

Inspection of Figure 4.3 shows that sums over the velocity vectors can be written as

$$\sum_i e_{i\alpha} = 0 + 0,$$

$$\sum_i e_{i\alpha} e_{i\beta} = 2c^2 \delta_{\alpha\beta} + 4c^2 \delta_{\alpha\beta},$$

$$\sum_i e_{i\alpha} e_{i\beta} e_{i\gamma} = 0 + 0,$$

$$\sum_i e_{i\alpha} e_{i\beta} e_{i\gamma} e_{i\delta} = 2c^4 \delta_{\alpha\beta\gamma\delta} + \left[4c^4 \left(\delta_{\alpha\beta} \delta_{\gamma\delta} + \delta_{\alpha\gamma} \delta_{\beta\delta} + \delta_{\alpha\delta} \delta_{\beta\gamma} \right) - 8c^4 \delta_{\alpha\beta\gamma\delta} \right],$$ (4.33)

where, to help see where the expressions come from, we have written the sums over first and second neighbors separately.

The coefficients in the expansion of the equilibrium distribution functions in Equations 4.31 and 4.32 can be fixed by imposing the relations 4.25–4.30 and then using Equation 4.33. We demonstrate how the calculation of the coefficients appearing in the expansion of f_i^0 proceeds. Using Equation 4.31 to sum the left-hand-sides of Equations 4.25, 4.26, and 4.28 gives

$$\sum_i f_i^0 = A^0 + 4A^1 + 4A^2 + \left(C^0 + 4C^1 + 4C^2\right)u^2 + 2c^2\left(D^1 + 2D^2\right)u^2$$

$$+ 2c^2\left(G_{\alpha\beta}^1 + 2G_{\alpha\beta}^2\right)\delta_{\alpha\beta}$$

$$\equiv n, \tag{4.34}$$

$$\sum_i f_i^0 e_{i\alpha} = \left(2B^1 + 4B^2\right)c^2 u_\alpha$$

$$\equiv n u_\alpha, \tag{4.35}$$

$$\sum_i f_i^0 e_{i\alpha} e_{i\beta} = \left[\left(2A^1 + 4A^2\right)c^2 + \left(2C^1 + 4C^2\right)c^2 u^2 + 4c^4\left(D^2 u^2 + G_{\gamma\delta}^2 \delta_{\gamma\delta}\right)\right]\delta_{\alpha\beta}$$

$$+ 8c^4\left(D^2 u_\alpha u_\beta + G_{\alpha\beta}^2\right) + \left[\left(2D^1 - 8D^2\right)u_\gamma u_\sigma\right.$$

$$\left. + \left(2G_{\gamma\sigma}^1 - 8G_{\gamma\sigma}^2\right)\right]c^4 \delta_{\alpha\beta\gamma\sigma}$$

$$\equiv P_{\alpha\beta} + n u_\alpha u_\beta. \tag{4.36}$$

One way of satisfying these equations is by choosing

$$A^0 = n - 20A^2, \qquad A^1 = 4A^2, \qquad A^2 = \frac{P_{\alpha\beta}\delta_{\alpha\beta}}{24c^2}, \tag{4.37}$$

$$B^1 = 4B^2, \qquad B^2 = \frac{n}{12c^2}, \tag{4.38}$$

$$C^0 = -\frac{2n}{3c^2}, \qquad C^1 = 4C^2, \qquad C^2 = -\frac{n}{24c^2}, \tag{4.39}$$

$$D^1 = 4D^2, \qquad D^2 = \frac{n}{8c^4}, \tag{4.40}$$

$$G_{\alpha\beta}^1 = 4G_{\alpha\beta}^2, \qquad G_{\alpha\beta}^2 = \frac{P_{\alpha\beta} - P_{\sigma\sigma}\delta_{\alpha\beta}/2}{8c^4}. \tag{4.41}$$

The coefficients A and G depend on derivatives of the concentration. For a fluid with only isotropic diagonal terms in the pressure tensor, one sees that the coefficients G are zero; indeed, they do not appear in lattice Boltzmann model for simple fluids [9]. Obtaining the coefficients in the expansion of the distributions g_i^0 proceeds in an analogous way. It transpires that they can be simply obtained from the results 4.37–4.41 by making the formal substitutions $n \to \psi$ and $P_{\alpha\beta} \to c^2 \Gamma \mu \delta_{\alpha\beta}$.

4.2.4   THE CONTINUUM LIMIT

In Sections 4.2.2 and 4.2.3, we have defined a lattice Boltzmann algorithm for binary fluids. Now we shall find the continuum limit of the evolution equations 4.23 and 4.24 to check the assertion that the algorithm solves the hydrodynamic equations of motion 4.16–4.18 [18]. Consider first the total density distributions. We Taylor-expand the left-hand side of Equation 4.23 to obtain

$$\sum_{k=1}^{\infty} \frac{\Delta t^k}{k!} (\partial_t + e_{i\alpha}\partial_\alpha)^k f_i = -\frac{1}{\tau_f}(f_i - f_i^0). \tag{4.42}$$

Solving this equation recursively it follows that, to second order in derivatives,

$$(\partial_t + e_{i\alpha}\partial_\alpha)f_i^0 - \Delta t\left(\tau_f - \frac{1}{2}\right)(\partial_t^2 + 2e_{i\alpha}\partial_t\partial_\alpha + e_{i\alpha}e_{i\beta}\partial_\alpha\partial_\beta)f_i^0 + \mathcal{O}(\partial^3)$$

$$= -\frac{1}{\Delta t\,\tau_f}(f_i - f_i^0). \tag{4.43}$$

Summing both sides of Equation 4.43 over i and using Equations 4.25, 4.26 to identify the density and momentum,

$$\partial_t n + \partial_\alpha(nu_\alpha) - \Delta t\left(\tau_f - \frac{1}{2}\right)\left[\partial_t^2 n + 2\partial_t\partial_\alpha(nu_\alpha) + \partial_\alpha\partial_\beta\sum_i f_i^0 e_{i\alpha}e_{i\beta}\right]$$

$$+ \mathcal{O}(\partial^3) = 0. \tag{4.44}$$

Similarly, multiplying Equation 4.43 by $e_{i\beta}$ and summing over i leads to

$$\partial_t(nu_\beta) + \partial_\alpha\sum_i f_i^0 e_{i\alpha}e_{i\beta} - \Delta t\left(\tau_f - \frac{1}{2}\right)$$

$$\times\left[\partial_t^2(nu_\beta) + 2\partial_t\partial_\alpha\sum_i f_i^0 e_{i\alpha}e_{i\beta} + \partial_\alpha\partial_\gamma\sum_i f_i^0 e_{i\alpha}e_{i\beta}e_{i\gamma}\right] + \mathcal{O}(\partial^3) = 0. \tag{4.45}$$

It is not hard to show that Equation 4.44 is equivalent to the continuity equation. From Equations 4.44 and 4.45 respectively,

$$\partial_t n = -\partial_\alpha(nu_\alpha) + \mathcal{O}(\partial^2), \tag{4.46}$$

$$\partial_t(nu_\beta) = -\partial_\alpha\sum_i f_i^0 e_{i\alpha}e_{i\beta} + \mathcal{O}(\partial^2). \tag{4.47}$$

Substituting Equations 4.46 and 4.47 into the curly brackets in Equation 4.44 shows that this term is higher order, leaving us with the continuity equation

$$\partial_t n + \partial_\alpha(nu_\alpha) = 0 + \mathcal{O}(\partial^3). \tag{4.48}$$

With a little more work, Equation 4.45 can be shown to be equivalent to the Navier–Stokes equation. Using Equation 4.47 shows that the first term, together with half the second term, in the curly brackets in Equation 4.45 can be neglected as higher order. Of the remaining terms in Equation 4.45, it follows immediately from Equation 4.28 that

$$\partial_\alpha \sum_i f_i^0 e_{i\alpha} e_{i\beta} = \partial_\alpha P_{\alpha\beta} + \partial_\alpha (nu_\alpha u_\beta). \tag{4.49}$$

Using the expansion of f_i^0 in Equation 4.31, together with the properties of the lattice vectors in Equation 4.33 and the expressions in Equation 4.38, the final term in the curly brackets in Equation 4.45 may be written as

$$\partial_\alpha \partial_\gamma \sum_i f_i^0 e_{i\alpha} e_{i\beta} e_{i\gamma} = \frac{c^2}{3} \partial_\alpha \partial_\gamma (nu_\gamma \delta_{\alpha\beta} + nu_\beta \delta_{\alpha\gamma} + nu_\alpha \delta_{\beta\gamma}). \tag{4.50}$$

Moreover, using Equations 4.28, 4.12, 4.46, and 4.47, it is possible to show that

$$\partial_t \partial_\alpha \sum_i f_i^0 e_{i\alpha} e_{i\beta} = -\partial_\beta \frac{dp_B}{dn} \partial_\gamma (nu_\gamma) + \partial_t \partial_\alpha P_{\alpha\beta}^{\text{chem}} - \partial_\alpha (u_\alpha \partial_\gamma P_{\beta\gamma} + u_\beta \partial_\gamma P_{\alpha\gamma})$$

$$- \partial_\gamma \partial_\alpha (nu_\alpha u_\beta u_\gamma). \tag{4.51}$$

We assume that all the terms except the first can be ignored, and the density can be removed from inside the derivatives in Equation 4.50. This is a good approximation close to the incompressible limit and works well in numerical simulations. Then, substituting Equations 4.49, 4.50, and 4.51 back into Equation 4.45, we obtain the Navier–Stokes equation

$$\partial_t (nu_\beta) + \partial_\alpha (nu_\alpha u_\beta) = -\partial_\alpha P_{\alpha\beta} + c^2 \Delta t \frac{(\tau_f - 1/2)}{3} \partial_\alpha$$

$$\times \left[\partial_\alpha (nu_\beta) + \left(2 - \frac{3}{c^2} \frac{dp_B}{dn} \right) \partial_\beta (nu_\alpha) \right]. \tag{4.52}$$

Here, we can identify the kinematic shear and bulk viscosities as

$$\nu = c^2 \Delta t \left(\frac{2\tau_f - 1}{6} \right),$$

$$\lambda = \nu \left(2 - \frac{3}{c^2} \frac{dp_B}{dn} \right), \tag{4.53}$$

respectively.

In a similar way, using Equations 4.27 and 4.29, one can write an equation analogous to Equation 4.44 for the concentration field

$$\partial_t \psi + \partial_\alpha(\psi u_\alpha) - \left(\tau_g - \frac{1}{2}\right)\Delta t \left[\partial_t^2 \psi + 2\partial_t\partial_\alpha(\psi u_\alpha) + \partial_\alpha\partial_\beta \sum_i g_i^0 e_{i\alpha}e_{i\beta}\right]$$
$$+ \mathcal{O}(\partial^3) = 0. \tag{4.54}$$

The first two terms on the left-hand side of this equation give

$$\partial_t \psi = -\partial_\alpha(\psi u_\alpha) + \mathcal{O}(\partial^2). \tag{4.55}$$

Thus the first term, together with half of the second term, in the square bracket of Equation 4.54 are of higher order and can be neglected. Using Equations 4.55, 4.47, and 4.46, one may write

$$\partial_\alpha\partial_t\psi u_\alpha = \partial_\alpha \left[u_\alpha\partial_t\psi + \frac{\psi}{n}\partial_t(nu_\alpha) - \frac{\psi u_\alpha}{n}\partial_t n\right]$$
$$= -\partial_\alpha \left[u_\alpha\partial_\beta(\psi u_\beta) + \frac{\psi}{n}\partial_\beta\left(\sum_i f_i^0 e_{i\alpha}e_{i\beta}\right) - \frac{\psi u_\alpha}{n}\partial_\beta(nu_\beta)\right]. \tag{4.56}$$

Inserting Equation 4.56 and the expression 4.30 in the remaining terms of the square bracket of Equation 4.54 and using Equation 4.28 gives the convection-diffusion equation 4.18 with diffusion constant

$$D = \Gamma\left(\tau_g - \frac{1}{2}\right)\Delta t\, c^2, \tag{4.57}$$

together with a higher-order derivative term proportional to $\partial_\alpha(\frac{\psi}{n}\partial_\beta P_{\alpha\beta})$.

4.3   PHASE ORDERING IN BINARY FLUIDS

In Section 1.5.2, we had discussed some aspects of phase-separation dynamics in binary fluid mixtures. In this section, we present a detailed discussion of this problem, along with results obtained from lattice Boltzmann simulations. As usual, we consider a binary fluid that is quenched from the disordered phase to below the critical temperature T_c. If the system is suddenly cooled into the spinodal region (see Figure 4.2), any fluctuation is unstable, and domains of the two phases form and grow. At early times, the velocity field is negligible, and the evolution of the concentration field ψ can be described by Equation 4.18 with $\vec{u} = 0$. Later the coupling between concentration and velocity becomes relevant, and both the convection-diffusion equation 4.18 and the Navier–Stokes equation 4.17 need to be considered.

Examples of the evolution of the order parameter in a two-dimensional binary fluid following a quench are shown in Figure 4.2. The average size of the domains grows with time, and one would like to characterize the growth. Just as for the many examples already considered in this book, typical length scales generally grow as a power law [19]

$$L \sim t^{\phi}. \tag{4.58}$$

The growth exponent ϕ depends on the physical mechanism operating during phase separation. A dynamical-scaling symmetry also holds that relates different spatial scales at different times. Recall the definition of the correlation function $C(\vec{r}, t)$ from Equation 1.88:

$$C(\vec{r}, t) \equiv \frac{1}{V} \int d\vec{R} \left[\langle \psi(\vec{R}, t) \psi(\vec{R} + \vec{r}, t) \rangle - \langle \psi(\vec{R}, t) \rangle \langle \psi(\vec{R} + \vec{r}, t) \rangle \right], \tag{4.59}$$

where the angular brackets indicate an average over different realizations of the system. The dynamical scaling property implies that

$$C(\vec{r}, t) = g\left(\frac{r}{L}\right), \tag{4.60}$$

and the structure factor $S(\vec{k}, t)$, corresponding with the Fourier transform of $C(\vec{r}, t)$, behaves as

$$S(\vec{k}, t) = L^d f(kL). \tag{4.61}$$

The rest of this section is organized as follows. We shall first explain how a simple scaling analysis can help to identify possible growth exponents for a binary fluid. These were briefly discussed in Section 1.5.2. In three dimensions, the existence of the different growth regimes has been confirmed by lattice Boltzmann solutions of the equations, and these results are discussed in Section 4.3.2. In Section 4.3.3, where we consider domain growth in two dimensions, we shall see that the simple scaling picture is not always confirmed and that more complex behavior can occur. Finally, we consider phase ordering in a sheared binary fluid, discussing the competition between the ordering tendency of the free energy and the disruption caused by the shear flow.

4.3.1 SCALING ANALYSIS

In many cases, binary mixtures can be considered incompressible so that Equations 4.17 and 4.18 reduce to

$$n(\partial_t u_\beta + u_\alpha \partial_\alpha u_\beta) = -\partial_\alpha P_{\alpha\beta} + \eta \partial_\alpha \partial_\alpha u_\beta, \tag{4.62}$$

$$\partial_t \psi + u_\alpha \partial_\alpha \psi = D \partial_\alpha \partial_\alpha \mu. \tag{4.63}$$

The existence of three different growth regimes can be deduced by analyzing these equations, assuming that there is only one relevant length scale.

4.3.1.1 Diffusive Growth

At early times after a quench, and in highly viscous fluids, the velocities are small, and it is sufficient to consider the nonlinear diffusion equation $\partial_t \psi = D\nabla^2 \mu$. The chemical potential behaves as $\mu \sim \sigma/L$, where σ is the surface tension [19]. Dimensional analysis then gives $L \sim (D\sigma t)^{1/3}$ and hence $\phi = 1/3$. The growth mechanism is the diffusion of molecules from smaller, higher-curvature domains to larger, lower-curvature ones. Diffusive growth has been experimentally observed in fluid systems [20–22]. It is also the mechanism for phase ordering in a binary alloy or an Ising model with Kawasaki dynamics, as discussed in Section 1.5.1.

4.3.1.2 Viscous Hydrodynamic Growth

In this regime, hydrodynamics becomes relevant, but velocities are still small and slowly varying. Therefore, the inertial terms on the left-hand side of Equation 4.62 are negligible, and interfacial forces balance the viscous term on the right-hand side. From Equation 4.15, one sees that $\partial_\alpha P_{\alpha\beta}$ includes the term $\psi\partial_\beta\mu$, and as $\mu \sim \sigma/L$, one has $\partial_\alpha P_{\alpha\beta} \sim \sigma/L^2$. Assuming that velocity gradients are also controlled by the same length scale L, and that $\vec{u}$ behaves like the velocity of interfaces $\dot{L}$, the viscous term scales as $\eta\dot{L}/L^2$. Balancing the two expressions gives $L \sim \sigma t/\eta$. The linear growth was first explained by Siggia [23] as driven by the Rayleigh instability. Tubes of fluid are unstable to fluctuations of their interfaces, and this results in pinch-off. The necks formed are quickly retracted into neighboring volumes of fluid, thus causing them to grow. Matching L between the diffusive and viscous regimes gives a crossover length and time $L_d \sim (D\eta)^{1/2}$, $t_d \sim \sqrt{D\eta^3/\sigma^2}$.

4.3.1.3 Inertial Hydrodynamic Growth

At higher velocities, the physics is determined by the balance between interfacial forces and the inertial term on the left-hand side of the Navier–Stokes equation [24]. Under the same assumptions as before, the inertial term scales as $n\ddot{L}$ so that the matching gives $L \sim (\sigma/n)^{1/3}t^{2/3}$. The crossover between the viscous and inertial regimes is expected to occur at $L_h \sim \eta^2/(n\sigma)$, $t_h \sim \eta^3/(n\sigma^2)$.

Interfacial forces drive both viscous and inertial hydrodynamic growth. They depend on curvature gradients, and their action can be described in terms of the Laplace–Young equation [25]. This equation says that the pressure difference between the inside and outside of domains is given by the surface tension times the sum of the inverse of the principal radii of curvature of the interface. The pressure is greater on the side where the interface is convex. Curvature differences induce tangential pressure gradients and hence a velocity field that favors the smoothing of protuberances and dimples from interfaces, thus driving domains spherical. An example is given in Figure 4.4. Note that once the domains are spherical, neither viscous nor inertial hydrodynamic growth can cause any further evolution of the domain pattern.

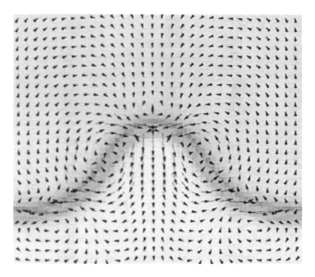

FIGURE 4.4   Flow fields around an interface, which act to equalize the curvature.

4.3.2 SPINODAL DECOMPOSITION IN THREE DIMENSIONS

Dimensional analysis has suggested the existence of three different growth regimes. It is not *a priori* obvious that they all occur in real fluids: there may be a physical constraint preventing a particular growth mechanism, or the assumption of a single length scale may break down. Therefore, we now present the results of Kendon et al. [26,27] who used a lattice Boltzmann algorithm to solve the equations of motion (4.17) and (4.18) with the aim of exploring the role of hydrodynamics in the phase separation of a three-dimensional symmetric (equal concentrations of A and B) mixture. Later, we will consider the two-dimensional case.

To obtain a quantitative measure of domain growth, a definition of the average domain size is needed. As discussed in Section 3.2, there are various possibilities. For example, different lengths can be defined as

$$L^{(n)} = 2\pi \left(\frac{\int dk S(k,t)}{\int dk k^n S(k,t)} \right)^{1/n}, \tag{4.64}$$

where $S(k,t)$ is the spherically averaged structure factor. This can be obtained by taking the average of the modulus squared of the Fourier transform $\psi(\vec{k},t)$ of the order parameter over a shell in $\vec{k}$-space at fixed $k = |\vec{k}|$. Lower order moments are more sensitive to larger structures but, if dynamical scaling holds, all the $L^{(n)}$ are equivalent. In Refs [26,27], there were no indications of violation of scaling, and the first moment $L = L^{(1)}$ of the structure factor was considered.

Dimensionless quantities are often useful in hydrodynamic problems to reduce the number of control parameters and to aid the comparison between different experiments. In the analysis in Refs [26,27], lengths and times were rescaled in terms of $L_0 = \eta^2/(n\sigma)$ and $t_0 = \eta^3/(n\sigma^2)$, respectively. These are the only quantities with the correct dimensions that can be constructed from n, η, and σ. Note that they coincide

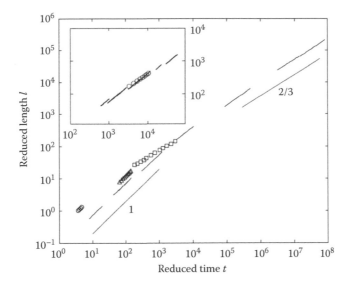

FIGURE 4.5 Reduced length plotted against reduced time from lattice Boltzmann simulations of phase ordering in a three-dimensional binary fluid. The lines denote the results of different runs and show a clear crossover from the viscous to the inertial regimes. Data from other numerical approaches is compared: □ Appert et al. (lattice gas simulations) [28], △ Laradji et al. (molecular dynamics) [29]. ○ Bastea and Lebowitz (direct simulation Monte Carlo combined with particle-in-cell methods) [30] Inset: ○ Jury et al. (dissipative particle dynamics) [31] compared with LBM data. (Reprinted with permission from Kendon, V.M. et al., *Phys. Rev. Lett.*, 83, 576, 1999; © 1999 The American Physical Society, http://link.aps.org/doi/10.1103/PhysRevLett.83.576)

with L_h and t_h, the crossover scales between the viscous and inertial growth regimes. Figure 4.5 is a plot, on a double-logarithmic scale, of L/L_0 against t/t_0 for runs with different dimensionless parameters. Only the data taken once a power law has been established, and before L is limited by the system size, are displayed.

Experience has shown that, in a single simulation run, it is generally not possible to access a range of times sufficiently long to observe the crossover between two different scaling regimes. Here, the use of dimensionless variables allows the data from several different runs to be displayed on the same plot. This gives a universal curve that tracks the domain growth across several decades in reduced time.

Clear evidence is seen in Figure 4.5 for the viscous and inertial growth exponents $\phi = 1$ and $\phi = 2/3$. The crossover is very broad and occurs for $t/t_0 \sim 10^2-10^6$, which is considerably larger than the expected value ~ 1. This delayed crossover may explain why, while the diffusive and viscous regimes have been observed [20–22], there are no experimental data showing inertial growth. At the end of the crossover region, the average size of domains for typical fluids should be of the order centimeters, where effects such as gravity and thermal convection hinder experiments.

The lattice Boltzmann algorithm used to obtain these results was a 15 velocity model on a cubic lattice. The largest system size considered was 256^3. A parallel code with domain decomposition and MPI library was used. Parallel implementation generally works well with lattice Boltzmann algorithms, as communication is only between neighboring sites of the mesh [32]. Typical runs were of 10^4 time steps and required about 3000 Cray T3D hours CPU.

Results obtained by different methods are also shown in Figure 4.5. These are particle-based methods that, primarily because they contain intrinsic noise, are harder to use for simulations of domain growth than the continuum equations we consider here. These data do not lie on the universal curve suggested by the lattice Boltzmann results. A likely reason for the discrepancies is that diffusive growth is still appreciable and introduces a second time scale in the problem. At present, a careful investigation of diffusive effects in three-dimensional fluids, and of the crossover between diffusive and hydrodynamic behavior, is not available. In two dimensions, however, the combined effects of hydrodynamic and diffusive transport have been investigated, and this will be the focus of the next section.

4.3.3 SPINODAL DECOMPOSITION IN TWO DIMENSIONS

We now consider phase ordering in a two-dimensional, symmetric, binary fluid. A major difference from three dimensions is that the viscous hydrodynamic regime is absent. This is because the Rayleigh instability does not occur in two-dimensional systems [33]. Thus, we are left with inertial hydrodynamic growth with $\phi = 2/3$, and diffusive growth with $\phi = 1/3$. Many of the interesting features of domain growth in two-dimensional fluids result from the simultaneous presence of the hydrodynamic (t_0) and diffusive $[T_d = \xi_b^3 \psi_{eq}^2/(D\sigma)]$ time scales. We discuss two phenomena related to the interplay between the two growth modes.

The sequence of configurations in Figure 4.6, obtained by two-dimensional lattice Boltzmann simulations, clearly shows that the domain structure is not scale-invariant [34]. At later times, there are a large number of isolated circular domains, which suggest the presence of a second length scale. This is emphasized in Figure 4.6b, where portions of the snapshots of Figure 4.6a are enlarged so that each corresponds with the same length scale.

Results from the simulations show that the inverse first moment of the structure factor $L^{(1)}$, defined in Equation 4.64, scales with a growth exponent 2/3. However, it is also possible to define two real-space measures of length [35]

$$l_I = \frac{L_s^2}{L_I(t)},$$

$$l_N = \frac{L_s}{\sqrt{N_C(t)}},$$

(4.65)

where $L_I(t)$, $N_C(t)$, and L_s are the total interface length, the number of disconnected domains, and the size of the system, respectively. The simulations suggest that $l_I \sim t^{2/3}$ but $l_N \sim t^{1/3}$. This emphasizes that care must be taken when measuring length scales in domain growth problems. $L^{(1)}$ preferentially weights longer length scales and can miss important physics. Moreover, real-space measures can be quicker to implement numerically.

FIGURE 4.6 (a) Phase ordering in a low-viscosity, two-dimensional binary fluid. (b) As (a) but showing a section of each frame magnified to the same length scale. Note the lack of scale-invariance. Hydrodynamics drives the domains circular, but, once they are circular, they can only grow by diffusion. (Reprinted with permission from Wagner, A.J. and Yeomans, J.M., *Phys. Rev. Lett.*, 80, 1429, 1998; © 1998 The American Physical Society, http://link.aps.org/doi/10.1103/PhysRevLett.80.1429)

The simultaneous existence of two coexisting length scales can be explained as follows [34]: Inertial hydrodynamic flow is driven by the difference in Laplace pressure between points of different curvature on the boundary of a domain. Hence it provides an efficient mechanism whereby domains can become circular. Once they are circular, there is no hydrodynamic driving force, and further coarsening can only occur by diffusion. Smaller domains become circular first, and therefore diffusion is operative at small length scales at the same time as hydrodynamics is driving the growth at larger length scales, with the crossover length between the two regimes increasing with time.

Another interesting phenomenon is double phase separation shown, for a mixture of ϵ-caprolactone oligomer (OCL) and styrene oligomer (OS), in Figure 4.1c [2]. This is observed when hydrodynamics is dominant and drives the domains circular long before diffusion has allowed the order parameter to reach the equilibrium value in each phase. If domains are sufficiently large that interaction between interfaces is small, and the value of the order parameter within each domain lies inside the spinodal region, further spinodal decomposition is possible. This leads to a topology of nested circular domains. Such patterns have been seen in simulations using both a lattice Boltzmann algorithm [34] and a direct discretization of the equations of motion [2].

We briefly mention domain growth in asymmetric binary fluids. As the volume of one component, say A, increases relative to that of B, the bicontinuous structure is lost, and isolated drops of the minority phase form in a background of the majority phase (see Figure 1.11). The larger drops then grow at the expense of the smaller ones. The typical evolution for off-critical binary fluids is shown in Figure 4.2. Growth is probably driven by the Lifshitz–Slyozov (LS) mechanism, as for a binary alloy. This is the diffusion of atoms from smaller, dissolving domains to larger ones, driven by the difference in chemical potential induced by the different interface curvatures.

However, subtle effects of hydrodynamics have not been excluded. Kumaran [36] argued that the anisotropic flux at an interface due to diffusion from neighboring drops can cause the drops to move toward each other. Tanaka [37,38] considered the possibility of coalescence-induced coalescence: When two drops collide, they induce concentration and velocity fields that tend to attract any third drop in the neighborhood. Simulations have so far been inconclusive, hampered by the slow growth, and there is no clear picture of the role of flow (if any) on phase ordering in asymmetric binary mixtures.

4.3.4 Growth Under Shear in Binary Mixtures

Phase separation can also occur in the presence of applied flows. Pressure differences in channels or moving boundaries induce velocity profiles that affect the domain morphology and the growth laws in phase-separating systems. Applied shear is an important case, particularly relevant in material processing, food, or cosmetic industries [39].

In a simple fluid between parallel walls moving with opposite velocities, the steady velocity field is given by the linear profile $\vec{u} = \dot{\gamma} y \vec{e}_x$, where $\dot{\gamma}$ is the shear rate and $\vec{e}_x$ denotes the flow direction [40]. In phase-separating systems, the main effect of shear is to stretch the domains in the direction of the flow causing breakups of the bicontinuous network. Late time configurations observed in simulations and experiments [41–43] typically consist of almost parallel string-like domains, with the average transverse domain size depending on the shear rate. Studying the convection-diffusion equation (4.63), together with a fixed linear velocity profile, gives relevant information on the behavior of the system [44]. However, velocity fluctuations, induced by the pressure tensor in the Navier–Stokes equation (4.62), are essential to fully understand the dynamics. Therefore, the lattice Boltzmann method has proved to be a convenient algorithm to study this problem [45,46].

The phenomenology of phase-separating systems under shear is very rich, as discussed by Onuki [47]. Here we only mention a basic question concerning the existence of a non-equilibrium steady state in quenched binary mixtures: Coarsening could continue indefinitely, as is the case without shear, or a steady state with finite domain size could be reached asymptotically as the result of the balance between the tendency of domains to grow isotropically and the stretching and breakup caused by the shear. Theoretical analysis of the convection-diffusion equation (in the case of an n-vector order parameter with $n \to \infty$) predicts indefinite coarsening and anisotropic growth with $L_x \sim \dot{\gamma} t L_y$ [44,48]. This scaling has been observed in experiments, and the diffusive exponent $1/3$ for L_y has been measured at early times [49], as also suggested by a simple scaling theory. On the other hand, hydrodynamic scaling arguments balancing interfacial with inertial or viscous forces predict a saturation of domain size.

It is not easy to explore the effects of hydrodynamics because the fast growth in the flow direction makes finite-size effects in a sheared phase-separating system important both in simulations and in experiments. However, recently, large-scale lattice Boltzmann simulations [46], using an algorithm similar to that described in Section 4.2.2, have produced evidence of domain size saturation in a regime where finite-size effects cannot be relevant, thus providing evidence for the occurrence of a non-equilibrium steady state.

4.4   PHASE ORDERING IN COMPLEX FLUIDS

Domain growth in simple binary fluid mixtures is rather well understood, and the focus is now on phase separation in more complex fluids [50]. There is an enormous range of these: nematic, cholesteric, and smectic liquid crystals, colloids, polymer melts and solutions, and ternary mixtures of oil, water, and surfactants are among the most common. Changing the external parameters such as temperature, mixing different complex fluids, adding microscopic particles, and introducing surfaces are all ways of changing the phase behavior of the systems. The kinetics of transitions between different phases can lead to new microstructures and hence materials with novel properties (i.e., see Figure 4.1d and e).

With such a wide range of materials, we must necessarily be somewhat selective in our discussion. Therefore, most of this section is devoted to "simpler" complex fluids where numerical results are available. We shall concentrate on results obtained using the lattice Boltzmann approach. In Section 4.4.1, we consider lamellar ordering such as that found in diblock copolymers. This has been investigated by lattice Boltzmann solutions of equations similar to Equations 4.17, 4.18 but with a different free energy. This subject will be discussed further in Chapter 7. In Section 4.4.2, we discuss extensions of the lattice Boltzmann algorithm to liquid crystals, where the equations of motion must describe the non-Newtonian properties of the flow.

4.4.1 LAMELLAR ORDERING

Fluids with competing interactions can order in lamellar structures comprising alternate layers of different composition. Lamellar order is found in diblock copolymer melts, where the steric constraints on chains of type A and B bonded in pairs cause them to line up at low temperatures to form a stack of lamellae [51], ternary mixtures where surfactants form interfaces between oil and water [52], and lipidic biological systems [53]. Examples of lamellar patterning are shown in Figure 4.1d.

The simplest model that can describe systems with lamellar order was introduced some years ago by Brazovskii [54]. It is currently used to model copolymer systems close to the disorder-lamellar transition. The model is based on the free energy

$$F = \int d\vec{r} \left[\frac{1}{3} n \ln n + \frac{a}{2} \psi^2 + \frac{b}{4} \psi^4 + \frac{K}{2} (\nabla \psi)^2 + \frac{d}{2} (\nabla^2 \psi)^2 \right], \qquad (4.66)$$

where we have included a term depending on the total density n of the system. This term gives a positive background pressure and does not affect the phase behavior. The scalar order parameter ψ represents (as before) the density difference between the two components of the mixture. The parameters b and d are taken strictly positive to ensure stability. For $a > 0$, the fluid is disordered; for $a < 0$ and $K > 0$, two homogeneous phases with $\psi = \pm\sqrt{-a/b}$ coexist as in the case of binary mixtures. A negative K favors the presence of interfaces, and a transition into a lamellar phase can occur. In the simplest approximation, assuming a profile $\sim A \sin(k_0 x)$ for the direction transverse to the lamellae, the transition ($|a| = b$) occurs at $a \simeq -1.11 K^2/d$, where

$$k_0 = \sqrt{\frac{-K}{2d}}, \quad A^2 = \frac{4}{3} \left(1 + \frac{K^2}{4db} \right). \qquad (4.67)$$

Expressions for the pressure tensor and chemical potential that follow from the free energy in Equation 4.66 are reported in Refs [55,56].

The dynamical equations 4.17 and 4.18 for this system have been solved using lattice Boltzmann methods in two and three dimensions [55–57]. Lamellar ordering was studied by following the kinetics of systems, with free-energy parameters corresponding with a stable lamellar phase, evolving from an initially disordered state. The relevant questions are similar to those discussed for binary mixtures: Are there universal exponents characterizing the ordering of lamellar domains? Does dynamical scaling hold? What are the effects of hydrodynamics?

The numerical studies have shown that the behavior is considerably more complex than for binary mixtures, and it is not yet fully understood. At early stages of segregation, domains of the two phases form and grow until the equilibrium wavelength of the lamellae is reached. At later times, at very high viscosity, when hydrodynamic

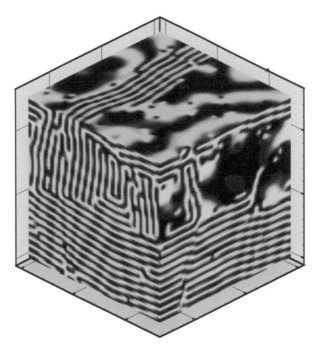

FIGURE 4.7   A late-time configuration from three-dimensional lattice Boltzmann simulations of lamellar ordering, showing the presence of grain boundaries between domains of lamellae oriented on mutually perpendicular planes. (From Xu, A., et al., *Europhys. Lett.*, 71, 651, 2005. With permission.)

effects are not relevant, the system remains frozen in an entangled state of intertwined lamellae. At lower viscosities, the system continues to order [55,56], but, still, at late times, grain boundaries between domains of differently oriented lamellae slow down or stop further evolution as shown in Figure 4.7. It is interesting that an applied shear flow facilitates the elimination of all defects and makes the ordering much faster [55,56,58], as needed in many applications.

Love and Coveney [59] have observed a similar slowing down in lamellar systems using a different lattice Boltzmann approach. They model the surfactant explicitly and, rather than describing the equilibrium state of the fluid in terms of a free energy, drive the ordering using coarse-grained interparticle interactions. Using this model, they have also explored the self-assembly of more complicated mesophases formed from amphiphilic sheets, in particular the gyroid cubic mesophase [60].

4.4.2 LIQUID CRYSTALS

We next consider ordering in liquid-crystalline fluids [61,62]. These are typically made up of highly anisotropic molecules. Subtle energy-entropy balances can lead to many different ordered phases. The simplest of these is the nematic phase where the molecules exhibit long-range orientational order—they line up, on average, along a preferred direction—but no long-range positional order. The order parameter describing the nematic is a symmetric traceless tensor $\mathbf{Q}$, which encodes both the direction and the magnitude of the orientational order. The free energy, written in terms of an expansion in $\mathbf{Q}$, is [61]

$$F_{lc} = \int d\vec{r} \left[A_2 \mathbf{Q}^2 - A_3 \mathbf{Q}^3 + A_4 \mathbf{Q}^4 + L_0 (\nabla \mathbf{Q})^2 \right]. \tag{4.68}$$

The first three terms in Equation 4.68 are bulk terms that describe a weakly first-order transition from the isotropic to the nematic phase. The derivative term models the energy associated with elastic distortions of the liquid-crystal order parameter.

As liquid-crystal molecules are extended in space, they couple with a shear flow leading to non-Newtonian hydrodynamics. The convection-diffusion equation for the binary fluid order parameter ψ is replaced by a similar equation for $\mathbf{Q}$, except for an additional term that allows the molecules to rotate in the flow. The pressure tensor in the Navier–Stokes equation becomes considerably more complicated because any rotation of the molecules in turn induces a backflow. The hydrodynamic equations are described by Beris and Edwards [63]. They can be solved using an

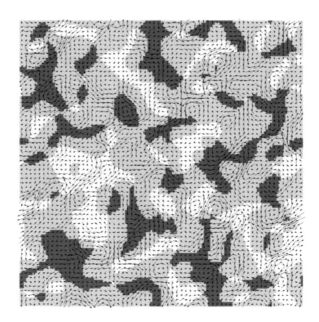

FIGURE 4.8 Lattice Boltzmann simulations of phase ordering for a two-dimensional liquid crystal quenched from the isotropic to the nematic phase. The shaded regions are the Schlieren pattern that would be observed if the liquid crystal was viewed between crossed polarizers. Regions of different shading correspond with aligned nematic order, but with the molecules pointing in different directions. Topological defects of strength $\pm 1/2$ are marked by junctions between black and white regions. The arrows denote the velocity field. It is apparent that the main features of the flow are the vortices associated with the moving defects. (Reprinted with permission from Denniston, C., et al., *Phys. Rev. E*, 64, 021701, 2001; © 2001 The American Physical Society, http://link.aps.org/doi/10.1103/PhysRevE.64.021701)

extension of the lattice Boltzmann algorithm described in Section 4.2.2. In particular, the scalar distribution functions describing the evolution of the concentration field must be replaced by a set of traceless symmetric tensors describing the motion of the liquid-crystal order parameter $\mathbf{Q}$. Details of the algorithm can be found in Ref. [64].

Figure 4.8 shows results from a lattice Boltzmann simulation of a liquid crystal quenched into the nematic phase [65]. Local nematic order is quickly established, but domains with the molecules pointing in different directions are separated by topological defects [66]. These anneal out much more slowly with a growth exponent $\phi = 1/2$. Hydrodynamics speeds up the growth but does not change the value of the growth exponent. However, flow does have an interesting effect on defect motion. As a defect moves it sets up a flow field, the exact form of which depends on its topological charge. Hence defects of different strengths move with different speeds, and a pair of annihilating defects will not meet at the mid-point of the line joining their centers [67].

There is considerable scope for further research on phase ordering in liquid crystals. For example, recent experimental work shows that a rich structure of dendritic fingers forms as a cholesteric phase grows into a nematic [68]. Another question of interest is the extent to which the stability of the elusive blue phases, which lie between the isotropic and cholesteric states, depends on the kinetic pathways to the cholesteric [69]. Mixtures of liquid crystals and colloids are also now attracting interest. For example, experiments by Vollmer et al. [70] show that the colloidal particles form networks as the liquid crystal is slowly cooled from the isotropic to the nematic phase. Many other examples of phase ordering in liquid crystals are given in the book by Dierking [4].

Liquid-crystal emulsions are binary mixtures of a nematic and isotropic phase. Araki and Tanaka [71] have solved the hydrodynamic equations of motion for a nematic emulsion using an explicit Euler method for the order parameter equation and a fast Fourier transform for the velocity equation. They argue that the coupling between the order parameter and the velocity field in the liquid-crystal phase leads to a break up of the bicontinuous structure seen for an isotropic mixture, with the liquid crystal forming isolated domains within a matrix of the isotropic component. Moreover, both experiments and simulations have shown that the preferred orientation of the director at nematic-isotropic interfaces can strongly affect the domain shapes during ordering [72,73].

4.4.3 OTHER APPLICATIONS

We conclude this chapter with a brief mention of a wide variety of recent work, both experimental and numerical, which emphasizes the breadth of phenomena to be explored as fluids order.

Aarts et al. [74] have reported experiments on fluid-fluid phase separation in a colloid-polymer mixture, which can be followed in great detail as the interfacial tension, and hence interface velocities, are very low. Domains coarsen through viscous hydrodynamic growth, and the pinch-off of necks is explicitly observed. At later times, gravity causes the domain structure to collapse. The minority phase, the liquid, breaks up and becomes discontinuous. Finally, a macroscopic interface between the two phases is formed and grows upwards to give complete ordering.

Stratford et al. [75] have performed large-scale lattice Boltzmann simulations following the behavior of neutrally wetting colloidal particles in a phase-ordering binary fluid. The colloids prefer to lie at the interface and, as the system coarsens, they form an increasingly densely packed layer that becomes jammed, inhibiting further domain growth. Thus one is left with a microstructure comprising a bicontinuous fluid phase pinned by a multiply-connected colloidal layer.

A promising approach to creating microstructures on well-defined length scales is to use reversible photochemical reactions to induce phase separation in a binary polymer mixture. For example, Tran-Cong-Miyata et al. [76] have used blends of trans-stilbene labeled polystyrene and poly(vinyl methyl ether) (PSS/PVME). Upon irradiation, the stilbene moieties on the PSS chains undergo a reversible trans-cis photoisomerization. Once the reaction reaches a certain threshold, the mixture undergoes phase separation because the cis-labeled polystyrene and PVME are immiscible. Thus, phase separation favoring ordering, and a reversible chemical reaction favoring mixing, take place simultaneously within the binary blend. By tuning the forward and reverse rates of the chemical reaction, hexagonal, cylindrical, or lamellar structures can be obtained. At higher light intensity, the rate of reaction is increased, the domains become more intermixed, and the characteristic domain size is smaller.

A very different coarsening phenomenon is observed in lipid vesicles. Artificial cell-sized lipid vesicles with a diameter of about 10 μm are often used as models for living cellular structures. If unsaturated and saturated phospholipids are present in a vesicle, and if cholesterol is also added, clusters of cholesterol and of the saturated lipid form and grow with time. These clusters move in the bilayer membrane and are called *rafts*. It has been suggested that, in living membranes, the rafts are the platforms on which proteins are attached for intermembrane exchanges [77]. Rafts are observed to coalesce [78], and an interesting question is whether the growth laws discussed in Section 4.3 are appropriate for these systems. Simulations have suggested that the scaling laws can change due to the fact that phase separation occurs on a curved manifold [79]. However, further experimental data and models are required to properly describe and understand the dynamics of rafts in real membranes.

REFERENCES

1. Jinnai, H., Hashimoto, T., Lee, D., and Chen, S.-H., Morphological characterization of bicontinuous phase-separated polymer blends and one-phase microemulsions, *Macromolecules*, 30, 130, 1997.
2. Tanaka, H., Hydrodynamic interface quench effects on spinodal decomposition for symmetric binary fluid mixtures, *Phys. Rev. E*, 51, 1313, 1995.
3. Park, M., Harrison, C., Chaikin, P.M., Register, R.A., and Adamson, D.H., Block copolymer lithography: Periodic arrays of $\sim 10^{11}$ holes in 1 square centimeter, *Science*, 276, 1401, 1997.
4. Dierking, I., *Textures of Liquid Crystals*, Wiley-VCH, Weinheim, 2003.

5. Wagner, A.J., *Theory and Application of the Lattice Boltzmann Method*, D. Phil. Thesis, University of Oxford, 1998.
6. Reichl, L.E., *A Modern Course in Statistical Physics*, Second Edition, John Wiley & Sons, New York, 1998.
7. Rowlinson, J.S. and Widom, B., *Molecular Theory of Capillarity*, Clarendon, Oxford, 1982.
8. Evans, R., The nature of the liquid-vapour interface and other topics in the statistical mechanics of non-uniform, classical fluids, *Adv. Phys.*, 28, 143, 1979.
9. Succi, S., *The Lattice Boltzmann Equation for Fluid Dynamics and Beyond*, Clarendon Press, Oxford, 2001.
10. Landau, L.D. and Lifshitz, E.M., *Fluid Mechanics*, Second Edition, Reed Educational and Professional Publishing Ltd., 1959.
11. De Groot, S.R. and Mazur, P., *Non-equilibrium Thermodynamics*, Dover Publications, New York, 1984.
12. Higuera, F.J., Succi, S., and Benzi, R., Lattice gas dynamics with enhanced collisions, *Europhys. Lett.*, 9, 345, 1989.
13. Benzi, R., Succi, S., and Vergassola, M., The lattice Boltzmann equation: theory and applications, *Phys. Rep.*, 222, 145, 1992.
14. McNamara, G. and Zanetti, G., Use of the Boltzmann equation to simulate lattice-gas automata, *Phys. Rev. Lett.*, 61, 2332, 1998.
15. Chen, S. and Doolen, G.D., Lattice Boltzmann method for fluid flows, *Annu. Rev. Fluid Mech.*, 30, 329, 1998.
16. Yeomans, J.M., Mesoscale simulations: Lattice Boltzmann and particle algorithms, *Physica A*, 369, 159, 2006.
17. Bhatnagar, P.L., Gross, E.P., and Krook, M., A Model for collision processes in gases. I. Small amplitude processes in charged and neutral one-component systems, *Phys. Rev.*, 94, 511, 1954.
18. Swift, M.R., Orlandini, E., Osborn, W.R., and Yeomans, J.M., Lattice Boltzmann simulations of liquid-gas and binary fluid systems, *Phys. Rev. E*, 54, 5041, 1996.
19. Bray, A.J., Theory of phase ordering kinetics, *Adv. Phys.*, 43, 357, 1994.
20. Wong, N.-C. and Knobler, C.M., Light scattering studies of phase separation in isobutyric acid + water mixtures: Hydrodynamic effects, *Phys. Rev. A*, 24, 3205, 1981.
21. Chou, Y.C. and Goldburg, W.I., Phase separation and coalescence in critically quenched isobutyric-acid-water and 2,6-lutidine-water mixtures, *Phys. Rev. A*, 20, 2105, 1979;
22. Bates, F.S. and Wiltzius, P., Spinodal decomposition of a symmetric critical mixture of deuterated and protonated polymer, *J. Chem. Phys.*, 91, 3258, 1988.
23. Siggia, E.D., Late stages of spinodal decomposition in binary mixtures, *Phys. Rev. A*, 20, 595, 1979.
24. Furukawa, H., Effect of inertia on droplet growth in a fluid, *Phys. Rev. A*, 31, 1103, 1985.
25. Yang, A.J.M, Fleming, P.D., and Gibbs, J.H., Molecular theory of surface tension, *J. Chem. Phys.*, 64, 3732, 1976.
26. Kendon, V.M., Desplat, J.-C., Bladon, P., and Cates, M.E., 3D spinodal decomposition in the inertial regime, *Phys. Rev. Lett.*, 83, 576, 1999.
27. Kendon, V.M., Cates, M.E., Desplat, J.-C., Pagonabarraga, I., and Bladon, P., Inertial effects in three-dimensional spinodal decomposition of a symmetric binary fluid mixture: a lattice Boltzmann study, *J. Fluid Mech.*, 440, 147, 2001.
28. Appert, C., Olson, J.F., Rothman, D.H., and Zaleski, S., Phase separation in a three-dimensional, two-phase, hydrodynamic lattice gas, *J. Stat. Phys.*, 81, 181, 1995.

29. Laradji, M., Toxvaerd, S., and Mouritsen, O.G., Molecular dynamics simulation of spinodal decomposition in three-dimensional binary fluids, *Phys. Rev. Lett.*, 77, 2253, 1996.

30. Bastea, S. and Lebowitz, J.L., Spinodal decomposition in binary gases, *Phys. Rev. Lett.*, 78, 3499, 1997.

31. Jury, S.I., Bladon, P., Krishna, S., and Cates, M.E., Tests of dynamical scaling in three-dimensional spinodal decomposition, *Phys. Rev. E*, 59, 2535, 1999.

32. Desplat, J.-C., Pagonabarraga, I., and Bladon, P., LUDWIG: A parallel lattice-Boltzmann code for complex fluids, *Comp. Phys. Comm.*, 134, 273, 2001.

33. San Miguel, M., Grant, M., and Gunton, J.D., Phase separation in two-dimensional binary fluids, *Phys. Rev. A*, 31, 1001, 1985.

34. Wagner, A.J. and Yeomans, J.M., Breakdown of scale invariance in the coarsening of phase-separating binary fluids, *Phys. Rev. Lett.*, 80, 1429, 1998.

35. Mecke, K.R. and Sofonea, V., Morphology of spinodal decomposition, *Phys. Rev. E*, 56, 3761, 1997.

36. Kumaran, V., Droplet interaction in the spinodal decomposition of a fluid, *J. Chem. Phys.*, 109, 7644, 1998.

37. Tanaka, H., New coarsening mechanisms for spinodal decomposition having droplet pattern in binary fluid mixture: Collision-induced collisions, *Phys. Rev. Lett.*, 72, 1702, 1994.

38. Tanaka, H., A new coarsening mechanism of droplet spinodal decomposition, *J. Chem. Phys.*, 103, 2361, 1995.

39. Larson, R.G., *The Structure and Rheology of Complex Fluids*, Oxford University Press, New York, 1999.

40. Acheson, D.J., *Elementary Fluid Dynamics*, Oxford University Press, Oxford, 1990.

41. Hashimoto, T., Matsuzaka, K., Moses, E., and Onuki, A., String phase in phase-separating fluids under shear flow, *Phys. Rev. Lett.*, 74, 126, 1995.

42. Ohta, T., Nozaki, H., and Doi, M., Computer simulations of domain growth under steady shear flow, *J. Chem. Phys.*, 93, 2664, 1990.

43. Corberi, F., Gonnella, G., and Lamura, A., Two-scale competition in phaseseparation with shear, *Phys. Rev. Lett.*, 83, 4057, 1999.

44. Corberi, F., Gonnella, G., and Lamura, A., Spinodal decomposition of binary mixtures in uniform shear flow, *Phys. Rev. Lett.*, 81, 3852, 1998.

45. Wagner, A.J. and Yeomans, J.M., Phase separation under shear in two-dimensional binary fluids, *Phys. Rev. E*, 59, 4366, 1999.

46. Stansell, P., Stratford, K., Desplat, J.-C., Adhikari, R., and Cates, M.E., Nonequilibrium steady states in sheared binary fluids, *Phys. Rev. Lett.*, 96, 085701, 2006.

47. Onuki, A., Phase transitions of fluids in shear flow, *J. Phys. Condens. Matter*, 9, 6119, 1997.

48. Rapapa, N.P. and Bray, A.J., Dynamics of phase separation under shear: A soluble model, *Phys. Rev. Lett.*, 83, 3856, 1999.

49. Läuger, J., Laubner, C., and Gronski, W., Correlation between shear viscosity and anisotropic domain growth during spinodal decomposition under shear flow, *Phys. Rev. Lett.*, 75, 3576, 1995.

50. Jones, R.A.L., *Soft Condensed Matter*, Oxford University Press, Oxford, 2002.

51. Bates, F.S. and Fredrickson, G.H., Block copolymer thermodynamics: theory and experiment, *Ann. Rev. Phys. Chem.*, 41, 525, 1990.

52. Gompper, G. and Schick, M., Self-assembling amphiphilic systems, in *Phase Transitions and Critical Phenomena*, Vol. 16, Domb, C. and Lebowitz, J.L., Eds., Academic Press, New York, 1994, 1.

53. Nelson, D., Piran, T., and Weinberg, S., Eds., *Statistical Mechanics of Membranes and Surfaces*, World Scientific, Singapore, 1989.
54. Brazovskii, S.A., Phase transition of an isotropic system to a nonuniform state, *Sov. Phys. JETP*, 41, 85, 1975.
55. Gonnella, G., Orlandini, E., and Yeomans, J.M., Spinodal decomposition to a lamellar phase: Effects of hydrodynamic flow, *Phys. Rev. Lett.*, 78, 1695, 1997.
56. Gonnella, G., Orlandini, E., and Yeomans, J.M., Lattice Boltzmann simulations of lamellar and droplet phases, *Phys. Rev. E*, 58, 480, 1998.
57. Xu, A., Gonnella, G., Lamura, A., Amati, G., and Massaioli, F., Scaling and hydrodynamic effects in lamellar ordering, *Europhys. Lett.*, 71, 651, 2005.
58. Xu, A., Gonnella, G., and Lamura, A., Morphologies and flow patterns in quenching of lamellar systems with shear, *Phys. Rev. E*, 74, 011505, 2006.
59. Love, P.J. and Coveney, P.V., Three-dimensional hydrodynamic lattice-gas simulations of ternary amphiphilic fluids under shear flow, *Phil. Trans. R. Soc. A*, 360, 357, 2002.
60. Gonzalez-Segredo, N. and Coveney, P.V., Self-assembly of the gyroid cubic mesophase: Lattice-Boltzmann simulations, *Europhys. Lett.*, 65, 795, 2004.
61. de Gennes, P.G. and Prost, J., *The Physics of Liquid Crystals*, Second Edition, Clarendon Press, Oxford, 1993.
62. Chandrasekhar, S., *Liquid Crystals*, Second Edition, Cambridge University Press, Cambridge, 1992.
63. Beris, A.N. and Edwards, B.J., *Thermodynamics of Flowing Systems*, Oxford University Press, Oxford, 1994.
64. Denniston, C., Marenduzzo, D., Orlandini, E., and Yeomans, J.M., Lattice Boltzmann simulations of three-dimensional liquid crystal hydrodynamics, *Phil. Trans. Royal Soc. A*, 362, 1745, 2004.
65. Denniston, C., Orlandini, E., and Yeomans, J.M., Phase ordering in nematic liquid crystals, *Phys. Rev. E*, 64, 021701, 2001.
66. Mermin, N.D., The topological theory of defects in ordered media, *Rev. Mod. Phys.*, 51, 591, 1979.
67. Toth, G., Denniston, C., and Yeomans, J.M., Hydrodynamics of topological defects in nematic liquid crystals, *Phys. Rev. Lett.*, 88, 105504, 2002.
68. Oswald, P., Baudry, J., and Rondepierre, T., Growth below and above the spinodal limit: The cholesteric-nematic front, *Phys. Rev. E*, 70, 041702, 2004.
69. Coles, H.J. and Pivnenko, M.N., Liquid crystal blue phases with a wide temperature range, *Nature*, 436, 997, 2005.
70. Vollmer, D., Hinze, G., Ullrich, B., Poon, W.C.K., Cates, M.E., and Schofield, A.B., Formation of self-supporting reversible cellular networks in suspensions of colloids and liquid crystals, *Langmuir*, 21, 4921, 2005.
71. Araki, T. and Tanaka, H., Nematohydrodynamic effects on the phase separation of a symmetric mixture of an isotropic liquid and a liquid crystal, *Phys. Rev. Lett.*, 93, 015702, 2004.
72. Lettinga, M.P., Kang, K., Imbof, A., Derks, D., and Dhont, J.K.G., Kinetic pathways of the nematic-isotropic phase transition as studied by confocal microscopy on rod-like viruses, *J. Phys. Condens. Matter*, 17, S3609, 2005.
73. Sulaiman, N.A., Mesoscopic modelling of nematic emulsions, D. Phil. Thesis, University of Oxford, 2006.
74. Aarts, D.G.A.L., Schmidt, M., and Lekkerkerker, H.N.W., Direct visual observation of thermal capillary waves, *Science*, 304, 847, 2004.

75. Stratford, K., Adhikari, R., Pagonabarraga, I., Desplat, J.-C., and Cates, M.E., Colloidal jamming at interfaces: A route to fluid-bicontinuous gels, *Science*, 309, 2198, 2005.
76. Tran-Cong-Miyata, Q., Nishigami, S., Ito, T., Komatsu, S., and Norisuye, T., Controlling the morphology of polymer blends using periodic irradiation, *Nature Materials*, 3, 448, 2004.
77. Simons, K. and Ikonen, E., Functional rafts in cell membranes, *Nature*, 387, 569, 1997.
78. Saeki, D., Hamada, T., and Yoshikawa, K., Domain-growth kinetics in a cell-sized liposome, *J. Phys. Soc. Jpn.*, 75, 013602, 2006.
79. Laradji, M. and Kumar, P.B.S., Dynamics of domain growth in self-assembled fluid vesicles, *Phys. Rev. Lett.*, 93, 198105, 2004.

5 Aging in Domain Growth

Marco Zannetti

CONTENTS

5.1 INTRODUCTION

Slow relaxation and aging are ubiquitous in the low-temperature behavior of complex systems, notably structural and spin glasses [1–3]. The challenge for nonequilibrium statistical mechanics is to formulate a theory for these phenomena. In such a context, domain growth is of particular importance as a paradigm where the mechanisms leading to slow relaxation and aging are believed to be well understood. In this chapter, we will review our understanding of aging in domain growth problems.

As discussed in Chapter 1, domain growth occurs in homogeneous mixtures that have been rendered thermodynamically unstable by (say) a rapid quench to a temperature $T_F < T_c$, where T_c is the critical temperature. However, in order to have a comprehensive view of the problem, it is useful to let the final quench temperature T_F span the whole temperature range, from above to below the critical temperature. More precisely, consider a typical phase diagram in the dimensionality-temperature plane, as schematically drawn in Figure 5.1. The critical temperature depends on the dimensionality d of the system. Although the specific form of the function $T_c(d)$ varies from model to model, some generic features are as follows. There exists a lower critical dimensionality d_L, such that $T_c(d) = 0$ for $d \leq d_L$, and there is a monotonic increase of $T_c(d)$ for $d > d_L$. The value of d_L depends on the symmetry of the model, with $d_L = 1$ for the case with discrete symmetry and $d_L = 2$ for the case with continuous symmetry. The disordered and ordered phases are above and below the critical line $T_c(d)$, respectively.

Let us first see how the slow relaxation arises. The system is initially in equilibrium at some temperature T_I well above the critical line. For simplicity, take $T_I = \infty$ corresponding with an initial state with vanishing correlation length. For a certain value of the dimensionality $d > d_L$ (dashed vertical line in Figure 5.1), the system may be quenched to a temperature T_F greater than, equal to, or lower than T_c. The relaxation involves the growth of the time-dependent correlation length $\xi(t)$ from the initial value $\xi_I = 0$ to a final value $\xi_F > 0$, which depends on T_F. Specifically, $0 < \xi_F < \infty$ for $T_F > T_c$, and $\xi_F = \infty$ for $T_F \leq T_c$ if the system is infinite. (It will be explained below why $\xi_F = \infty$ when $T_F < T_c$.) Therefore, for $T_F > T_c$, there is a finite equilibration time t_{eq}. For $T_F \leq T_c$, because $\xi(t)$ typically grows with a power

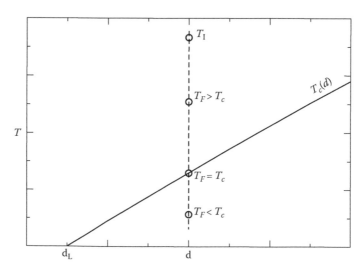

FIGURE 5.1   Generic phase diagram of a phase-ordering system in the (d, T) plane. d_L is the lower critical dimensionality.

law $\xi(t) \sim t^{1/z}$ [4], the system is always out-of-equilibrium, no matter for how long it is observed. Hence, for phase-ordering systems, the mechanism responsible for the slow relaxation in the quench to or below T_c is clear—it is the growth of correlated regions of arbitrarily large size.

The next step is to focus on the features of the relaxation, like aging, which depend on where the system is quenched to in the (d, T)-plane. Assuming that the instantaneous quench takes place at $t = 0$, aging is manifested through the behavior of two-time observables, such as the order-parameter *autocorrelation function* $C(t, t_w)$ and the *autoresponse* function $R(t, t_w)$. The shortest time after the quench ($t_w \geq 0$) and the longest time ($t \geq t_w$) are conventionally called the *waiting time* and the *observation time*. Regarding t_w as the age of the system, aging usually means that older systems relax slower and younger ones faster [5]. However, this needs to be explained in greater detail, because a rich phenomenology goes under the heading of aging.

The first relevant feature is the separation of the timescales [3]. This means that when t_w is sufficiently large, the range of $\tau = t - t_w$ is divided into the short-time ($\tau \ll t_w$) and the long-time ($\tau \gg t_w$) separations, and that quite different behaviors are observed in the two regimes. For short τ, the system appears equilibrated. The two-time quantities are *time-translation-invariant* (TTI) and exhibit the same behavior as if equilibrium at the final temperature of the quench had been reached:

$$C(t, t_w) = C_{eq}(\tau, T_F),$$
$$R(t, t_w) = R_{eq}(\tau, T_F). \tag{5.1}$$

This is the quasi-equilibrium regime, where the system is ageless. Conversely, for large τ, there is a genuine off-equilibrium behavior obeying a scaling form called simple aging [1–3]:

$$C(t, t_w) = C_{ag}(t, t_w) = t_w^{-b} h_C\left(\frac{t}{t_w}\right), \tag{5.2}$$

$$R(t, t_w) = R_{ag}(t, t_w) = t_w^{-(1+a)} h_R\left(\frac{t}{t_w}\right), \tag{5.3}$$

where a, b are non-negative exponents and h_C, h_R are scaling functions. Here, the additional important features are that (a) the system does not realize that it is off-equilibrium until $\tau \sim t_w$, and (b) once the off-equilibrium relaxation gets started, the timescale of the relaxation is fixed by t_w itself. Hence, the system ages.

Aging is common to all quenches: above, to, and below T_c. In the first case, where the equilibration time is finite, for aging to be observable it is necessary that t_{eq} is sufficiently large to allow for both a large t_w and $t_w \ll t_{eq}$, so that the separation of the timescales is possible. Eventually, when $t_w \sim t_{eq}$, equilibrium is reached and aging is interrupted.

The phenomenology outlined above suggests the existence of *fast* and *slow* degrees of freedom [6–8]. The separation of the timescales means that, for sufficiently large t_w, the fast degrees of freedom have already thermalized, and the slow degrees of freedom are still out-of-equilibrium. This picture is easy to visualize in the case of domain growth. For example, in a ferromagnetic system, the fast degrees of freedom are responsible for the thermal fluctuations within ordered domains, and the slow ones are the labels of domains, like the spontaneous magnetization within each domain. The label fluctuates because a given site may belong to different domains at different times. For a given t_w, the typical size of a domain is $L(t_w) \sim t_w^{1/z}$ and it takes an interval of time $\tau \sim t_w$ for a domain wall to sweep the whole domain. Hence, in the short-time regime, the slow degrees of freedom are frozen and only the fast degrees of freedom contribute to the decay of the two-time quantities, yielding the behavior in Equation 5.1. Conversely, in the long-time regime, the evolution is dominated by the motion of domain walls, producing the off-equilibrium behavior of Equations 5.2 and 5.3.

The picture outlined above is simple and intuitive enough to suggest that aging in domain growth is well understood. Indeed, in Section 5.5.1, an example will be presented where the construction of the fast and slow degrees of freedom can be carried out exactly. However, though the picture works well for quenches below T_c, it cannot be so easily extended to quenches to T_c, where the interpretation in terms of fast and slow degrees of freedom remains less clear. Despite the common features in the phenomenology, a qualitative difference between these two instances of aging arises in the way the matching between the stationary and the aging behavior is implemented. In the quench to T_c, stationary and aging behaviors match *multiplicatively*, whereas in the quench below T_c, the matching is *additive*.

The existence of these two different realizations of aging poses quite an interesting problem for what happens in the quench to the final state located at $(d = d_L, T_F = 0)$. Looking at Figure 5.1, it is clear that such a state can be regarded as a limit state reached either along the critical line or from the ordered region, as $d \to d_L$. However, the two points of view are not equivalent, because the structure of the two-time quantities should be multiplicative in one case and additive in the other. As we shall see later, the second alternative is the correct one. Nonetheless, in the quench to $(d = d_L, T_F = 0)$, there are peculiar features that make it a case apart from both the quenches to and below T_c with $d > d_L$. As a matter of fact, the processes listed above can be hierarchically organized in terms of increasing degree of deviation from equilibrium. At the bottom, there is the quench to $T_F > T_c$, where t_{eq} is finite. Immediately above, there is the quench to T_c, where it takes an infinite time to reach equilibrium. Further above, there is the quench to $T_F < T_c$ with $d > d_L$, where equilibrium is not reached even in an infinite time. However, this is shown only by the autocorrelation function without appearing in the dynamic susceptibility, which behaves as if equilibrium was reached. Lastly, at the top of the hierarchy, there is the quench to $(d = d_L, T_F = 0)$, where the dynamic susceptibility also displays out-of-equilibrium behavior over all timescales.

For the computation of the aging properties, namely the exponents a, b and the scaling functions h_C, h_R, the systematic expansion methods of field theory, like the ϵ-expansion, can be successfully used in the quench to T_c [9–11]. However, perturbative methods are useless in the quench below T_c, where the best theoretical tool remains the uncontrolled *Gaussian auxiliary field* (GAF) approximation of the Ohta–Jasnow–Kawasaki type [12] in its various formulations [13–18]. The GAF approach was discussed at length in Section 1.4. With methods of this type, a good understanding of the autocorrelation function $C(t, t_w)$ has been achieved (cf. Equations 1.130 and 1.135) [19]. However, the computation of the autoresponse function $R(t, t_w)$, in particular of the exponent a, remains an open problem. For this reason, the investigation of aging in quenches below T_c relies heavily on numerical simulations, although the accurate numerical computation of $R(t, t_w)$ is a difficult problem of its own [20–23]. In order to complete the theoretical perspective, the *local scale invariance hypothesis* [24,25] must be mentioned. This is a conjecture according to which the response function transforms covariantly under the group of local scale transformations, both in quenches to and below T_c. However, the predictions of this theory are affected by discrepancies with renormalization-group calculations at T_c [11] and with numerical simulations at and below T_c [26,27], which indicate that the local scale invariance hypothesis is akin to a Gaussian approximation. Nonetheless, a definite assessment of this approach cannot yet be made, as work is in progress with the proposal of modified versions of the theory [28].

In the following sections, an overview of aging properties in the various quenches mentioned above will be presented, first in general and then through analytical and numerical results for specific models. As a prelude to this discussion, we present a short summary of static properties.

5.2   STATICS

In general, order-parameter configurations will be denoted by $[\varphi(\vec{x})]$ and the Hamiltonian of the system by $H[\varphi(\vec{x})]$. The variable φ may be a scalar or a vector, and $\vec{x}$ may denote either the points in a continuous region or the sites of a lattice. The set of all possible configurations forms the phase space Ω. Symmetries of the system are ther groups of transformations of Ω onto itself, which leave $H[\varphi(\vec{x})]$ invariant. The equilibrium state at the temperature T is the Gibbs state with probability distribution

$$P_G[\varphi(\vec{x})] = \frac{1}{Z} \exp\{-H[\varphi(\vec{x})]/T\},$$

$$Z = \sum_{[\varphi(\vec{x})]} \exp\{-H[\varphi(\vec{x})]/T\}, \tag{5.4}$$

where the Boltzmann constant is taken to be unity ($k_B = 1$). It is evident that $P_G[\varphi(\vec{x})]$ shares the symmetries of the Hamiltonian.

The models considered are characterized by the existence of a critical temperature T_c and a phase diagram as in Figure 5.1. For $T \geq T_c$, the symmetry is not broken and $P_G[\varphi(\vec{x})]$ is a pure state. For $T < T_c$, the symmetry is broken and $P_G[\varphi(\vec{x})]$ is a mixture

$$P_G[\varphi(\vec{x})] = \sum_{\alpha} w_\alpha P_\alpha[\varphi(\vec{x})]. \tag{5.5}$$

Here, the broken-symmetry pure states $P_\alpha[\varphi(\vec{x})]$ are invariant under a subgroup of the symmetry group of the Hamiltonian, and the weights $w_\alpha \geq 0$ satisfy $\sum_\alpha w_\alpha = 1$. In the Gibbs state, due to symmetry, all the weights are equal,

$$w_\alpha = w \; \forall \alpha. \tag{5.6}$$

Thus, for example, in the case of the Ising model, where the order-parameter configurations are spin configurations ($S_i = \pm 1$) on a lattice, there are two pure states below T_c. These are labeled by $\alpha = \pm$, and transform one into the other under spin inversion

$$P_+[S_i] = P_-[-S_i], \tag{5.7}$$

and remain invariant under the trivial subgroup of the identity. The Gibbs state is given by

$$P_G[S_i] = w_+ P_+[S_i] + w_- P_-[S_i], \tag{5.8}$$

with $w_+ = w_- = w = 1/2$. With a vector order parameter and $H[\vec{\varphi}(\vec{x})]$ invariant under rotations, the labels of the pure states are the unit vectors $\hat{\alpha}$ in the order-parameter space, and each broken-symmetry state $P_{\hat{\alpha}}[\vec{\varphi}(\vec{x})]$ is invariant under the subgroup of rotations around the $\hat{\alpha}$-axis.

5.2.1 MAGNETIZATION

We shall assume throughout, except when it is explicitly stated otherwise, that all expectation values are space translation invariant. Then, the equilibrium magnetization is given by

$$m_{\mathrm{eq}} = \langle \varphi(\vec{x}) \rangle_G, \tag{5.9}$$

where $\langle \cdot \rangle_G$ denotes averages taken with respect to $P_G[\varphi(\vec{x})]$. Because $\varphi(\vec{x})$ is not symmetrical, m_{eq} must vanish for all temperatures. Thus, the existence of a phase transition cannot be detected from the magnetization in the Gibbs state. Below T_c, using Equations 5.5 and 5.6,

$$m_{\mathrm{eq}} = w \sum_{\alpha} m_{\alpha} = 0, \tag{5.10}$$

where $m_{\alpha} = \langle \varphi(\vec{x}) \rangle_{\alpha} \neq 0$ is the spontaneous magnetization in the pure state $P_{\alpha}[\varphi(\vec{x})]$. The quantity that vanishes below T_c is the sum of all possible values of the spontaneous magnetization, but separately each contribution is not zero.

5.2.2 CORRELATION FUNCTION

Conversely, the behavior of the order-parameter correlation function (cf. Equation 1.88),

$$C_{\mathrm{eq}}(\vec{r}, T) = \langle \varphi(\vec{x} + \vec{r}) \varphi(\vec{x}) \rangle_G - \langle \varphi(\vec{x} + \vec{r}) \rangle_G \langle \varphi(\vec{x}) \rangle_G, \tag{5.11}$$

allows one to detect the existence of the phase transition, even in the symmetrical Gibbs state. The scaling behavior of C_{eq} for $T \geq T_c$,

$$C_{\mathrm{eq}}(\vec{r}, T) = r^{-(d-2+\eta)} f_{\mathrm{eq}}\left(\frac{r}{\xi}\right), \tag{5.12}$$

where ξ is the correlation length, and $f_{eq}(x)$ is a rapidly vanishing scaling function, gives rise to the clustering property:

$$\lim_{r \to \infty} C_{eq}(\vec{r}, T) = 0. \tag{5.13}$$

However, for $T < T_c$, we have from Equations 5.5 and 5.11

$$C_{eq}(\vec{r}, T) = \sum_{\alpha} w_{\alpha} C_{\alpha}(\vec{r}, T) + q_{eq}. \tag{5.14}$$

Here,

$$C_{\alpha}(\vec{r}, T) = \langle \varphi(\vec{x} + \vec{r}) \varphi(\vec{x}) \rangle_{\alpha} - \langle \varphi(\vec{x} + \vec{r}) \rangle_{\alpha} \langle \varphi(\vec{x}) \rangle_{\alpha} \tag{5.15}$$

is the correlation function in the αth broken-symmetry state, and

$$q_{eq} = \sum_{\alpha} w_{\alpha} m_{\alpha}^2 - \left[\sum_{\alpha} w_{\alpha} m_{\alpha} \right]^2 \tag{5.16}$$

is the variance of the spontaneous magnetization in the Gibbs state. In the context of spin glasses, this quantity is the *Edwards–Anderson order parameter* [29]. Because $C_{\alpha}(\vec{r}, T)$ and m_{α}^2 are independent of α, it is convenient to introduce the notation $G_{eq}(\vec{r}, T) = C_{\alpha}(\vec{r}, T)$, $M^2 = m_{\alpha}^2$ and to rewrite

$$C_{eq}(\vec{r}, T) = G_{eq}(\vec{r}, T) + M^2. \tag{5.17}$$

Here, $G_{eq}(\vec{r}, T)$ has a form similar to Equation 5.12,

$$G_{eq}(\vec{r}, T) - r^{-(d-2+\eta)} g_{eq}\left(\frac{r}{\xi}\right). \tag{5.18}$$

The appearance of a nonzero value of M^2 upon crossing T_c,

$$\lim_{r \to \infty} C_{eq}(\vec{r}, T) = M^2 \neq 0, \tag{5.19}$$

signals the occurrence of the phase transition and the breaking of ergodicity [30]. Notice that $C_{eq}(\vec{r}, T)$ is a connected correlation function, and its decay to a non-vanishing value at large distances implies that the correlation length ξ_G in the Gibbs state is divergent for $T < T_c$, as opposed to ξ in the pure states, which is finite for $T < T_c$.

5.2.3 SPLITTING OF THE ORDER PARAMETER

The structure of the correlation function in Equation 5.17 can be viewed as arising from the splitting of the order parameter into the sum of two statistically independent contributions:

$$\varphi(\vec{x}) = \psi(\vec{x}) + \sigma, \tag{5.20}$$

each with zero mean

$$\langle \psi(\vec{x}) \rangle = \langle \sigma \rangle = 0, \tag{5.21}$$

and such that

$$G_{eq}(\vec{r}, T) = \langle \psi(\vec{x} + \vec{r})\psi(\vec{x}) \rangle, \tag{5.22}$$

$$M^2 = \langle \sigma^2 \rangle. \tag{5.23}$$

The first contribution represents the thermal fluctuations in any of the pure states and is obtained by shifting the order parameter by its mean

$$\psi(\vec{x}) = \varphi(\vec{x}) - m_\alpha. \tag{5.24}$$

This quantity averages to zero by construction, and the probability distribution

$$P[\psi(\vec{x})] = P_\alpha[\psi(\vec{x}) + m_\alpha] \tag{5.25}$$

is independent of α, because the deviations from the mean are equally distributed in all pure states. The second contribution σ fluctuates over the possible values of the spontaneous magnetization, taking the values m_α with probabilities $P(m_\alpha) = w_\alpha$. Then, from Equations 5.10 and 5.16, we have

$$\langle \sigma \rangle = 0, \quad M^2 = \langle \sigma^2 \rangle. \tag{5.26}$$

5.2.4 Static Susceptibility

The introduction of a field $h(\vec{x})$ conjugate to the order parameter modifies the Hamiltonian:

$$H_h[\varphi(\vec{x})] = H[\varphi(\vec{x})] - \sum_{\vec{x}} h(\vec{x})\varphi(\vec{x}), \tag{5.27}$$

and the corresponding Gibbs state is

$$P_{G,h}[\varphi(\vec{x})] = \frac{1}{Z_h} \exp\{-H_h[\varphi(\vec{x})]/T\}. \tag{5.28}$$

With an $\vec{x}$-dependent external field, averages are no more space-translation invariant. If the field is small, the magnetization at the site $\vec{x}$ is given by

$$\langle \varphi(\vec{x}) \rangle_{G,h} = \langle \varphi(\vec{x}) \rangle_0 + \int d\vec{y} \, \chi_{st}(\vec{x} - \vec{y}, T)h(\vec{y}) + \mathcal{O}(h^2), \tag{5.29}$$

where $\langle \varphi(\vec{x}) \rangle_0$ is the magnetization in the state

$$P_0[\varphi(\vec{x})] = \lim_{h \to 0} P_{G,h}[\varphi(\vec{x})] = \begin{cases} P_G[\varphi(\vec{x})] & \text{for } T \geq T_c, \\ P_\alpha[\varphi(\vec{x})] & \text{for } T < T_c. \end{cases} \tag{5.30}$$

Here, $P_\alpha[\varphi(\vec{x})]$ stands for the particular broken-symmetry state selected when the external field is switched off. The site-dependent static susceptibility is given by

$$\chi_{st}(\vec{x} - \vec{y}, T) = \frac{\delta \langle \varphi(\vec{x}) \rangle_{G,h}}{\delta h(\vec{y})} \bigg|_{h=0}. \tag{5.31}$$

Making an expansion up to first order in $h(\vec{x})$ also in the definition,

$$\langle \varphi(\vec{x}) \rangle_{G,h} = \sum_{[\varphi(\vec{x}')]} \varphi(\vec{x}) P_{G,h}[\varphi(\vec{x}')] \tag{5.32}$$

and comparing this with Equation 5.29, it is straightforward to derive the *fluctuation-response theorem*:

$$\chi_{st}(\vec{r}, T) = \frac{1}{T} C_0(\vec{r}, T), \tag{5.33}$$

where $C_0(\vec{r}, T)$ is the correlation function in the state $P_0[\varphi(\vec{x})]$.

5.3 DYNAMICS

Let us now examine the time-dependent properties in the various quenches described in the introductory section. The order-parameter expectation value and correlator in the infinite-temperature initial state $P_G[\varphi, T_I]$ are given by

$$\langle \varphi(\vec{r}) \rangle_I = 0, \tag{5.34}$$

$$\langle \varphi(\vec{r}) \varphi(\vec{r}') \rangle_I = \Delta \delta(\vec{r} - \vec{r}'). \tag{5.35}$$

Modeling the dynamics by a Markov stochastic process, the time-dependent probability distribution obeys the master equation (cf. Equation 1.22):

$$\frac{\partial}{\partial t} P[\varphi, t] = \sum_{[\varphi']} W([\varphi']|[\varphi]) P[\varphi', t], \tag{5.36}$$

where $W([\varphi']|[\varphi])$ is the transition probability per unit time from $[\varphi']$ to $[\varphi]$. As discussed in Section 1.3, the transition probability must satisfy the detailed-balance condition with the Gibbs state at the final temperature T_F:

$$W([\varphi']|[\varphi]) P_G[\varphi', T_F] = W([\varphi]|[\varphi']) P_G[\varphi, T_F]. \tag{5.37}$$

Therefore, the dynamics preserves the symmetry of the Hamiltonian and, *if* equilibrium is reached, then *necessarily* $P[\varphi, t]$ must go over to $P_G[\varphi, T_F]$. The crucial question, of course, is whether the system equilibrates or not. As anticipated in the introductory section, four qualitatively different relaxation processes arise, depending on the values of d and T_F:

1. Quench to $T_F > T_c$: there is a finite equilibration time t_{eq}.
2. Quench to $T_F = T_c$: equilibrium is reached in an infinite time.
3. Quench to $(d > d_L, T_F < T_c)$: $C(t, t_w)$ does not equilibrate, whereas the dynamic susceptibility $\chi(t, t_w)$ equilibrates.
4. Quench to $(d = d_L, T_F = 0)$: neither $C(t, t_w)$ nor $\chi(t, t_w)$ equilibrate.

The dynamic susceptibility $\chi(t, t_w)$ will be defined more precisely in Section 5.4 as the zero-field-cooled susceptibility.

5.3.1 Equilibration

Because $P[\varphi, t]$ remains symmetrical during the relaxation, no information on the equilibration process can be extracted from the time-dependent magnetization, which vanishes throughout: $m(t) = \langle \varphi(\vec{r}, t) \rangle \equiv 0$. One must turn to the time-dependent correlation function

$$C(\vec{r}, t, t_w) = \langle \varphi(\vec{x} + \vec{r}, t) \varphi(\vec{x}, t_w) \rangle - \langle \varphi(\vec{x} + \vec{r}, t) \rangle \langle \varphi(\vec{x}, t_w) \rangle, \qquad (5.38)$$

where the angular brackets denote an average over the initial condition and the thermal noise. This quantity is the two-time generalization of the correlation function defined in Equation 1.88. For the subsequent discussion, it is sufficient to consider the autocorrelation function $C(t, t_w) = C(\vec{r} = 0, t, t_w)$.

If there exists a finite equilibration time t_{eq}, then for $t_w > t_{eq}$, the dynamics becomes TTI with

$$C(t, t_w) = C_{eq}(\tau, T_F). \qquad (5.39)$$

So, if it is not known how large t_{eq} is (or if it exists at all), in order to ascertain whether equilibration has occurred or not, it is necessary to look at the large-t_w behavior. The $t_w \to \infty$ limit requires us to also specify how $t \to \infty$. This is done by rewriting $C(t, t_w)$ in terms of the new pairs of variables (τ, t_w) and $(x = t/t_w, t_w)$,

$$C(t, t_w) = \widetilde{C}(\tau, t_w) = \widehat{C}(x, t_w), \qquad (5.40)$$

and by taking the limit $t_w \to \infty$ while keeping either τ or x fixed. The short-time $(\tau \ll t_w)$ and large-time $(\tau \gg t_w)$ regimes are explored, respectively, in the first and second cases. From the relation $x = \tau/t_w + 1$, it follows that when using the x variable, the short-time regime gets all compressed into $x = 1$. Assuming that the limits exist, one has

$$\lim_{t_w \to \infty} \widetilde{C}(\tau, t_w) = K_C(\tau), \qquad (5.41)$$

and

$$\lim_{t_w \to \infty} \widehat{C}(x, t_w) = C(x). \qquad (5.42)$$

In general, these two functions are not related to each other. However, if there exists an equilibration time t_{eq} and Equation 5.39 holds on all timescales, then $K_C(\tau) = C_{eq}(\tau, T_F)$ and for $t_w > t_{eq}$,

$$\widehat{C}(x, t_w) = C_{eq}[(x-1)t_w, T_F], \tag{5.43}$$

which implies the singular limit

$$C(x) = \begin{cases} C_{eq}(0, T_F) & \text{for } x = 1, \\ C_{eq}(\infty, T_F) & \text{for } x > 1. \end{cases} \tag{5.44}$$

This is a necessary condition for equilibration. To also be a sufficient condition, it should hold for all two-time observables. As we shall see below, it may hold for some quantities, but not for others.

5.3.2 GENERIC PROPERTIES OF $C(t, t_w)$

In the quench to $T_F > T_c$, because there is a finite equilibration time, Equation 5.44 necessarily holds. The interesting cases are those of quenches to and below T_c, where t_{eq} diverges.

In a quench to T_c, the generic form of $C(t, t_w)$ displays a multiplicative combination of the stationary and aging contributions [11,31]:

$$C(t, t_w) = (\tau + t_0)^{-b_c} g_C\left(\frac{t}{t_w}\right), \tag{5.45}$$

where

$$b_c = \frac{d - 2 + \eta}{z_c}. \tag{5.46}$$

Here, η is the usual static exponent, z_c is the dynamical critical exponent [32], and t_0 is a microscopic time* needed to regularize the equal-time autocorrelation function. Taking the short-time limit in Equation 5.41, $K_C(\tau)$ is found to coincide with the autocorrelation function of equilibrium critical dynamics:

$$C_{eq}(\tau, T_c) = (\tau + t_0)^{-b_c} g_C(1), \tag{5.47}$$

where $C_{eq}(0, T_c) = t_0^{-b_c} g_C(1) = \langle \varphi^2(\vec{x}) \rangle_G$. In order to explore the large-time regime, Equation 5.45 can be rewritten in the simple aging form:

$$C(t, t_w) = t_w^{-b_c} f_C(x, y), \tag{5.48}$$

with $y = t_0/t_w$. The scaling function,

$$f_C(x, y) = (x - 1 + y)^{-b_c} g_C(x), \tag{5.49}$$

---

*For instance, t_0 might be the time it takes for a domain wall to advance by one lattice spacing.

decreases asymptotically with the power law

$$f_C(x, y) = A_C x^{-\lambda_c/z_c}, \tag{5.50}$$

where λ_c is the critical autocorrelation exponent [33]. Then, taking the $t_w \to \infty$ limit, one finds

$$C(x) = \begin{cases} t_0^{-b_c} g_C(1) & \text{for } x = 1, \\ 0 & \text{for } x > 1, \end{cases} \tag{5.51}$$

in agreement with Equation 5.44. Therefore, although it is clear that equilibrium is not reached in any finite time, the autocorrelation function behaves as if equilibrium was reached in an infinite time.

As an illustration, consider the behavior of $C(t, t_w)$ in the quench to T_c of the kinetic Glauber–Ising model [34], which was introduced in Section 1.3.2. In Figure 5.2, $\widetilde{C}(\tau, t_w)$ is plotted against τ for increasing values of t_w, showing the separation of the timescales, as the curves collapse in the short-time regime and spread out as the large-time regime is entered. Furthermore, the collapse improves with increasing t_w, showing the convergence of $\widetilde{C}(\tau, t_w)$ toward $C_{eq}(\tau, T_c)$. The multiplicative structure of $C(t, t_w)$ is demonstrated in Figure 5.3, where a plot of $\tau^{b_c} C(t, t_w)$ versus x (for different values of t_w) shows the data collapse required by Equation 5.45, as in the chosen t_w-range, t_0 is negligible. This plot has been made using $b_c = 0.115$, obtained from Equation 5.46 with $\eta = 1/4$ and $z_c = 2.167$ [35].

When the quench is made to $T_F < T_c$, in the short-time regime the $t_w \to \infty$ limit gives the equilibrium behavior

$$K_C(\tau) = C_{eq}(\tau, T_F) = G_{eq}(\tau, T_F) + M^2. \tag{5.52}$$

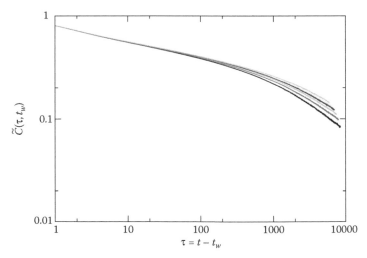

FIGURE 5.2 **(See color insert following page 180.)** Plot of $\widetilde{C}(\tau, t_w)$ vs. τ for the $d = 2$ Glauber–Ising model quenched to T_c. The waiting time t_w increases in steps of 500 from 1500 up to 4000 (bottom to top). In this, and in the following figures, times are always measured in units of Monte Carlo steps.

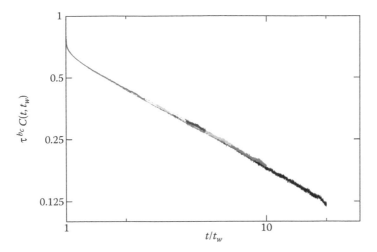

FIGURE 5.3 (See color insert following page 180.) Plot of $\tau^{b_c} C(t, t_w)$ vs. $x = t/t_w$ for the $d = 2$ Glauber–Ising model quenched to T_c and for the same values of t_w as in Figure 5.2. The collapse of the different curves demonstrates the multiplicative structure of Equation 5.45.

However, for large times, one finds

$$C(x) = \begin{cases} C_{\text{eq}}(0, T_F) & \text{for } x = 1, \\ h_C(x) & \text{for } x > 1. \end{cases} \tag{5.53}$$

Here, $h_C(x)$ is a monotonically decaying function with the limiting behaviors:

$$h_C(x) = \begin{cases} M^2 & \text{for } x = 1, \\ x^{-\lambda/z} & \text{for } x \gg 1, \end{cases} \tag{5.54}$$

and λ is the autocorrelation exponent below T_c [36]. Therefore, (a) the necessary condition (5.44) for equilibration is violated, and (b) the above result, together with Equation 5.52, implies the noncommutativity of the limits $t_w \to \infty$ and $t \to \infty$ with

$$\lim_{\tau \to \infty} \lim_{t_w \to \infty} C(\tau + t_w, t_w) = M^2, \tag{5.55}$$

and

$$\lim_{t_w \to \infty} \lim_{t \to \infty} C(t, t_w) = 0. \tag{5.56}$$

This is the phenomenon of *weak ergodicity breaking* [37], as ergodicity appears broken in the short-time regime but not in the large-time separations. The behaviors described above are illustrated in Figures 5.4 and 5.5, where $C(t, t_w)$ is plotted against τ and x for the Glauber–Ising model quenched to $T_F = 0.66T_c$. Both plots highlight the separation of timescales, with collapse of the data either in the short-time or in the large-time regime. Moreover, the curves fall below the Edwards–Anderson plateau at

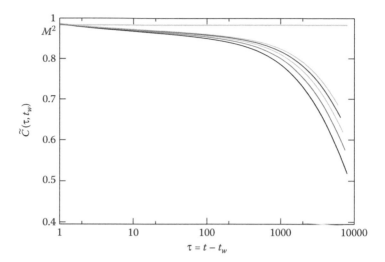

FIGURE 5.4 (See color insert following page 180.) Plot of $\tilde{C}(\tau, t_w)$ vs. τ for the $d = 2$ Glauber–Ising model quenched to $T_F = 0.66T_c$. The waiting time t_w increases in steps of 500 from 2000 up to 4000 (bottom to top). The horizontal line indicates the Edwards–Anderson order parameter $M^2 = 0.97$.

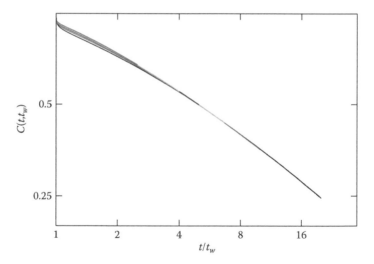

FIGURE 5.5 (See color insert following page 180.) Plot of $C(t, t_w)$ vs. $x = t/t_w$ for the $d = 2$ Glauber–Ising model quenched to $T_F = 0.66T_c$. The values of t_w are the same as in Figure 5.4.

$M^2 = 0.97$, revealing that the system keeps on decorrelating for arbitrarily large time-scales. The curve in Figure 5.5 for $x > 3$, where the collapse is good, is essentially a plot of $h_C(x)$.

Weak ergodicity breaking is incompatible with the multiplicative form in Equation 5.45. Then, in order to put together the separation of the timescales and weak ergodicity breaking, $C(t, t_w)$ must have the additive structure:

$$C(t, t_w) = G_{eq}(\tau, T_F) + C_{ag}(t, t_w), \tag{5.57}$$

where $C_{ag}(t, t_w)$ obeys the simple aging form in Equation 5.2 with the scaling function $h_C(x)$ and $b = 0$. Notice that this value of b is of geometrical origin, as it is related through the multi-time scaling of Furukawa [38],

$$C_{ag}(\vec{r}, t, t_w) = \tilde{g}\left(\frac{r}{t^{1/z}}, \frac{t}{t_w}\right), \tag{5.58}$$

to the scaling of the equal-time correlation function in Equation 1.88:

$$C_{ag}(\vec{r}, t) = g\left(\frac{r}{t^{1/z}}\right) = \tilde{g}\left(\frac{r}{t^{1/z}}, 1\right). \tag{5.59}$$

This scaling, in turn, is a consequence of the compact nature of the ordered domains.* Here, z is the phase-ordering growth exponent with the value $z = 2$ for dynamics with nonconserved order parameter (see Section 1.4).

5.3.3 SPLITTING OF THE ORDER PARAMETER

The additive structure in Equation 5.57 leads, for large time, to the generalization of the splitting in Equation 5.20 of the order parameter into the sum of two time-dependent components:

$$\varphi(\vec{x}, t) = \psi(\vec{x}, t) + \sigma(\vec{x}, t). \tag{5.60}$$

The two contributions are still statistically independent, have zero mean, and the respective autocorrelation functions are given by

$$\langle \psi(\vec{x}, t)\psi(\vec{x}, t_w)\rangle = G_{eq}(\tau, T_F), \tag{5.61}$$

$$\langle \sigma(\vec{x}, t)\sigma(\vec{x}, t_w)\rangle = C_{ag}(t, t_w). \tag{5.62}$$

The variables $\psi(\vec{x}, t)$ and $\sigma(\vec{x}, t)$ are associated with the fast and slow degrees of freedom [6–8]. Keeping in mind the domain structure of the configurations, the local ordering variable is defined by

$$\sigma(\vec{x}, t) = m_{\alpha(\vec{x}, t)}, \tag{5.63}$$

where $\alpha(\vec{x}, t)$ is the value of α selected by the domain to which the site $\vec{x}$ belongs at the time t. Hence, $\sigma(\vec{x}, t)$ represents the local equilibrium magnetization within domains, and $\psi(\vec{x}, t)$ is the field of thermal fluctuations, as in Equation 5.20. The off-equilibrium character of the dynamics enters only through the ordering variable $\sigma(\vec{x}, t)$. In fact,

---

*In the usual treatment of phase-ordering kinetics [4], the thermal contribution G_{eq} is either negligible or absent when quenches to $T_F = 0$ are considered.

while $\psi(\vec{x}, t)$ executes equilibrium thermal fluctuations, $\sigma(\vec{x}, t)$ can change only if a defect (or an interface) goes by the site $\vec{x}$ at the time t. In other words, the evolution of $\sigma(\vec{x}, t)$ is strictly related to the existence of defects in the system as discussed in Section 1.4—this is precisely what keeps the system out-of-equilibrium. As we shall see in Section 5.5.1, in the case of the large-n model, which was briefly discussed in Section 1.4.2, the construction in Equation 5.60 can be carried out exactly.

5.4   LINEAR RESPONSE FUNCTION

As previously stated, the quench of phase-ordering systems offers the full spectrum of off-equilibrium phenomena, with increasing degree of deviation from equilibrium as T_F is lowered from above to below T_c. The survey of aging, so far, has been conducted using the order-parameter autocorrelation function as a probe. However, once the basic features of the phenomena involved have been brought into focus, it is relevant to look jointly at the autocorrelation and autoresponse functions. In particular, the deviations from the *fluctuation-dissipation theorem* (FDT) have proved to be a most effective tool of investigation [1–3]. This approach to the study of out-of-equilibrium dynamics has been formulated by Cugliandolo and Kurchan in their work on mean-field models of spin glasses [39,40].

Let us begin by defining the time-dependent linear response function. If a small space- and time-dependent external field $h(\vec{x}, t)$ is switched on in the time interval (t_1, t_2) after the quench, then the magnetization at the time $t \geq t_2$ is given by

$$\langle \varphi(\vec{x}, t) \rangle_h = \langle \varphi(\vec{x}, t) \rangle + \int d\vec{y} \int_{t_1}^{t_2} dt'\, R(\vec{x} - \vec{y}, t, t') h(\vec{y}, t') + \mathcal{O}(h^2). \qquad (5.64)$$

Here, $\langle \varphi(\vec{x}, t) \rangle$ is the magnetization in the absence of the field, and

$$R(\vec{x} - \vec{y}, t, t_w) = \left. \frac{\delta \langle \varphi(\vec{x}, t) \rangle_h}{\delta h(\vec{y}, t_w)} \right|_{h=0} \qquad (5.65)$$

is the space- and time-dependent linear response function. The autoresponse function $R(t, t_w)$ is obtained by setting $\vec{x} = \vec{y}$.

With a time-independent external field, Equation 5.64 takes the form

$$\langle \varphi(\vec{x}, t) \rangle_h = \langle \varphi(\vec{x}, t) \rangle + \int d\vec{y}\, \zeta(\vec{x} - \vec{y}, t, t_1, t_2) h(\vec{y}) + \mathcal{O}(h^2), \qquad (5.66)$$

where

$$\zeta(\vec{r}, t, t_2, t_1) = \int_{t_1}^{t_2} dt'\, R(\vec{r}, t, t') \qquad (5.67)$$

is the integrated linear response function. Particular cases, frequently encountered in the literature, are those of the *thermoremanent magnetization* (TRM) corresponding with the protocol $t_1 = 0$, $t_2 = t_w$,

$$\rho(\vec{r}, t, t_w) = \zeta(\vec{r}, t, t_w, 0), \qquad (5.68)$$

and of the *zero-field-cooled* (ZFC) susceptibility corresponding with $t_1 = t_w$, $t_2 = t$,

$$\chi(\vec{r}, t, t_w) = \zeta(\vec{r}, t, t, t_w). \tag{5.69}$$

5.4.1 Fluctuation–Dissipation Theorem

For convenience, let us briefly derive the FDT. We assume that the small external field has been applied from a time so distant in the past that equilibrium in the field is established at the time t_w, and that it is switched off for $t > t_w$. Then, the magnetization is given by

$$\langle \varphi(\vec{x}, t) \rangle_h = \sum_{[\varphi],[\varphi']} \varphi(\vec{x}) Q([\varphi', t_w] | [\varphi, t]) P_{G,h}[\varphi'], \tag{5.70}$$

where $Q([\varphi, t] | [\varphi', t_w])$ is the conditional probability in the absence of the field, as $t > t_w$. Recalling that $P_{G,h}[\varphi]$ is given by Equations 5.27 and 5.28, and expanding up to first order in the field, one finds

$$\langle \varphi(\vec{x}, t) \rangle_h = \langle \varphi(\vec{x}, t) \rangle + \frac{1}{T_F} \int d\vec{y} \, C_0(\vec{x} - \vec{y}, t - t_w, T_F) h(\vec{y}), \tag{5.71}$$

where $C_0(\vec{x} - \vec{y}, t - t_w, T_F)$ is the equilibrium, unperturbed correlation function in the stationary state in Equation 5.30. Hence, comparing with Equation 5.66,

$$\frac{1}{T_F} C_0(\vec{r}, t - t_w, T_F) = \int_{-\infty}^{t_w} dt' R(\vec{r}, t, t'), \tag{5.72}$$

and, differentiating with respect to t_w, the FDT is obtained as follows:

$$R_{\text{eq}}(\vec{r}, \tau, T_F) = -\frac{1}{T_F} \frac{\partial}{\partial \tau} C_0(\vec{r}, \tau, T_F), \tag{5.73}$$

where $R_{\text{eq}}(\vec{r}, \tau, T_F)$ is the equilibrium response function. From this, it is straightforward to derive the integrated form of the FDT in terms of the equilibrium ZFC susceptibility:

$$\chi_{\text{eq}}(\vec{r}, \tau, T_F) = \frac{1}{T_F} \left[C_0(\vec{r}, T_F) - C_0(\vec{r}, \tau, T_F) \right]. \tag{5.74}$$

Using $\lim_{\tau \to \infty} C_0(\vec{r}, \tau, T_F) = 0$, this gives the identification via Equation 5.33, of the large-time limit of the equilibrium ZFC susceptibility with the static susceptibility:

$$\lim_{\tau \to \infty} \chi_{\text{eq}}(\vec{r}, \tau, T_F) = \chi_{\text{st}}(\vec{r}, T_F). \tag{5.75}$$

5.4.2 GENERIC PROPERTIES OF $R(t, t_w)$

Before exploring the deviations from the FDT when the system is not in equilibrium, it is convenient to go over the generic properties of $R(t, t_w)$, as was done for $C(t, t_w)$ in Section 5.3.2. Apart from the few cases where analytical results are available, $R(t, t_w)$ is less known than $C(t, t_w)$ because it is much more difficult to measure numerically. Actually, until very recently, $R(t, t_w)$ was numerically accessible only indirectly through the measurement of the integrated response functions. This situation has partially changed after the introduction of new and more efficient algorithms [20–23]. In any case, whenever TTI holds, as in equilibrium or in the short-time sector, the task is easier because the form of $R(t, t_w)$ is related to that of $C(t, t_w)$ via the FDT. When TTI does not hold, scaling arguments will be used.

In the quench to T_c, the analog of the multiplicative form in Equation 5.45 is

$$R(t, t_w) = (\tau + t_0)^{-(1+a_c)} g_R\left(\frac{t}{t_w}\right). \tag{5.76}$$

The short-time limit of this quantity, $K_R(\tau)$, coincides with the equilibrium response function:

$$R_{eq}(\tau, T_c) = (\tau + t_0)^{-(1+a_c)} g_R(1), \tag{5.77}$$

and should be related to $C_{eq}(\tau, T_c)$ by the FDT. This yields the constraints

$$a_c = b_c, \tag{5.78}$$

and

$$T_c \, g_R(1) = b_c \, g_C(1). \tag{5.79}$$

Switching to the (x, t_w)-variables, $R(t, t_w)$ is rewritten in the simple aging form

$$R(t, t_w) = t_w^{-(1+a_c)} f_R(x, y). \tag{5.80}$$

The scaling function,

$$f_R(x, y) = (x - 1 + y)^{-(1+a_c)} g_R(x), \tag{5.81}$$

decreases for large x with the same power law as $f_C(x, y)$ [9,10]:

$$f_R(x, y) = A_R x^{-\lambda_c/z_c}. \tag{5.82}$$

Then, taking the $t_w \to \infty$ limit, the analog of Equation 5.51 is obtained as

$$\mathcal{R}(x) = \begin{cases} t_0^{-(1+b_c)} g_R(1) & \text{for } x = 1, \\ 0 & \text{for } x > 1. \end{cases} \tag{5.83}$$

Here, $\mathcal{R}(x)$ satisfies the necessary condition for equilibration:

$$\mathcal{R}(x) = \begin{cases} R_{eq}(0, T_F) & \text{for } x = 1, \\ R_{eq}(\infty, T_F) & \text{for } x > 1, \end{cases} \tag{5.84}$$

in the same way as the autocorrelation function.

The above behavior is well illustrated by the data for $R(t, t_w)$, obtained with the algorithm of Ref. [22], in the quench to T_c of the $d = 2$ kinetic Ising model. In Figure 5.6, $R(t, t_w)$ is plotted against τ, displaying (as in Figure 5.2) the separation of the timescales, with the collapse and spread of curves in the short-time and large-time regimes, respectively. The collapse of the curves in Figure 5.7, obtained by plotting $\tau^{1+a_c} R(t, t_w)$ against x, demonstrates (as in Figure 5.3) the multiplicative structure in Equation 5.76. The validity of the FDT in the short-time regime is illustrated by Figure 5.8.

In the quench below T_c, the additivity of $C(t, t_w)$ induces the corresponding structure in the response function,

$$R(t, t_w) = R_{eq}(\tau, T_F) + R_{ag}(t, t_w), \tag{5.85}$$

in the following way. The stationary response of the fast degrees of freedom $R_{eq}(\tau, T_F)$ is defined by requiring the FDT to hold in the short-time regime:

$$R_{eq}(\tau, T_F) = -\frac{1}{T_F} \frac{\partial}{\partial \tau} G_{eq}(\tau, T_F), \tag{5.86}$$

where $G_{eq}(\tau, T_F)$ is the stationary contribution entering Equation 5.57. The aging component associated with the slow degrees of freedom remains defined by

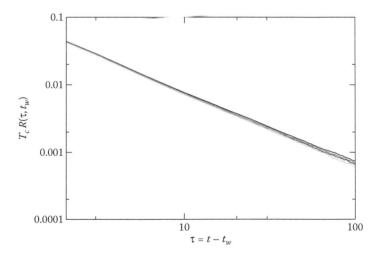

FIGURE 5.6 **(See color insert following page 180.)** Plot of $T_c R(\tau, t_w)$ vs. τ for the $d = 2$ Glauber–Ising model quenched to T_c. The waiting time t_w increases in steps of 500 from 1500 up to 4000 (bottom to top).

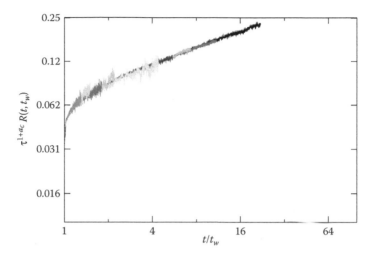

FIGURE 5.7 (**See color insert following page 180.**) Plot of $\tau^{1+a_c}R(t,t_w)$ vs. $x = t/t_w$ for the $d = 2$ Glauber–Ising model quenched to T_c and for the same values of t_w as in Figure 5.6. The collapse of the different curves demonstrates the multiplicative structure of Equation 5.76.

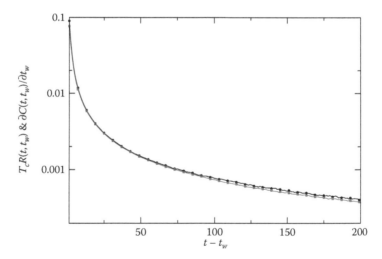

FIGURE 5.8 (**See color insert following page 180.**) Plot of $T_c R(t,t_w)$ and $\partial C(t,t_w)/\partial t_w$ for the $d = 2$ Glauber–Ising model quenched to T_c, with $t_w = 1000$. The superposition of the curves for $\tau \ll t_w$ demonstrates the validity of the FDT in the short-time regime.

Equation 5.85 as the difference $R_{\mathrm{ag}}(t,t_w) = R(t,t_w) - R_{\mathrm{eq}}(\tau,T_F)$. It obeys the simple aging form in Equation 5.3, where the scaling function decays asymptotically with the same power law as $h_C(x)$ [31]:

$$h_R(x) \sim x^{-\lambda/z}. \tag{5.87}$$

Because in the short-time regime the FDT must also be satisfied by the full $R(t, t_w)$, this implies that $R_{ag}(t, t_w)$ vanishes for short times. Conversely, because $R_{eq}(\tau, T_F)$ is typically a rapidly decaying function, in the large-time regime only $R_{ag}(t, t_w)$ survives. Then, the behavior of $R(x, t_w)$ for large t_w is given by

$$R(x, t_w) = \begin{cases} R_{eq}(0, T_F) & \text{for } x = 1, \\ t_w^{-(1+a)} h_R(x) & \text{for } x > 1. \end{cases} \quad (5.88)$$

From this, it is easy to see that the condition in Equation 5.84 is satisfied for $d > d_L$, after taking into account that $a > 0$ (as will be explained shortly) and that $R_{eq}(\infty, T_F) = 0$. Therefore, in contrast with the case of the autocorrelation function, the large-t_w behavior of $R(t, t_w)$ does not reveal that the system remains out-of-equilibrium in the large-time regime.

Regarding the exponent a, there is a major difference with respect to the case of the quench to T_c. Because the FDT in Equation 5.86 relates only the stationary components $G_{eq}(\tau, T_F)$ and $R_{eq}(\tau, T_F)$, there is now no constraint relating a to b. (For the T_c-quench, the equality in Equation 5.78 was enforced by the multiplicative structure.) As mentioned in the introduction, the determination of the exponent a is a difficult problem, which will be discussed at the end of this chapter. In any case, although the actual value of a has not been definitely established, yet there is general consensus that $a > 0$ for $d > d_L$. The case of $d = d_L$ stands apart, and will be discussed in Section 5.4.6.

5.4.3 ZFC SUSCEPTIBILITY BELOW T_c

It is clear that the additive structure of the response function generates the analogous form of the ZFC susceptibility:

$$\chi(t, t_w) = \chi_{eq}(\tau, T_F) + \chi_{ag}(t, t_w), \quad (5.89)$$

where the stationary component satisfies, by construction, the equilibrium FDT in Equation 5.74. Inserting the scaling form (5.3) of R_{ag} in the definition of the aging component,

$$\chi_{ag}(t, t_w) = \int_{t_w}^t dt' R_{ag}(t, t'), \quad (5.90)$$

one finds

$$\chi_{ag}(t, t_w) = t_w^{-a} h_\chi(x, y). \quad (5.91)$$

Here,

$$h_\chi(x, y) = x^{-a} \mathcal{I}(x, y), \quad (5.92)$$

and

$$\mathcal{I}(x, y) = \int_{1/x}^1 dz \, z^{-(1+a)} h_R\left(z^{-1}, z^{-1} y/x\right). \quad (5.93)$$

Here, we make an observation that turns out to be quite important in general, when the data for χ_{ag} from the simulations is used to measure the exponent a [41]. If one seeks to determine a from Equation 5.91 by examining the behavior of χ_{ag} as t_w is varied and x is kept fixed, one must be aware that the t_w-dependence coming from $y = t_0/t_w$ may play a role. In other words, t_0 may act as a dangerous irrelevant variable through a mechanism quite similar to the one causing the breakdown of hyperscaling in static critical phenomena above the upper critical dimensionality d_U [42]. In those cases where analytical calculations can be carried out with arbitrary d [8,43,44], there exists a value d^* of the dimensionality such that the $y \to 0$ limit of the integral $\mathcal{I}(x, y)$ is finite for $d < d^*$, whereas for $d > d^*$ there is a singularity of the type

$$\mathcal{I}(x, y) = (y/x)^{-c}\, \widehat{\mathcal{I}}(x). \tag{5.94}$$

Here, $c > 0$, which becomes logarithmic ($c = 0$) for $d = d^*$. Hence, d^* plays the same role as d_U in critical phenomena but is clearly not related to d_U. The assumption is that this is a generic feature of the relaxation below T_c. Then, Equation 5.91 can be rewritten as

$$\chi_{ag}(t, t_w) = t_w^{-a_\chi}\widehat{h}_\chi(x), \tag{5.95}$$

with

$$a_\chi = \begin{cases} a & \text{for } d < d^*, \\ a & \text{with log-corrections for } d = d^*, \\ a - c & \text{for } d > d^*. \end{cases} \tag{5.96}$$

For large x,

$$\widehat{h}_\chi(x) \sim x^{-a_\chi}, \tag{5.97}$$

which implies

$$\chi_{ag}(t, t_w) \sim t^{-a_\chi} \tag{5.98}$$

for large t. The analogy with the breaking of hyperscaling is quite close because, as we shall see, there is a dependence of a_χ on d for $d < d^*$, which disappears for $d > d^*$.

Our second observation concerns the value of a_χ and the asymptotic behavior of $\chi(t, t_w)$. From Equation 5.89,

$$\lim_{t \to \infty} \chi(t, t_w) = \lim_{\tau \to \infty} \chi_{eq}(\tau, T_F) + \lim_{t \to \infty} \chi_{ag}(t, t_w). \tag{5.99}$$

Recalling Equations 5.75 and 5.98, this gives

$$\lim_{t \to \infty} \chi(t, t_w) = \chi_{st}(T_F) + \chi^*, \tag{5.100}$$

with

$$\chi^* = \begin{cases} 0 & \text{for } a_\chi > 0, \\ \lim_{x \to \infty} \widehat{h}_\chi(x) & \text{for } a_\chi = 0. \end{cases} \tag{5.101}$$

Hence, the ZFC susceptibility reaches the equilibrium value for $a_\chi > 0$ but not for $a_\chi = 0$. In other words, for $a_\chi > 0$ the contribution of the slow degrees of freedom disappears asymptotically, and for $a_\chi = 0$ there remains an extra contribution χ^* over the equilibrium one. This is a very interesting phenomenon. Rewriting Equation 5.33 as

$$T_F \chi_{st}(T_F) = C_{eq}(0, T_F) - M^2, \tag{5.102}$$

and recalling that M^2 plays the role of the Edwards–Anderson order parameter, there is a formal similarity with what happens in the mean-field theory of spin glasses. There, the large-time limit of the ZFC susceptibility can be written [45,46] exactly as in Equation 5.100. The substantial difference is that, in the spin glass case, χ^* is an equilibrium quantity whose appearance is due to *replica symmetry breaking* [29]. As a matter of fact, the observation of $\chi^* > 0$ in the simulations of finite-dimensional spin glasses is taken as evidence of replica symmetry breaking [47]. Here, χ^* is the difference between the large-time limit of the ZFC susceptibility and the same quantity computed from equilibrium statistical mechanics. It is a quantity of purely dynamical origin that appears in the quench to $(d_L, T_F = 0)$, revealing the strong out-of-equilibrium nature of the relaxation.

5.4.4 FLUCTUATION–DISSIPATION RATIO

If the system is not in equilibrium, the FDT does not hold and the violation of the theorem can be used as a measure of the deviation from equilibrium. This idea was implemented by Cugliandolo and Kurchan [39] through the introduction of the *fluctuation–dissipation ratio* (FDR):

$$X(t, t_w) = \frac{T_F R(t, t_w)}{\partial_{t_w} C(t, t_w)}, \tag{5.103}$$

which satisfies $X(t, t_w) = 1$ in equilibrium and $X(t, t_w) \neq 1$ off-equilibrium. Formally, the FDR allows us to define the temperature-like quantity

$$T_{eff}(t, t_w) = \frac{T_F}{X(t, t_w)}, \tag{5.104}$$

whose interpretation as an effective temperature, however, requires some care [48].

As we have seen, the characterization of systems that remain out-of-equilibrium for arbitrary long times requires the exploration of the various asymptotic regimes reached as $t_w \to \infty$. An efficient way of doing this is through the reparameterization of the time t in terms of the autocorrelation function. For fixed t_w, $C(t, t_w)$ is a monotonically decreasing function of t. Hence, inverting with respect to t, the function $\widehat{X}(C, t_w) = X(t(C, t_w), t_w)$ is obtained. The limit of this function for fixed C,

$$\lim_{t_w \to \infty} \widehat{X}(C, t_w) = \mathcal{X}(C), \tag{5.105}$$

defines the limit FDR in the time-sector characterized by the chosen value of C. The associated effective temperature is given by

$$\mathcal{T}(C) = \frac{T_F}{\mathcal{X}(C)}. \tag{5.106}$$

The correspondence between values of C and the short-time and long-time regimes will be clarified below.

Inserting $X(t, t_w)$ into the definition of the ZFC susceptibility in Equation 5.69, we have

$$T_F \chi(t, t_w) = \int_{t_w}^{t} dt' \, \widehat{X}(C(t, t'), t') \frac{\partial}{\partial t'} C(t, t'). \tag{5.107}$$

For values of t_w so large that Equation 5.105 can be used under the integral, the parametric representation of χ is obtained:

$$T_F \widehat{\chi}(C(t, t_w)) = \int_{C(t, t_w)}^{C(t, t)} dC' \, \mathcal{X}(C'). \tag{5.108}$$

Differentiating with respect to C, this gives

$$-T_F \frac{d\widehat{\chi}(C)}{dC} = \mathcal{X}(C), \tag{5.109}$$

which relates the limit FDR to the slope of $\widehat{\chi}(C)$. This is the most commonly used way of estimating the FDR, due to the relative ease of computing $\widehat{\chi}(C)$ numerically. Notice that the integrated form in Equation 5.74 of the FDT is recovered from Equation 5.108 when $\mathcal{X}(C) \equiv 1$. In that case, the plot of $T_F \widehat{\chi}(C)$ is a straight line with slope -1, the so-called trivial plot, which is the hallmark of equilibrium. Off-equilibrium behavior is conveniently detected through deviations from the trivial plot.

5.4.5 PARAMETRIC PLOTS

The shape of the parametric plots (see Figure 5.9) can be derived from general considerations. In the quench to above T_c, the system equilibrates in a finite time, and it is straightforward to obtain

$$\mathcal{X}(C) \equiv 1, \tag{5.110}$$

as $X(t, t_w) = 1$ for $t_w > t_{eq}$. In the case of the quench to T_c, the outcome is almost the same, but the derivation is less straightforward. From Equations 5.48, 5.78, 5.80, one finds

$$X(t, t_w) = F(x, y), \tag{5.111}$$

where

$$F(x, y) = -T_c \frac{f_R(x, y)}{f_{\partial C}(x, y)}, \tag{5.112}$$

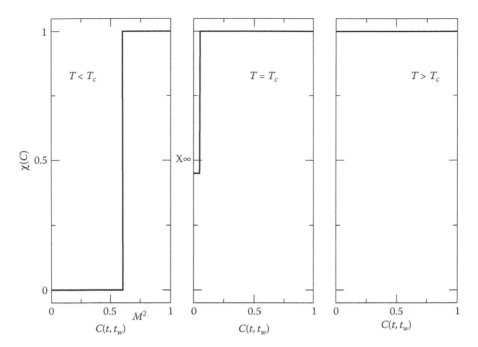

FIGURE 5.9 The limit FDR $\mathcal{X}(C)$.

and

$$f_{\partial C}(x, y) = b_c f_C(x, y) + \left[x \frac{\partial}{\partial x} f_C(x, y) + y \frac{\partial}{\partial y} f_C(x, y) \right]. \qquad (5.113)$$

Inverting the form of the autocorrelation function in Equation 5.51 with respect to x,

$$x(C) = \begin{cases} \infty & \text{for } C = 0, \\ 1 & \text{for } 0 < C \leq C_{eq}(0, T_c). \end{cases} \qquad (5.114)$$

Inserting this into Equation 5.111, the limit FDR is obtained

$$\mathcal{X}(C) = \begin{cases} \mathcal{X}_\infty & \text{for } C = 0, \\ 1 & \text{for } 0 < C \leq C_{eq}(0, T_c), \end{cases} \qquad (5.115)$$

where

$$\mathcal{X}_\infty = \lim_{y \to 0} \lim_{x \to \infty} F(x, y) = \frac{T_c A_R}{A_C(\lambda/z_c - b_c)} \qquad (5.116)$$

is a new universal quantity characteristic of the critical relaxation [31,49]. The second line of Equation 5.115 comes from $F(1, 0) = T_c g_R(1)/b_c g_C(1)$, together with the FDT requirement in Equation 5.77. The above result illustrates the utility of the FDR, and the parametric plot, in the precise characterization of off-equilibrium relaxation. Equation 5.114 shows that, in the quench to T_c, all values of $C > 0$ correspond with the short-time regime, whereas the large-time regime corresponds with $C = 0$. In the latter regime, the system remains off-equilibrium because, in general, $X_\infty < 1$ (see center panel of Figure 5.9).

In the quench below T_c, from the additive structures of C and R, it follows that

$$X(t, t_w) = T_F \left[\frac{R_{\mathrm{eq}}(\tau, T_F)}{\partial_{t_w} G_{\mathrm{eq}}(\tau, T_F) + \partial_{t_w} C_{\mathrm{ag}}(t, t_w)} + \frac{R_{\mathrm{ag}}(t, t_w)}{\partial_{t_w} G_{\mathrm{eq}}(\tau, T_F) + \partial_{t_w} C_{\mathrm{ag}}(t, t_w)} \right].$$
(5.117)

Recalling that R_{eq} and R_{ag} are not simultaneously different from zero when t_w is large, the first term in the brackets contributes in the short-time regime and the second one in the large-time regime, yielding

$$X(t, t_w) = \begin{cases} 1 & \text{for } x = 1, \\ t_w^{-a} H(x, y) & \text{for } x > 1. \end{cases}$$
(5.118)

Here,

$$H(x, y) = -T_F \frac{h_R(x, y)}{h_{\partial C}(x, y)},$$
(5.119)

with

$$h_{\partial C}(x, y) = \left[x \frac{\partial}{\partial x} h_C(x, y) + y \frac{\partial}{\partial y} h_C(x, y) \right].$$
(5.120)

Again, inverting the autocorrelation function in Equation 5.53 with respect to x,

$$x(C) = \begin{cases} h_C^{-1}(C) & \text{for } C < M^2, \\ 1 & \text{for } M^2 \leq C \leq C_{\mathrm{eq}}(0, T_F). \end{cases}$$
(5.121)

Inserting into Equation 5.118, one finds

$$X(C, t_w) = \begin{cases} t_w^{-a} H(h_C^{-1}(C), y) & \text{for } C < M^2, \\ 1 & \text{for } M^2 \leq C \leq C_{\mathrm{eq}}(0, T_F), \end{cases}$$
(5.122)

whose $t_w \to \infty$ limit for $d > d_L$ (that is, when $a > 0$) gives

$$X(C) = \begin{cases} 0 & \text{for } C < M^2, \\ 1 & \text{for } M^2 \leq C \leq C_{\mathrm{eq}}(0, T_F). \end{cases}$$
(5.123)

Here, the off-equilibrium character of the relaxation is expanded and enhanced, with respect to the T_c-quench (see left panel of Figure 5.9). The quasi-equilibrium behavior in the short-time regime is limited to values of C above the Edwards–Anderson plateau, and the deviation from equilibrium for large times is more pronounced, as the FDR vanishes.

The corresponding parametric representations of the ZFC susceptibility are easily obtained (see Figure 5.10) by integration. In the quenches above and to T_c, both Equations 5.110 and 5.115 yield the same trivial plot:

$$T_c \hat{\chi}(C) = C_{\mathrm{eq}}(0, T_c) - C.$$
(5.124)

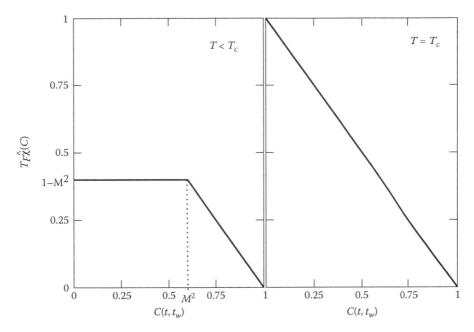

FIGURE 5.10 Parametric plots of the ZFC susceptibility $\hat{\chi}(C)$ with $C_{eq}(0, T_F) = 1$.

Notice that, if one goes back differentiating with respect to C, one finds identically $\mathcal{X}(C) = 1$ for all values of C, above and at T_c. This clarifies that, in order to uncover the existence of the non-trivial FDR $\mathcal{X}_\infty$, the order of the limits in Equation 5.116 is crucial. In the quench below T_c, instead, the departure from the trivial behavior is most evident:

$$T_F \hat{\chi}(C) = \begin{cases} C_{eq}(0, T_F) - M^2 & \text{for } C < M^2, \\ C_{eq}(0, T_F) - C & \text{for } M^2 \le C \le C_{eq}(0, T_F), \end{cases} \tag{5.125}$$

as the plot is flat for $C < M^2$. This plot clearly shows that the susceptibility equilibrates, whereas the autocorrelation function does not. The rise from zero to $[C_{eq}(0, T_F) - M^2]$ in the left panel of Figure 5.10 shows the saturation of $\hat{\chi}(C)$ to the static value $T_F \chi_{st}(T_F) = G_{eq}(0, T_F)$, as C decays to the Edwards–Anderson plateau. For larger times, the susceptibility remains fixed at the equilibrium value in the flat portion of the plot, and C falls below the plateau, according to the weak-ergodicity-breaking scenario.

The results described above are universal, as all the non-universal features of $F(x, y)$ and $H(x, y)$ have been eliminated in the $t_w \to \infty$ limit. Therefore, in all quenches to T_c, the parametric plots of $\mathcal{X}(C)$ and $T_c \hat{\chi}(C)$ are trivial, except for the value of $\mathcal{X}_\infty$. In all quenches below T_c, the deviation from the trivial plot takes the form of the flat behavior below the Edwards–Anderson plateau. For the parametric plots in the quench to (d_L, T_F), it is not possible to make statements of such generality. Some comments will be made in the next section.

Finally, let us say a few words about the effective temperature. From Equations 5.106 and 5.110, it follows that $\mathcal{T}(C)$ coincides with the temperature of the thermal bath T_F, when the plot of $T_F \hat{\chi}(C)$ is trivial. For $T_F > T_c$, this happens for all values of C. For $T_F = T_c$, this happens for all values of C except $C = 0$, where $\mathcal{T}(0) = T_F / \mathcal{X}_\infty > T_F$. Finally, for $T_F < T_c$,

$$\mathcal{T}(C) = \begin{cases} \infty & \text{for } C < M^2, \\ T_F & \text{for } M^2 \le C \le C_{eq}(0, T_F). \end{cases} \tag{5.126}$$

The latter result indicates that, in the quench below T_c, whereas the fast degrees of freedom thermalize, the slow ones do not interact with the thermal bath and retain the temperature T_I of the initial condition.

5.4.6 A SPECIAL CASE: THE QUENCH TO (d_L, $T_F = 0$)

An interesting and non-trivial situation arises if the system is at the lower critical dimensionality d_L and the quench is made to $T_F = 0$. As discussed in the introduction, this process can be regarded as the $d \to d_L$ limit, either of quenches to T_c or of quenches below T_c. In the first case, from Equation 5.115,

$$\mathcal{X}(C) = \begin{cases} 0 & \text{for } C = 0, \\ 1 & \text{for } 0 < C \le C_{eq}(0, T_c), \end{cases} \tag{5.127}$$

as $\mathcal{X}_\infty = 0$. In the second case, from Equation 5.123,

$$\mathcal{X}(C) = \begin{cases} 0 & \text{for } C < C_{eq}(0, T_F), \\ 1 & \text{for } C = C_{eq}(0, T_F), \end{cases} \tag{5.128}$$

as $C_{eq}(0, T_F) = M^2$ at $T_F = 0$. Hence, two very different results are obtained, and one has to identify which is the correct one. What can be stated on general grounds is that the quench to $(d_L, T_F = 0)$ is akin to a quench below T_c, as $G_{eq}(\tau, T_F) = 0$ at $T_F = 0$. Therefore, $C(t, t_w)$ must necessarily have an additive structure, otherwise it would vanish identically. Hence, Equation 5.127 can be discarded. However, from the analytical [8,44,50,51] and numerical [52,53] evidence accumulated so far, $\mathcal{X}(C)$ does not obey Equation 5.128 either. Rather, $\mathcal{X}(C)$ appears to be a non-trivial and non-universal smooth function. This behavior is compatible with Equation 5.122 if $a = 0$, which yields the smooth behavior

$$\mathcal{X}(C) = H(h_C^{-1}(C), 0), \tag{5.129}$$

preserving all the non-universal features of $h_C(x, 0)$ and $H(x, 0)$. Therefore, the quench to $(d_L, T_F = 0)$ belongs to the class of quenches below T_c, but there are peculiarities induced by the vanishing of the exponent a.

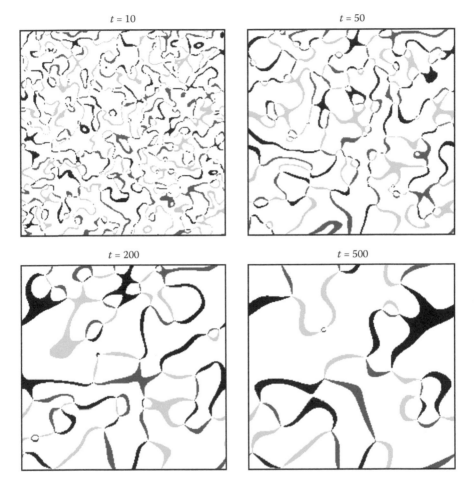

FIGURE 1.9 Evolution of the dimensionless XY model from a disordered initial condition. The pictures were obtained from a Euler-discretized version of Equation 1.164 with $n = 2$ and $\epsilon = 0$. The numerical details are the same as for Figure 1.4. The snapshots show regions of constant phase $\theta_\psi = \tan^{-1}(\psi_2/\psi_1)$, measured in radians, with the following color coding: $\theta_\psi \in [1.85, 2.15]$ (black); $\theta_\psi \in [3.85, 4.15]$ (red); $\theta_\psi \in [5.85, 6.15]$ (green). Typically, a meeting point of the three colors denotes a vortex defect.

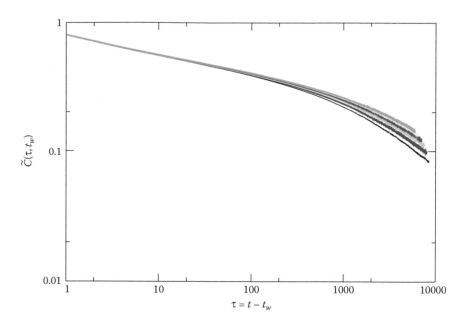

FIGURE 5.2 Plot of $\widetilde{C}(\tau, t_w)$ vs. τ for the $d = 2$ Glauber–Ising model quenched to T_c. The waiting time t_w increases in steps of 500 from 1500 up to 4000 (bottom to top). In this, and in the following figures, times are always measured in units of Monte Carlo steps.

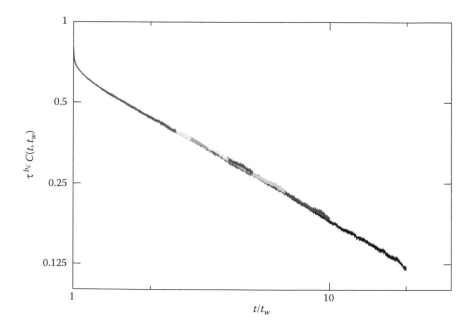

FIGURE 5.3 Plot of $\tau^{b_c} C(t, t_w)$ vs. $x = t/t_w$ for the $d = 2$ Glauber–Ising model quenched to T_c and for the same values of t_w as in Figure 5.2. The collapse of the different curves demonstrates the multiplicative structure of Equation 5.45.

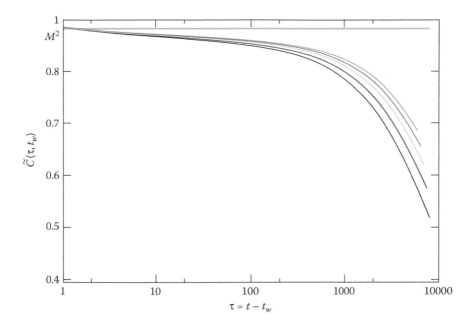

FIGURE 5.4 Plot of $\tilde{C}(\tau, t_w)$ vs. τ for the $d = 2$ Glauber–Ising model quenched to $T_F = 0.66T_c$. The waiting time t_w increases in steps of 500 from 2000 up to 4000 (bottom to top). The horizontal line indicates the Edwards–Anderson order parameter $M^2 = 0.97$.

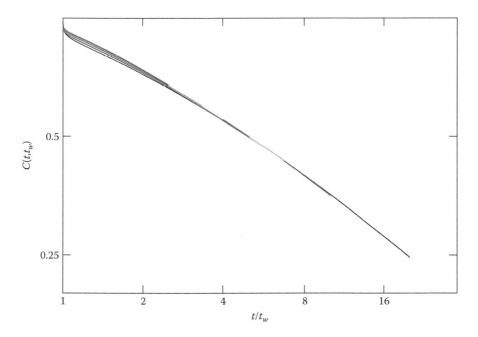

FIGURE 5.5 Plot of $C(t, t_w)$ vs. $x = t/t_w$ for the $d = 2$ Glauber–Ising model quenched to $T_F = 0.66T_c$. The values of t_w are the same as in Figure 5.4.

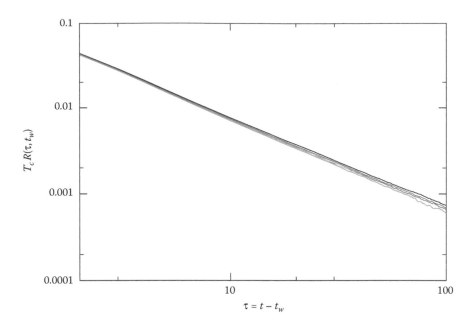

FIGURE 5.6 Plot of $T_c R(\tau, t_w)$ vs. τ for the $d = 2$ Glauber–Ising model quenched to T_c. The waiting time t_w increases in steps of 500 from 1500 up to 4000 (bottom to top).

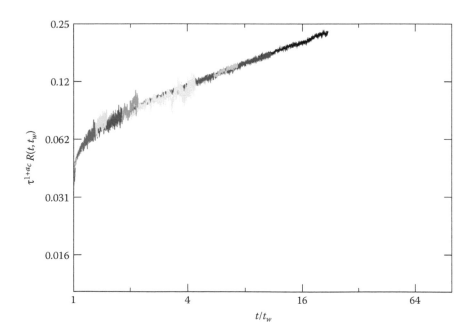

FIGURE 5.7 Plot of $\tau^{1+a_c} R(t, t_w)$ vs. $x = t/t_w$ for the $d = 2$ Glauber–Ising model quenched to T_c and for the same values of t_w as in Figure 5.6. The collapse of the different curves demonstrates the multiplicative structure of Equation 5.76.

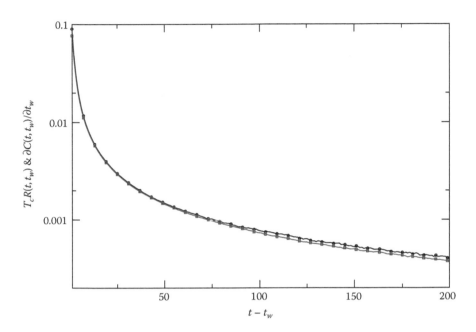

FIGURE 5.8 Plot of $T_c R(t, t_w)$ and $\partial C(t, t_w)/\partial t_w$ for the $d = 2$ Glauber–Ising model quenched to T_c, with $t_w = 1000$. The superposition of the curves for $\tau \ll t_w$ demonstrates the validity of the FDT in the short-time regime.

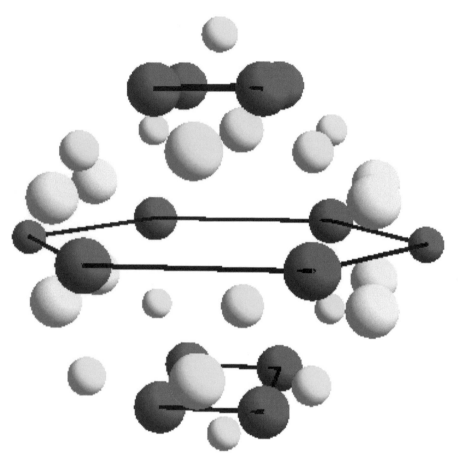

FIGURE 7.4 Bragg spots of the gyroid structure. The size of the Bragg spot is proportional to the intensity.

FIGURE 7.10 Three successive layers of the perforated lamellar structure. The location of holes is indicated by the dark areas.

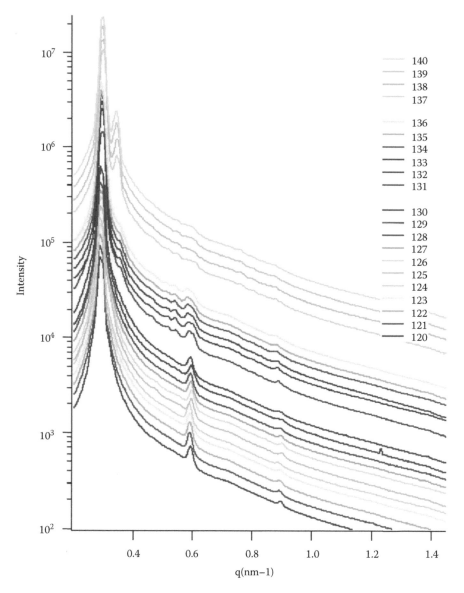

	140
	139
	138
	137
	136
	135
	134
	133
	132
	131
	130
	129
	128
	127
	126
	125
	124
	123
	122
	121
	120

FIGURE 7.14 Small-angle X-ray scattering intensities for different temperatures [91]. The material is poly(styrene-b-isoprene) with PS/PI = 36.2/63.8 vol/vol, $M_n = 2.64 \times 10^4$, and $M_w/M_n = 1.02$.

5.5 MODELS

The general concepts introduced in the previous sections will now be illustrated through analytical and numerical results for specific models, limiting the discussion to the case of nonconserved order parameter. The models considered are as follows:

- The Ising model with the Hamiltonian

$$H\{S_i\} = -J \sum_{\langle ij \rangle} S_i S_j, \quad S_i = \pm 1, \qquad (5.130)$$

where the sum runs over the pairs $\langle ij \rangle$ of the nearest-neighbor spins with the ferromagnetic coupling $J > 0$. The time-evolution with Glauber spin-flip dynamics [55] is governed by the master Equation 1.22. We reproduce the master equation here for ease of reference:

$$\frac{d}{dt} P(\{S_i\}, t) = - \sum_{j=1}^{N} W(S_1, \ldots S_j, \ldots S_N | S_1, \ldots - S_j, \ldots S_N) P(\{S_i\}, t)$$

$$+ \sum_{j=1}^{N} W(S_1, \ldots - S_j, \ldots S_N | S_1, \ldots S_j, \ldots S_N) P(\{S_i'\}, t),$$

(5.131)

where $P(\{S_i\}, t)$ is the probability of realization of the spin configuration $\{S_i\}$ at time t. The transition probability is of the form in Equation 1.30:

$$W(\{S_i\} | \{S_i'\}) = \frac{\lambda}{2} \left[1 - S_j \tanh \left(\beta J \sum_{L_j} S_{L_j} + \beta h \right) \right], \qquad (5.132)$$

where L_j denotes the neighbors of site j.
- The continuous Ginzburg–Landau–Wilson (GLW) model with the free energy

$$G[\vec{\phi}(\vec{x})] = \int d\vec{x} \left[\frac{1}{2} \left(\nabla \vec{\phi} \right)^2 + \frac{\mu}{2} \vec{\phi}^2 + \frac{u}{4n} \left(\vec{\phi}^2 \right)^2 \right], \qquad (5.133)$$

where $\vec{\phi}(\vec{x}) = (\phi_1, \ldots, \phi_n)$ is an n-component vector order parameter, as discussed in Section 1.4.2, and $\mu < 0$, $u > 0$ are constants. The purely relaxational dynamics (Model A in the classification of Hohenberg–Halperin [32]) is governed by the vector generalization of the Langevin Equation 1.59 [56,57]:

$$\frac{\partial}{\partial t} \vec{\phi}(\vec{x}, t) = -\frac{\delta G[\vec{\phi}]}{\delta \vec{\phi}} + \vec{\theta}(\vec{x}, t). \qquad (5.134)$$

Here, $\vec{\theta}(\vec{x}, t)$ is a Gaussian white noise that satisfies

$$\left\langle \vec{\theta}(\vec{x}, t) \right\rangle = 0, \tag{5.135}$$

and

$$\left\langle \theta_\alpha(\vec{x}, t)\theta_\beta(\vec{x}', t') \right\rangle = 2T\delta_{\alpha\beta}\delta(\vec{x} - \vec{x}')\delta(t - t'). \tag{5.136}$$

These models can be solved exactly only in a limited number of cases: in $d = 1$ for the Ising model, and in arbitrary d for the vector GLW model, after taking the $n \to \infty$ (large-n) limit [44,49,57–59]. Otherwise, one must resort either to numerical simulations or to approximation methods, as discussed in Chapter 1.

5.5.1 LARGE-n MODEL

When the number n of order-parameter components goes to infinity, the mean-field-like linearization of the GLW Hamiltonian, obtained by the replacement $(\vec{\phi}^2)^2 \to 2\langle\vec{\phi}^2\rangle\vec{\phi}^2$ in Equation 5.133, becomes exact. This holds for both statics and dynamics, with the proviso that the average $\langle\vec{\phi}^2\rangle$ must be computed self-consistently.

Therefore, in the large-n limit, the equation of motion for the Fourier transform of the order parameter,

$$\vec{\phi}(\vec{k}, t) = \int d\vec{x} \, e^{i\vec{k}\cdot\vec{x}}\vec{\phi}(\vec{x}, t), \tag{5.137}$$

takes the linear form

$$\frac{\partial}{\partial t}\vec{\phi}(\vec{k}, t) = -[k^2 + I(t)]\vec{\phi}(\vec{k}, t) + \vec{\theta}(\vec{k}, t). \tag{5.138}$$

Here,

$$I(t) = \mu + \frac{u}{n}\left\langle \vec{\phi}^2(\vec{x}, t) \right\rangle, \tag{5.139}$$

and the average is taken over the noise ensemble, which satisfies

$$\left\langle \vec{\theta}(\vec{k}, t) \right\rangle = 0,$$

$$\left\langle \theta_\alpha(\vec{k}, t)\theta_\beta(\vec{k}', t') \right\rangle = 2T_F\delta_{\alpha\beta}V\delta_{\vec{k}+\vec{k}',0}\delta(t - t'), \tag{5.140}$$

and the initial condition, which satisfies

$$\left\langle \vec{\phi}(\vec{k}) \right\rangle_I = 0,$$

$$\left\langle \phi_\alpha(\vec{k})\phi_\beta(\vec{k}') \right\rangle_I = \Delta\delta_{\alpha\beta}V\delta_{\vec{k}+\vec{k}',0}. \tag{5.141}$$

5.5.1.1 Statics

If the volume V is kept finite, the system equilibrates in a finite time t_{eq} and the order-parameter distribution reaches the Gibbs state:

$$P_G[\vec{\phi}(\vec{k})] = \frac{1}{Z}\exp\left[-\frac{1}{2T_F V}\sum_{\vec{k}}(k^2 + \xi_F^{-2})\vec{\phi}(\vec{k})\cdot\vec{\phi}(-\vec{k})\right], \qquad (5.142)$$

where ξ_F is the correlation length defined by the static self-consistency condition

$$\xi_F^{-2} = \mu + \frac{u}{n}\left\langle\vec{\phi}^2(\vec{x})\right\rangle_G. \qquad (5.143)$$

In order to analyze the properties of $P_G[\vec{\phi}(\vec{k})]$, it is necessary to extract the dependence of ξ_F on T_F and V. Evaluating the average, the above equation takes the form

$$\xi_F^{-2} = \mu + \frac{u}{V}\sum_{\vec{k}}\frac{T_F}{k^2 + \xi_F^{-2}}, \qquad (5.144)$$

whose solution is well known [44,61]. There exists a critical temperature

$$T_c = -\frac{\mu}{u}(4\pi)^{d/2}\Lambda^{2-d}\frac{(d-2)}{2}, \qquad (5.145)$$

where Λ is a high-momentum cutoff. For $T_F > T_c$, the solution of Equation 5.144 is independent of the volume. For $T_F \leq T_c$, it depends on the volume. The required solution is

$$\xi_F \begin{cases} \sim [(T_F - T_c)/T_c]^{-\nu} & \text{for } T_F > T_c, \\ \sim V^{1/d} & \text{for } T_F = T_c, \\ = \sqrt{M^2 V/T_F} & \text{for } T_F < T_c, \end{cases} \qquad (5.146)$$

where

$$M^2 = M_0^2\left(\frac{T_c - T_F}{T_c}\right) \qquad (5.147)$$

is the square of the spontaneous magnetization ($M_0^2 = -\mu/u$), and $\nu = 1/(d-2)$. Notice that, from Equation 5.145, it follows that the critical line $T_c(d)$ in Figure 5.1 is a straight line and that $d_L = 2$.

Let us now see what the implications are for the equilibrium state. As Equation 5.142 shows, the individual Fourier components are independent random variables, Gaussian-distributed with zero average for all temperatures. The variance of each mode is given by

$$\frac{1}{n}\left\langle\vec{\phi}(\vec{k})\cdot\vec{\phi}(-\vec{k})\right\rangle_G = VC_{eq}(\vec{k}), \qquad (5.148)$$

where

$$C_{eq}(\vec{k}) = \frac{T_F}{k^2 + \xi_F^{-2}} \tag{5.149}$$

is the equilibrium structure factor. For $T_F > T_c$, all $\vec{k}$-modes behave in the same way, with the variance growing linearly with the volume. For $T_F \leq T_c$, instead, ξ_F^{-2} is negligible with respect to k^2 except at $\vec{k} = 0$, yielding

$$C_{eq}(\vec{k}) = \begin{cases} \left(1 - \delta_{\vec{k},0}\right) T_c/k^2 + \kappa V^{2/d}\delta_{\vec{k},0} & \text{for } T_F = T_c, \\ \left(1 - \delta_{\vec{k},0}\right) T_F/k^2 + M^2 V\delta_{\vec{k},0} & \text{for } T_F < T_c, \end{cases} \tag{5.150}$$

where κ is a constant. Therefore, for $T_F \leq T_c$, the $\vec{k} = 0$ mode behaves differently from all the other modes, as the variance grows faster than linear with the volume. In particular, for $T_F < T_c$, the Gibbs state takes the form

$$P_G[\vec{\phi}(\vec{k})] = \frac{1}{Z} \exp\left[-\frac{\vec{\phi}^2(0)}{2M^2V^2}\right] \exp\left[-\frac{1}{2T_FV} \sum_{\vec{k}} k^2\vec{\phi}(\vec{k}) \cdot \vec{\phi}(-\vec{k})\right]. \tag{5.151}$$

Therefore, crossing T_c, there is a transition from the usual disordered high-temperature phase to a low-temperature phase. The latter phase, instead of being a mixture of broken-symmetry states, is characterized by a macroscopic variance in the Gaussian distribution of the $\vec{k} = 0$ mode. In place of the transition from disorder to order, it is more appropriate to speak of the condensation of fluctuations in the $\vec{k} = 0$ mode. The distinction between the condensed phase and the mixture of pure states has been discussed in detail in Ref. [62].

Despite the difference in the mechanism of the transition, from Equations 5.149 and 5.150 it is easy to see that the correlation function follows the same pattern outlined in general in Equations 5.12, 5.17, 5.18, with $\eta = 0$ and M^2 given by Equation 5.147. Furthermore, the splitting of the order parameter in Equation 5.20,

$$\vec{\phi}(\vec{x}) = \vec{\psi}(\vec{x}) + \vec{\sigma}, \tag{5.152}$$

can now be carried out explicitly, taking

$$\vec{\sigma} = \frac{1}{V} \vec{\phi}(\vec{k} = 0) \tag{5.153}$$

and

$$\vec{\psi}(\vec{x}) = \frac{1}{V} \sum_{\vec{k} \neq 0} \vec{\phi}(\vec{k}) e^{i\vec{k} \cdot \vec{x}}. \tag{5.154}$$

Then, rewriting the Gibbs state as

$$P_G[\vec{\phi}(\vec{x})] = P(\vec{\sigma})P[\vec{\psi}(\vec{x})], \tag{5.155}$$

with

$$P(\vec{\sigma}) = \frac{1}{(2\pi M^2)^{n/2}} \exp\left(-\frac{\vec{\sigma}^2}{2M^2}\right), \tag{5.156}$$

and

$$P[\vec{\psi}(\vec{x})] = \frac{1}{Z} \exp\left[-\frac{1}{2T_F} \int d\vec{x} \; (\nabla\vec{\psi})^2\right], \tag{5.157}$$

the two contributions in Equation 5.17 are given by

$$G_{eq}(\vec{x} - \vec{x}', T_F) = \frac{1}{n} \left\langle \vec{\psi}(\vec{x}) \cdot \vec{\psi}(\vec{x}') \right\rangle_G, \tag{5.158}$$

and

$$M^2 = \frac{1}{n} \langle \vec{\sigma} \cdot \vec{\sigma} \rangle_G. \tag{5.159}$$

5.5.1.2 Dynamics

Taking advantage of the rotational symmetry, and of the effective decoupling of the vector components, from now on we shall drop vectors and refer to the generic order-parameter component. The formal solution of the equation of motion (5.138) reads

$$\phi(\vec{k}, t) = R(\vec{k}, t, 0)\phi_0(\vec{k}) + \int_0^t dt' R(\vec{k}, t, t')\theta(\vec{k}, t'), \tag{5.160}$$

where

$$R(\vec{k}, t, t') = \frac{Y(t')}{Y(t)} e^{-k^2(t-t')} \tag{5.161}$$

is the response function. Furthermore, $\phi_0(\vec{k}) = \phi(\vec{k}, 0)$ is the initial value of the order parameter, and

$$Y(t) = \exp\left[\int_0^t ds \, I(s)\right] \tag{5.162}$$

is the key quantity in the exact solution of the model. In order to find $Y(t)$, notice that it follows from the definition of Y that

$$\frac{dY^2(t)}{dt} = 2\left[\mu + u\left\langle\phi^2(\vec{x}, t)\right\rangle\right] Y^2(t). \tag{5.163}$$

We can write $\left\langle\phi^2(\vec{x}, t)\right\rangle$ in terms of the structure factor,

$$\left\langle\phi^2(\vec{x}, t)\right\rangle = \int \frac{d\vec{k}}{(2\pi)^d} S(\vec{k}, t) \, e^{-k^2/\Lambda^2}, \tag{5.164}$$

and use Equation 5.160 to evaluate $S(\vec{k}, t)$:

$$S(\vec{k}, t) = R^2(k, t, 0)\Delta + 2T_F \int_0^t dt' R^2(\vec{k}, t, t'). \tag{5.165}$$

Then, from Equation 5.163, one obtains the integro-differential equation

$$\frac{dY^2(t)}{dt} = 2\mu Y^2(t) + 2u\Delta J \left(t + \frac{1}{2\Lambda^2}\right) + 4uT_F \int_0^t dt' J \left(t - t' + \frac{1}{2\Lambda^2}\right) Y^2(t'), \tag{5.166}$$

where

$$J(x) = \int \frac{d\vec{k}}{(2\pi)^d} e^{-2k^2 x} = (8\pi x)^{-d/2}. \tag{5.167}$$

Solving Equation 5.166 by Laplace transformation [49,63], the leading behavior of $Y(t)$ for large times is given by [44,49]

$$Y(t) = \begin{cases} A_a \, e^{t/\xi_F} & \text{for } T_F > T_c, \\ A_c \, t^{(d-4)/4} & \text{for } T_F = T_c, \\ A_b \, t^{-d/4} & \text{for } T_F < T_c, \end{cases} \tag{5.168}$$

where A_a, A_b, A_c are constants.

5.5.1.3   Splitting of the Order Parameter

The solution of the model will now be used to show how the splitting in Equation 5.60 of the order parameter into the sum of two independent contributions, with the properties (5.61) and (5.62), can be explicitly carried out. Furthermore, we will also give a derivation of the properties of $C(t, t_w)$ and $R(t, t_w)$, which have been stated in general in the previous sections. From the multiplicative property of the response function,

$$R(\vec{k}, t, t')R(\vec{k}, t', t^*) = R(\vec{k}, t, t^*), \tag{5.169}$$

for any ordered triplet of times $t^* < t' < t$, it is easy to show that the solution in Equation 5.160 can be rewritten as the sum of two statistically independent contributions $\phi(\vec{k}, t) = \psi(\vec{k}, t) + \sigma(\vec{k}, t)$. Here,

$$\sigma(\vec{k}, t) = R(\vec{k}, t, t^*)\phi(\vec{k}, t^*), \tag{5.170}$$

and

$$\psi(\vec{k}, t) = \int_{t^*}^t dt' R(\vec{k}, t, t')\theta(\vec{k}, t'), \tag{5.171}$$

because for $0 \le t^* < t$, $\phi(\vec{k}, t^*)$ and $\theta(\vec{k}, t)$ are independent by causality. In other words, the order parameter at the time t is split into the sum of a component $\sigma(\vec{k}, t)$, driven by the fluctuations of the order parameter at the earlier time t^*, and a component

$\psi(\vec{k},t)$, driven by the thermal history between t^* and t. Recall that t^* can be chosen arbitrarily between the initial time of the quench ($t = 0$) and the observation time t. With the particular choice $t^* = 0$, the component $\sigma(\vec{k},t)$ is driven by the fluctuations in the initial condition described by Equation 5.141. The ψ-component describes fluctuations of thermal origin, and the σ-component (as will be clear below) describes the local condensation of the order parameter if t^* is chosen sufficiently large.

From Equations 5.140 and 5.141, it follows that $\langle \sigma(\vec{k},t) \rangle = \langle \psi(\vec{k},t) \rangle = 0$, and the two-time structure factor separates into the sum

$$S(\vec{k},t,t_w) = S_\sigma(\vec{k},t,t_w,t^*) + S_\psi(\vec{k},t,t_w,t^*). \tag{5.172}$$

Here,

$$S_\sigma(\vec{k},t,t_w,t^*) = R(\vec{k},t,t^*)R(\vec{k},t_w,t^*)S(\vec{k},t^*), \tag{5.173}$$

and

$$S_\psi(\vec{k},t,t_w,t^*) = 2T_F \int_{t^*}^{t_w} dt'' R(\vec{k},t,t'') \, R(\vec{k},t_w,t''). \tag{5.174}$$

Of course, the t^*-dependence of the two contributions cancels out in the sum. Then, going to real space, the autocorrelation function can be rewritten as

$$C(t,t_w) = C_\sigma(t,t_w,t^*) + C_\psi(t,t_w,t^*), \tag{5.175}$$

with

$$C_\psi(t,t_w,t^*) = \frac{2T_F}{Y(t)Y(t_w)} \int_{t^*}^{t_w} dt'' J\left(\frac{t+t_w}{2} - t'' + \frac{1}{2\Lambda^2}\right) Y^2(t''), \tag{5.176}$$

and

$$C_\sigma(t,t_w,t^*) = \frac{Y^2(t^*)}{Y(t)Y(t_w)} \int \frac{d\vec{k}}{(2\pi)^d} e^{-k^2(t+t_w-2t^*+1/\Lambda^2)} S(\vec{k},t^*). \tag{5.177}$$

Assuming that t and t_w are sufficiently larger than t^*, and *a fortiori* of the microscopic time $t_0 = \Lambda^{-2}$, the above integral is dominated by the $\vec{k} = 0$ contribution, yielding

$$C_\sigma(t,t_w,t^*) = \frac{Y^2(t^*)}{Y(t)Y(t_w)} J\left(\frac{t+t_w}{2}\right) C^*, \tag{5.178}$$

where $C^* = S(\vec{k} = 0, t^*)$. This can be rewritten as

$$C_\sigma(t,t_w,t^*) = t_w^{-b_\sigma} f_\sigma(x, t^*), \tag{5.179}$$

where for $T_F \leq T_c$

$$f_\sigma(x, t^*) = \Upsilon(t^*) x^{-\omega/2}(x + 1)^{-d/2}, \tag{5.180}$$

$$\omega = \begin{cases} d/2 - 2 & \text{for } T_F = T_c, \\ -d/2 & \text{for } T_F < T_c, \end{cases} \tag{5.181}$$

$$b_\sigma = \begin{cases} d - 2 & \text{for } T_F = T_c, \\ 0 & \text{for } T_F < T_c, \end{cases} \tag{5.182}$$

and $\Upsilon(t^*)$ is a t^*-dependent constant to be determined.

With a simple change of the integration variable, the contribution of the thermal fluctuations in Equation 5.176 can be rewritten in the scaling form:

$$C_\psi(t, t_w, t^*) = t_w^{-b_\psi} f_\psi\left(x, y, \frac{t^*}{t_w}\right). \tag{5.183}$$

Here,

$$f_\psi\left(x, y, \frac{t^*}{t_w}\right) = \frac{2T_F}{(4\pi)^{d/2}} x^{-\omega/2} \int_{t^*/t_w}^1 dz \, z^\omega (x + 1 - 2z + y)^{-d/2}, \tag{5.184}$$

and

$$b_\psi = \frac{(d - 2)}{2}. \tag{5.185}$$

The autoresponse function is obtained by integrating Equation 5.161 over $\vec{k}$:

$$R(t, t_w) = t_w^{-(1+a)} f_R(x, y), \tag{5.186}$$

where

$$a = \frac{(d - 2)}{2}, \tag{5.187}$$

and the scaling function is given by

$$f_R(x, y) = (4\pi)^{-d/2}(x - 1 + y)^{-(1+a)} x^{-\omega/2}. \tag{5.188}$$

5.5.1.3.1   Quench to $T_F = T_c$

In this case, $b_\sigma = 2b_\psi$ and C_σ becomes negligible with respect to C_ψ, when t_w is sufficiently large, both for short-time and large-time separations. Hence, Equation 5.175 can be rewritten as

$$C(t, t_w) = C_\psi(t, t_w), \tag{5.189}$$

where $C_\psi(t, t_w)$ is obtained by letting $t^*/t_w \to 0$ in Equation 5.184, as the integral is well-behaved at the lower limit of integration. Setting $f_C(x, y) = f_\psi(x, y, 0)$, and rewriting the scaling function in the form

$$f_C(x, y) = (x - 1 + y)^{-b_\psi} g_C(x, y), \tag{5.190}$$

with

$$g_C(x, y) = \frac{T_c}{(4\pi)^{d/2}} x^{-\omega/2} \int_0^{2/(x-1+y)} dz (1+z)^{-d/2} [1 - (x - 1 + y)z/2]^{\omega},$$

(5.191)

the autocorrelation function displays the multiplicative form in Equation 5.45. Furthermore, as x becomes large, $f_C(x, y) \sim x^{-\lambda_c/2}$ with $\lambda_c = 3d/2 - 2$, and from Equations 5.185, 5.187 it follows that $a_c = b_c = b_\psi$. This is in agreement with Equations 5.46 and 5.78 because, in the large-n model, $\eta = 0$ and $z_c = 2$.

5.5.1.3.2 Quench to $T_F < T_c$

In this case, the roles of $C_\sigma(t, t_w)$ and $C_\psi(t, t_w)$ are reversed, as now $b_\psi > b_\sigma$. In the short-time regime, both contributions are stationary with

$$C_\sigma(\tau) = 2^{-d/2} \Upsilon(t^*),$$

(5.192)

and

$$C_\psi(\tau) = (M_0^2 - M^2) \left[\left(\frac{2t^*}{t_0} \right)^{1-d/2} + \left(\frac{\tau}{t_0} + 1 \right)^{1-d/2} \right].$$

(5.193)

This has been obtained by accounting for the fact that the integral in Equation 5.184 now develops a singularity at the lower limit of integration as $t^*/t_w \to 0$. In order to determine $\Upsilon(t^*)$, notice that Equation 5.143, together with $\xi_F^{-2} = 0$ for $T_F < T_c$, requires $C_{eq}(\vec{r} = 0, T_F) = M_0^2$. Therefore, requiring that the equilibrium sum rule,

$$C_\sigma(\tau = 0) + C_\psi(\tau = 0) = M_0^2,$$

(5.194)

be satisfied, one gets

$$\Upsilon(t^*) = 2^{d/2} \left[M^2 - (M_0^2 - M^2) \left(\frac{2t^*}{t_0} \right)^{1-d/2} \right].$$

(5.195)

With the above expression for $\Upsilon(t^*)$, in the large-time regime one has

$$C_\sigma(t, t_w) = \left[M^2 - (M_0^2 - M^2) \left(\frac{2t^*}{t_0} \right)^{1-d/2} \right] \left[\frac{4x}{(x+1)^2} \right]^{d/4},$$

(5.196)

and

$$C_\psi(t, t_w) = (M_0^2 - M^2) \left(\frac{2t^*}{t_0} \right)^{1-d/2} \left[\frac{4x}{(x+1)^2} \right]^{d/4}.$$

(5.197)

Taking the limit $t^*/t_0 \to \infty$, the dependence on t^* is eliminated, yielding for short times:

$$C_\sigma(t, t_w) = M^2,$$

$$C_\psi(t, t_w) = (M_0^2 - M^2) \left(\frac{\tau}{t_0} + 1 \right)^{1-d/2},$$

(5.198)

and for large times:

$$C_\sigma(t, t_w) = M^2 \left[\frac{4x}{(x+1)^2} \right]^{d/4},$$

$$C_\psi(t, t_w) = 0, \tag{5.199}$$

which implies $\lambda = d/2$. Finally, comparing with Equations 5.57 and 5.54, the identifications

$$C_\sigma(t, t_w) = C_{\text{ag}}(t, t_w) \tag{5.200}$$

and

$$C_\psi(t, t_w) = G_{\text{eq}}(\tau, T_F) \tag{5.201}$$

are obtained. The power-law decay of the stationary component C_ψ in Equation 5.198 is a peculiarity of the large-n limit, because to lowest order in $1/n$ only the Goldstone modes contribute to the thermal fluctuations [64,65], yielding critical behavior for $T_F \leq T_c$.

Following the prescription outlined in Section 5.4.2 for the construction of the corresponding components of the response function, from Equations 5.86 and 5.201 one finds

$$R_{\text{eq}}(\tau, T_F) = (4\pi)^{-d/2}(\tau + t_0)^{-(1+a)}. \tag{5.202}$$

Furthermore,

$$R_{\text{ag}}(t, t_w) = R(t, t_w) - R_{\text{eq}}(\tau, T_F)$$

$$= t_w^{-(1+a)} h_R(x, y), \tag{5.203}$$

with

$$h_R(x, y) = (4\pi)^{-d/2} \frac{x^{d/4} - 1}{(x - 1 + y)^{1+a}}. \tag{5.204}$$

This completes the check that the analytical solution of the model fits into the generic pattern presented in Sections 5.3.2 and 5.4.2.

Summarizing, the additive structures of $C(t, t_w)$ and $R(t, t_w)$ have been derived from the exact solution of the large-n model. Furthermore, the pair of fields $[\sigma(\vec{x}, t), \psi(\vec{x}, t)]$, taking $t^* \gg t_0$, provide an explicit realization of the decomposition in Equation 5.60, satisfying the requirements in Equations 5.61–5.62.

5.5.1.4 ZFC Susceptibility below T_c

It is quite instructive to examine the behavior of the ZFC susceptibility in the quench below T_c, as this gives an opportunity to see a concrete realization of the general considerations in Section 5.4.3. Recalling that $\chi(t, t_w)$ is also the sum of two contributions, it is straightforward to find the stationary component by integration of Equation 5.202:

$$\chi_{eq}(\tau, T_F) = (4\pi)^{-d/2} \frac{2}{(d-2)} \left[t_0^{1-d/2} - (\tau + t_0)^{1-d/2} \right]. \qquad (5.205)$$

The long-time limit of this quantity gives the static susceptibility, as required by Equation 5.75, because it follows from $t_0 = \Lambda^{-2}$ and Equation 5.145 that

$$(4\pi)^{-d/2} \frac{2}{(d-2)} t_0^{1-d/2} = \frac{M_0^2 - M^2}{T_F} = \frac{M_0^2}{T_c}, \qquad (5.206)$$

which coincides with the definition of $\chi_{st}(T_F)$ in Equation 5.33, for $\vec{r} = 0$.

Turning to the aging component and using Equation 5.204, the integral in Equation 5.93 entering the scaling function $h_\chi(x, y)$ is given by

$$\mathcal{I}(x, y) = \frac{1}{(4\pi)^{d/2}} \int_{1/x}^{1} dz \frac{z^{-d/4} - 1}{(1 - z + y/x)^{1+a}}. \qquad (5.207)$$

As y/x becomes small, $\mathcal{I}(x, y)$ remains finite for $d < 4$, and for $d \geq 4$ there is a divergence with the leading behaviors

$$\mathcal{I}(x, y) \sim \begin{cases} \ln(x/y) & \text{for } d = 4, \\ (y/x)^{-(d-4)/2} & \text{for } d > 4. \end{cases} \qquad (5.208)$$

Hence, from the comparison with Equations 5.94–5.96, it follows that $d^* = 4$, $c = (d - 4)/2$, and

$$a_\chi = \begin{cases} a = (d-2)/2 & \text{for } d < d^*, \\ 1 & \text{with log-corrections for } d = d^*, \\ 1 & \text{for } d > d^*, \end{cases} \qquad (5.209)$$

showing that the scenario presented in Section 5.4.3 is verified. Thus, y does act as a dangerous irrelevant variable for $d \geq d^*$, producing a difference between the exponents a and a_χ, and a_χ becomes independent of the dimensionality for $d \geq d^*$. As explained in Section 5.4.3, d^* plays the role of an upper dimensionality, although the current context bears no relationship with critical phenomena. Therefore, the coincidence of the values $d^* = d_U$ must be regarded as fortuitous.

From the above result for a_χ and from Equation 5.100, it follows that the ZFC susceptibility reaches the equilibrium value for $d > 2$, and a non-vanishing value of χ^* is expected in the quench to $(d_L = 2, T_F = 0)$, where $a_\chi = 0$. As $d \rightarrow 2$, from Equation 5.205 one has

$$\chi_{eq}(\tau, T_F) = \frac{1}{4\pi} \ln \left(1 + \frac{\tau}{t_0} \right). \qquad (5.210)$$

This quantity diverges as $\tau \to \infty$, as it should because the static susceptibility in Equation 5.206 diverges at d_L, where $T_c = 0$.

Switching to the aging component, the evaluation of the integral in Equation 5.207 for $d = 2$ gives

$$
\mathcal{I}(x, y) = \frac{1}{4\pi} \left[\left(\frac{\sqrt{\kappa} + 1}{\sqrt{\kappa}} \right) \ln(\sqrt{\kappa} + 1) + \left(\frac{\sqrt{\kappa} - 1}{\sqrt{\kappa}} \right) \ln(\sqrt{\kappa} - 1) \right.
$$
$$
\left. + \frac{1}{\sqrt{\kappa}} \ln \left(\frac{\sqrt{\kappa} - \sqrt{1/x}}{\sqrt{\kappa} + \sqrt{1/x}} \right) - \ln(\kappa - 1/x) \right], \tag{5.211}
$$

where $\kappa = 1 + y/x$. Hence, taking the limit $y \to 0$, one finds

$$
\chi_{ag}(x) = \hat{h}_\chi(x) = \frac{1}{2\pi} \ln \left(\frac{2}{1 + x^{-1/2}} \right). \tag{5.212}
$$

Letting $x \to \infty$,

$$
\chi^* = \frac{1}{2\pi} \ln 2, \tag{5.213}
$$

which shows that the ZFC susceptibility does not equilibrate, although the effect is not observable because $\chi_{st}(T_F)$ diverges. As we shall see, in the $d = 1$ Ising model it is not so, as the effect is observable. The best way to visualize the formation of χ^* is through a parametric plot. Because for $T_F = 0$ the autocorrelation function is entirely given by $C_{ag}(t, t_w)$, eliminating x between Equations 5.199 and 5.212 yields

$$
\chi_{ag}(C) = \frac{1}{2\pi} \log \left[\frac{2}{1 + \frac{M_0^2}{C} \left(1 - \sqrt{1 - \frac{C^2}{M_0^4}} \right)} \right]. \tag{5.214}
$$

The plot of χ_{ag} vs. C is displayed in Figure 5.11 and is qualitatively similar to the one in Figure 5.12 for the $d = 1$ Glauber-Ising model. The value $\chi^* = 0.1103$ is given by the intercept with the vertical axis at $C = 0$.

5.5.2   Kinetic Ising Model in $d = 1$

In Section 1.3.2, we introduced the kinetic Ising model with Glauber spin-flip kinetics. The Glauber–Ising model in $d = 1$ is the other exactly solvable model [55] we consider here. As in the large-n model, it is possible to analytically derive [50,66,67] the relaxation properties in the quench to $(d_L, T_F = 0)$, because for discrete symmetry $d_L = 1$. As a matter of fact, one cannot straightforwardly set $T_F = 0$, because the linear response function $R(t, t_w)$ is not well-defined for hard spins when $T_F = 0$. This is due to the dependence on the temperature (and the external field) entering the dynamics through the transition rates in Equation 5.132 in the combination h/T_F, which does not allow for a linear response regime when $T_F \to 0$. There is no such problem with the Langevin Equation 5.134, where the temperature enters only through the

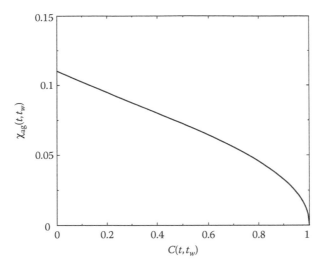

FIGURE 5.11 Parametric plot of $\chi_{ag}(t, t_w)$ in the quench of the large-n model to ($d = 2$, $T_F = 0$). (From Corberi, F., Lippiello, E., and Zannetti, M., *Phys. Rev. E*, 65, 046136, 2002. With permission.)

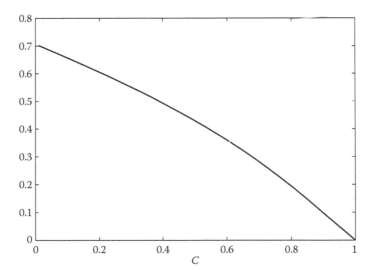

FIGURE 5.12 Plot of $T_F \chi$ against C for the $d = 1$ Glauber–Ising model, quenched to an arbitrary value of T_F with $J = \infty$.

noise, allowing one to deal with a small external field in the zero-temperature limit. The problem can be bypassed by making the quench to a non-zero temperature T_F, where there is a linear regime and aging for $t_w \ll t_{eq} = \xi_F^2$, with

$$\xi_F = -\left\{ \ln\left[\tanh\left(\frac{J}{T_F} \right) \right] \right\}^{-1}. \tag{5.215}$$

Hence, aging lasts longer as the temperature is lowered and t_{eq} increases [50]. Then, the behavior in the quench to $T_F = 0$ must be understood as the limit for $T_F \to 0$ of the off-equilibrium behavior observed in the quenches to non-zero T_F. Alternatively, as can be seen from Equation 5.215, t_{eq} diverges by letting the coupling constant $J \to \infty$ while keeping T_F finite and, thus, preserving the linear regime.

With this proviso, let us derive $R(t, t_w)$. From the exact solution of the model [50], the autocorrelation and autoresponse functions turn out to be related as

$$R(t, t_w) = \frac{1}{2T_F} \left[\frac{\partial}{\partial t_w} C(t, t_w) - \frac{\partial}{\partial t} C(t, t_w) \right]. \tag{5.216}$$

Because for $T_F = 0$, or equivalently for $J = \infty$, there are no thermal fluctuations and $G_{eq}(\tau, T_F)$ vanishes, the autocorrelation function is entirely given by the aging component [68,69]:

$$C(t, t_w) = \frac{2}{\pi} \sin^{-1} \sqrt{\frac{2}{1+x}}. \tag{5.217}$$

The same is also true for $R(t, t_w)$, because the stationary response $R_{eq}(\tau, T_F)$ vanishes at $T_F = 0$ and only the aging component gives a contribution. Inserting the above expression for $C(t, t_w)$ into Equation 5.216, $R(t, t_w)$ is found to obey the scaling form in Equation 5.3:

$$R(t, t_w) = \frac{1}{\sqrt{2\pi} T_F} t_w^{-1} (x-1)^{-1/2}, \tag{5.218}$$

which implies $a = 0$. Hence, as in the large-n model, it is an exact result that the exponent a vanishes at d_L. The consequence, according to the general analysis of Section 5.4.3, is that the ZFC susceptibility does not equilibrate because the aging contribution does not disappear in the $t_w \to \infty$ limit. This feature is much more conspicuous here than in the large-n model, because now $\chi_{eq}(\tau, T_F)$ vanishes identically. Thus, $\chi(t, t_w)$ has only the contribution of the aging component, obtained by integration of Equation 5.218:

$$T_F \chi(t, t_w) = \frac{1}{\sqrt{2\pi}} \left[\frac{\pi}{2} + \sin^{-1} \left(1 - \frac{2}{x} \right) \right], \tag{5.219}$$

whose $x \to \infty$ limit gives

$$T_F \chi^* = \frac{1}{\sqrt{2}}. \tag{5.220}$$

The parametric representation, obtained by eliminating x with $C(t, t_w)$ in Equation 5.217, is

$$T_F \widehat{\chi}(C) = \frac{\sqrt{2}}{\pi} \tan^{-1} \left[\sqrt{2} \cot \left(\frac{\pi}{2} C \right) \right]. \tag{5.221}$$

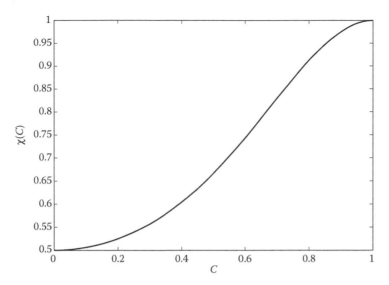

FIGURE 5.13 Parametric plot of the FDR for the $d = 1$ Glauber–Ising model, quenched to an arbitrary value of T_F with $J = \infty$.

In Figure 5.12, we plot $T_F \chi$ vs. C, which shows a qualitative similarity with the behavior of $\chi_{ag}(C)$ vs. C in Figure 5.11. Finally, the FDR is obtained by differentiating $T_F \widehat{\chi}(C)$ with respect to C:

$$\mathcal{X}(C) = \left[2 - \sin^2 \left(\frac{\pi}{2}C\right)\right]^{-1}, \tag{5.222}$$

providing see Figure 5.13 an example of the smooth and non-universal behavior mentioned in Section 5.4.6.

5.6 THE EXPONENT a

As we have seen above, from the exact results of the large-n model and the $d = 1$ Ising model, the exponent a vanishes in the quenches to $(d_L, T_F = 0)$. Although a_c, defined in Equation 5.46, vanishes as $d \to d_L$ along the critical line, this is not an adequate explanation because (as pointed out in Section 5.4.6) the quench to $(d_L, T_F = 0)$ is not a critical quench. Hence, the vanishing of a must be accounted for within the framework of the response function in quenches below the critical line. However, there is an additional complication due to a popular argument [70] identifying a, for $T_F < T_c$, with the exponent m/z in the time-dependence of the density of defects $\rho(t) \sim t^{-m/z}$, where $m = 1$ or $m = 2$ for scalar or vector order parameter [4]. In order to reproduce the argument, let us rewrite Equation 5.66 in the context of the ZFC susceptibility:

$$\langle \varphi(\vec{x}, t)\rangle_h = \langle \varphi(\vec{x}, t)\rangle + \int d\vec{y}\, \chi(\vec{x} - \vec{y}, t, t_w)h(\vec{y}) + \mathcal{O}(h^2). \tag{5.223}$$

Further, let us assume that $h(\vec{x})$ is an uncorrelated random field with expectations

$$\overline{h(\vec{x})} = 0, \tag{5.224}$$

and

$$\overline{h(\vec{x})h(\vec{y})} = h_0^2 \delta(\vec{x} - \vec{y}), \tag{5.225}$$

where the overbar denotes the average. Then, multiplying Equation 5.223 by $h(\vec{x})$ and taking the average over the field, one finds

$$\chi(t, t_w) = \frac{1}{h_0^2} \overline{\langle \varphi(\vec{x}, t)\rangle_h \, h(\vec{x})}, \tag{5.226}$$

which shows that the ZFC susceptibility is proportional to the correlation of the local magnetization with the external random field on the same site.

We can write the magnetization as the sum

$$\langle \varphi(\vec{x}, t)\rangle_h = \langle \varphi(\vec{x}, t)\rangle_{h,B} + \langle \varphi(\vec{x}, t)\rangle_{h,D}, \tag{5.227}$$

where the first contribution comes from the bulk of domains, where equilibrium has been established, and the second from the defects. Then, Equation 5.226 takes the form

$$\chi(t, t_w) = \frac{1}{h_0^2} \left[\overline{\langle \varphi(\vec{x}, t)\rangle_{h,B} \, h(\vec{x})} + \overline{\langle \varphi(\vec{x}, t)\rangle_{h,D} \, h(\vec{x})} \right]. \tag{5.228}$$

Associating the two terms on the right-hand side with χ_{eq} and χ_{ag}, respectively, and *assuming* that the defect contribution to the magnetization is proportional to the density of defects, $\langle \varphi(\vec{x}, t)\rangle_{h,D} \sim \rho(t)h(\vec{x})$, one finds

$$\chi_{ag}(t, t_w) \sim \rho(t). \tag{5.229}$$

Recalling Equation 5.98, this leads to the identification

$$a_\chi = \frac{m}{z}. \tag{5.230}$$

Because for $T_F < T_c$, the dynamical exponent z is independent of dimensionality, according to this argument a_χ is also independent of dimensionality, implying that there is no distinction between a_χ and a. Then, in the scalar case one should have $a = 1/2$, and in the vector case $a = 1$, independently from d and, therefore, also at d_L.

This simple and intuitive picture is contradicted by the exact result for the large-n model, which gives a_χ dependent on d for $d < d^*$, and by the vanishing of a in the $d = 1$ Glauber–Ising model. A point of contact with the prediction in Equation 5.230 is found only if one looks at a_χ in Equation 5.209 for $d > d^*$. Then, the question is whether the large-n model and the $d = 1$ Ising model are peculiar cases producing exceptions to Equation 5.230, or whether the lack of dimensionality-dependence in

Equation 5.230 is indicative that some important element of the response mechanism is missed in the intuitive argument when $d < d^*$.

In order to attempt an answer, one must enrich the phenomenology, necessarily resorting to approximate methods and to numerical simulations. Calculations of a in the scalar GLW model, with the GAF approximation [8,43] and an improved version of it [71], give

$$a = \frac{d-1}{2}.$$ (5.231)

This result shows a linear dependence on d as in the large-n model and reproduces the vanishing of a at $d_L = 1$ as in the Ising model. Furthermore, from the computation of the ZFC susceptibility [8,43], one finds for a_χ the same pattern as in Equation 5.209 for the large-n model:

$$a_\chi = \begin{cases} a = (d-1)/2 & \text{for } d < d^*, \\ 1/2 & \text{with log-corrections for } d = d^*, \\ 1/2 & \text{for } d > d^*, \end{cases}$$ (5.232)

except that now $d^* = 2$, instead of $d^* = 4$ in the large-n model. The details of the calculation show that the existence of the upper dimensionality d^* occurs through the same mechanism as in the large-n model, namely with the microscopic time t_0 acting as a dangerous irrelevant variable at and above d^*.

Therefore, we may conclude that all the analytical results presented so far, exact and approximate, follow the same pattern. This is summarized as

$$a = \frac{m}{z} \left(\frac{d - d_L}{d^* - d_L} \right),$$ (5.233)

and

$$a_\chi = \begin{cases} a & \text{for } d < d^*, \\ m/z & \text{with log-corrections for } d = d^*, \\ m/z & \text{for } d > d^*, \end{cases}$$ (5.234)

with $d^* > d_L$ and dependent on the model.

The next step is to see whether this pattern also holds for models accessible only through numerical simulations. The problem with simulations is that the measurement of the instantaneous response function $R(t, t_w)$ is numerically demanding. Although significant progress has been made recently [20–23], yet a large-scale survey of the d-dependence of a in different models is unrealizable, as of now. One way forward is to turn to the measurement of integrated response functions, such as the ZFC susceptibility, which are less noisy than $R(t, t_w)$ [41]. This program has been carried out through the measurement of a_χ in systems in different universality classes [53], that is, with scalar or vector order parameter, with and without conservation of the order parameter, and at different dimensionalities. The outcome fits into the pattern of Equation 5.234, provided one takes $d^* = 3$ and $d^* = 4$ for scalar and vector order parameter, respectively; and $z = 3$ or $z = 4$ for scalar conserved or vector conserved order parameter,

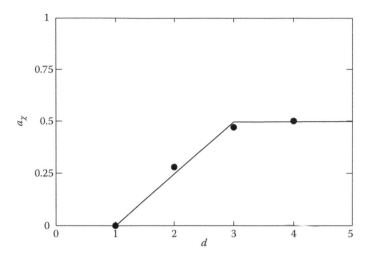

FIGURE 5.14 Plot of a_χ vs. d for the Glauber–Ising model. The dot for $d = 1$ denotes the exact value. The dots for $d = 2, 3, 4$ are the results of numerical simulations. The continuous line is the plot of Equation 5.234 with $m = 1, z = 2, d_L = 1, d^* = 3$. (From Corberi, F., Lippiello, E., and Zannetti, M., *Phys. Rev. E*, 68, 046131, 2003. With permission.)

respectively. As an example, the values of a_χ for the nonconserved Ising model with dimensionality varying from $d = 1$ to $d = 4$ are plotted in Figure 5.14. Similar plots for other models can be found in Ref. [53]. For convenience, the values of d_L, d^*, and a_χ in the two analytically solvable cases and in the Ising model are summarized in the following table:

Model	d_L	d^*	$a_\chi = a, d \le d^*$	$a_\chi, d > d^*$
Large-n	2	4	$(d - 2)/2$	1
GAF scalar	1	2	$(d - 1)/2$	1/2
Ising numerical	1	3	$(d - 1)/4$	1/2

Clearly, the expression for a_χ in the last row for the Ising case must be understood as a phenomenological formula.

Despite the evidence supporting Equations 5.233 and 5.234, the issue of the exponent a cannot be regarded as settled. The main reason is that a first-principles derivation is lacking. Some particularly challenging problems are as follows:

1. What produces the deviation of a_χ from m/z below d^*? Preliminary answers have been put forward [53,72] on the basis of the roughening of interfaces for scalar models, but much work remains to be done before a full understanding is reached.
2. In the scalar model treated in the GAF approximation, or its improved version [71], one finds $d^* = 2$ while the fit of the numerical data requires $d^* = 3$. This discrepancy reveals the strong non-perturbative nature of the exponent

a, for which even the best analytical tools currently available seem to be inadequate.

In conclusion, aging in domain growth is highly non-trivial and poses some hard problems to our understanding of phase-ordering kinetics.

REFERENCES

1. Bouchaud, J.-P., Cugliandolo, L.F., Kurchan, J., and Mezard, M., Out of equilibrium dynamics in spin-glasses and other glassy systems in *Spin Glasses and Random Fields*, Young, A.P., Ed., World Scientific, Singapore, 1997, chap. 6.
2. Crisanti, A. and Ritort, F., Violations of the fluctuation-dissipation theorem in glassy systems: Basic notions and the numerical evidence, *J. Phys. A Math. Gen.*, 36, R181, 2003.
3. Cugliandolo, L.F., Dynamics of glassy systems, in *Slow Relaxation and Non-Equilibrium Dynamics in Condensed Matter*, Barrat, J.-L., Dalibard, J., Kurchan, J., and Feigel'man, M.V., Eds., Springer-Verlag, Heidelberg, 2002.
4. Bray, A.J., Theory of phase ordering kinetics, *Adv. Phys.*, 43, 357, 1994.
5. Struick, L.C.E., *Physical Aging in Amorphous Polymers and Other Materials*, Elsevier, Houston, 1976.
6. Mazenko, G.F., Valls, O.T., and Zannetti, M., Field theory for growth kinetics, *Phys. Rev. B*, 38, 520, 1988.
7. Franz, S. and Virasoro, M.A., Quasi-equilibrium interpretation of aging dynamics, *J. Phys. A Math. Gen.*, 33, 891, 2000.
8. Corberi, F., Lippiello, E., and Zannetti, M., On the connection between off-equilibrium response and statics in non-disordered coarsening systems, *Eur. Phys. J. B*, 24, 359, 2001.
9. Janssen, H.K., Schaub, B., and Schmittmann, B., New universal short-time scaling behaviour of critical relaxation processes, *Z. Phys. B Cond. Matter*, 73, 539, 1989.
10. Janssen, H.K., On the renormalized field theory of nonlinear critical relaxation, in *From Phase Transitions to Chaos: Topics in Modern Statistical Physics*, Györgyi, G., Kondor, I., Sasvári, L., and Tel, T., Eds., World Scientific, Singapore, 1992, 68.
11. Calabrese, P. and Gambassi, A., Aging properties of critical systems, *J. Phys. A Math. Gen.*, 38, R133, 2005.
12. Ohta, T., Jasnow, D., and Kawasaki, K., Universal scaling in the motion of random interfaces, *Phys. Rev. Lett.*, 49, 1223, 1982.
13. Oono, Y. and Puri, S., Large wave number features of form factors for phase transition kinetics, *Mod. Phys. Lett. B*, 2, 861, 1988.
14. Mazenko, G.F., Theory of unstable thermodynamic systems, *Phys. Rev. Lett.*, 63, 1605, 1989.
15. Mazenko, G.F., Theory of unstable growth, *Phys. Rev. B*, 42, 4487, 1990.
16. Mazenko, G.F., Theory of unstable growth. II. Conserved order parameter, *Phys. Rev. B*, 43, 5747, 1991.
17. Bray, A.J. and Humayun, K., Towards a systematic calculation of the scaling functions for the ordering kinetics of nonconserved fields, *Phys. Rev. E*, 48, R1609, 1993.
18. De Siena, S. and Zannetti, M., Nature of the Gaussian approximations in phase-ordering kinetics, *Phys. Rev. E*, 50, 2621, 1994.
19. Liu, F. and Mazenko, G.F., Nonequilibrium autocorrelations in phase-ordering dynamics, *Phys. Rev. B*, 44, 9185, 1991.

20. Chatelain, C., A far-from-equilibrium fluctuation-dissipation relation for an Ising–Glauber-like model, *J. Phys. A Math. Gen.*, 36, 10739, 2003.

21. Ricci-Tersenghi, F., Measuring the fluctuation-dissipation ratio in glassy systems with no perturbing field, *Phys. Rev. E*, 68, 065104, 2003.

22. Lippiello, E., Corberi, F., and Zannetti, M., Off-equilibrium generalization of the fluctuation dissipation theorem for Ising spins and measurement of the linear response function, *Phys. Rev. E*, 71, 036104, 2005.

23. Corberi, F., Lippiello, E., and Zannetti, M., Correction to scaling in the response function of the two-dimensional kinetic Ising model, *Phys. Rev. E*, 72, 056103, 2005.

24. Henkel, M., Pleimling, M., Godrèche, C., and Luck, J.M., Aging, phase ordering, and conformal invariance, *Phys. Rev. Lett.*, 87, 265701, 2001.

25. Henkel, M., Phenomenology of local scale invariance: from conformal invariance to dynamical scaling, *Nucl. Phys. B*, 641, 405, 2002.

26. Pleimling, M. and Gambassi, A., Corrections to local scale invariance in the nonequilibrium dynamics of critical systems: Numerical evidences, *Phys. Rev. B*, 71, 180401, 2005.

27. Lippiello, E., Corberi, F., and Zannetti, M., Test of local scale invariance from the direct measurement of the response function in the Ising model quenched to and below T_c, *Phys. Rev. E*, 74, 041113, 2006.

28. Henkel, M. and Pleimling, M., On the scaling and aging behaviour of the alternating susceptibility in spin glasses and local scale invariance, *J. Phys. Condens. Matter*, 17, S1899, 2005.

29. Mezard, M., Parisi, G., and Virasoro, M., *Spin Glass Theory and Beyond*, World Scientific, Singapore, 1987.

30. Palmer, R.G., Broken ergodicity, *Adv. Phys.*, 31, 669, 1982.

31. Godrèche, C. and Luck, J.M., Nonequilibrium critical dynamics of ferromagnetic spin systems, *J. Phys. Condens. Matter*, 14, 1589, 2002.

32. Hohenberg, P.C. and Halperin, B.I., Theory of dynamic critical phenomena, *Rev. Mod. Phys.*, 49, 435, 1977.

33. Huse, D.A., Remanent magnetization decay at the spin-glass critical point: A new dynamic critical exponent for nonequilibrium autocorrelations, *Phys. Rev. B*, 40, 304, 1989.

34. Landau, D.P. and Binder, K., *A Guide to Monte Carlo Simulations in Statistical Physics*, Cambridge University Press, Cambridge, 2000.

35. Wang, F.C. and Hu, C.K., Universality in dynamic critical phenomena, *Phys. Rev. E*, 56, 2310, 1997.

36. Fisher, D.S. and Huse, D.A., Nonequilibrium dynamics of spin glasses, *Phys. Rev. B*, 38, 373, 1989.

37. Bouchaud, J.-P., Weak ergodicity breaking and aging in disordered systems, *J. Phys. I (France)*, 2, 1705, 1992.

38. Furukawa, H., Numerical study of multitime scaling in a solid system undergoing phase separation, *Phys. Rev. B*, 40, 2341, 1989.

39. Cugliandolo, L.F. and Kurchan, J., Analytical solution of the off-equilibrium dynamics of a long-range spin-glass model, *Phys. Rev. Lett.*, 71, 173, 1993.

40. Cugliandolo, L.F. and Kurchan, J., On the out-of-equilibrium relaxation of the Sherrington–Kirkpatrick model, *J. Phys. A Math. Gen.*, 27, 5749, 1994.

41. Corberi, F., Lippiello, E., and Zannetti, M., Scaling of the linear response function from zero-field-cooled and thermoremanent magnetization in phase-ordering kinetics, *Phys. Rev. E*, 68, 046131, 2003.

42. Fisher, M.E., *Scaling, Universality and Renormalization Group Theory*, Lecture Notes in Physics 186, Springer-Verlag, Berlin, 1986.

43. Berthier, L., Barrat, J.L., and Kurchan, J., Response function of coarsening systems, *Eur. Phys. J. B*, 11, 635, 1999.

44. Corberi, F., Lippiello, E., and Zannetti, M., Slow relaxation in the large-N model for phase ordering, *Phys. Rev. E*, 65, 046136, 2002.

45. Fischer, K.H. and Hertz, J.A., *Spin Glasses*, Cambridge University Press, Cambridge, 1991.

46. Yoshino, H., Hukushima, K., and Takayama, H., Extended droplet theory for aging in short-range spin glasses and a numerical examination, *Phys. Rev. B*, 66, 064431, 2002.

47. Parisi, G., Ricci-Tersenghi, F., and Ruiz-Lorenzo, J.J., Generalized off-equilibrium fluctuation-dissipation relations in random Ising systems, *Eur. Phys. J. B*, 11, 317, 1999.

48. Cugliandolo, L.F., Kurchan, J., and Peliti, L., Energy flow, partial equilibration, and effective temperatures in systems with slow dynamics, *Phys. Rev. E*, 55, 3898, 1997.

49. Godrèche, C. and Luck, J.M., Response of non-equilibrium systems at criticality: ferromagnetic models in dimension two and above, *J. Phys. A Math. Gen.*, 33, 9141, 2000.

50. Lippiello, E. and Zannetti, M., Fluctuation dissipation ratio in the one-dimensional kinetic Ising model, *Phys. Rev. E*, 61, 3369, 2000.

51. Godrèche, C. and Luck, J.M., Response of non-equilibrium systems at criticality: exact results for the Glauber–Ising chain, *J. Phys. A Math. Gen.*, 33, 1151, 2000.

52. Corberi, F., Castellano, C., Lippiello, E., and Zannetti, M., Universality of the off-equilibrium response function in the kinetic Ising chain, *Phys. Rev. E*, 65, 066114, 2002.

53. Corberi, F., Castellano, C., Lippiello, E., and Zannetti, M., Generic features of the fluctuation dissipation relation in coarsening systems, *Phys. Rev. E*, 70, 017103, 2004.

54. Glauber, R.J., Time-dependent statistics of the Ising model, *J. Math. Phys.*, 4, 294, 1963.

55. Goldenfeld, N., *Lectures on Phase Transitions and the Renormalization Group*, Addison-Wesley, Reading, MA, 1992.

56. Chaikin, P.M. and Lubensky, T.C., *Principles of Condensed Matter Physics*, Cambridge University Press, Cambridge, 1995.

57. Ma, S.K., *Modern Theory of Critical Phenomena*, Addison-Wesley, Reading, MA, 1976.

58. Cugliandolo, L.F. and Dean, D.S., Full dynamical solution for a spherical spin-glass model, *J. Phys. A Math. Gen.*, 28, 4213, 1995.

59. Chamon, C., Cugliandolo, L.F., and Yoshino, H., Fluctuations in the coarsening dynamics of the $O(N)$ model with $N \to \infty$: are they similar to those in glassy systems?, *JSTAT: Theory and Experiment*, stacks.iop.org/JSTAT/2006/P01006, 2006.

60. Baxter, R.J., *Exactly Solved Models in Statistical Mechanics*, Academic Press, New York, 1982.

61. Castellano, C., Corberi, F., and Zannetti, M., Condensation vs. phase ordering in the dynamics of first-order transitions, *Phys. Rev. E*, 56, 4973, 1997.

62. Newman, T.J. and Bray, A.J., Dynamic correlations in domain growth: a $1/n$ expansion, *J. Phys. A Math. Gen.*, 23, 4491, 1990.

63. Mazenko, G.F. and Zannetti, M., Growth of order in a system with continuous symmetry, *Phys. Rev. Lett.*, 53, 2106, 1984.

64. Mazenko, G.F. and Zannetti, M., Instability, spinodal decomposition, and nucleation in a system with continuous symmetry, *Phys. Rev. B*, 32, 4565, 1985.

65. Godrèche, C. and Luck, J.M., Response of non-equilibrium systems at criticality: exact results for the Glauber–Ising chain, *J. Phys. A Math. Gen.*, 33, 1151, 2000.

66. Mayer, P. and Sollich, P., General solutions for multispin two-time correlation and response functions in the Glauber–Ising chain, *J. Phys. A Math. Gen.*, 37, 9, 2004.

67. Bray, A.J., Universal scaling function for domain growth n the Glauber–Ising chain, *J. Phys. A Math. Gen.*, 22, L67, 1989.

68. Prados, A., Brey, J.J., and Sánchez-Rey, B., Aging in the one-dimensional Ising model with Glauber dynamics, *Europhys. Lett.*, 40, 13, 1997.

69. Barrat, A., Monte Carlo simulations of the violation of the fluctuation-dissipation theorem in domain growth processes, *Phys. Rev. E*, 57, 3629, 1998.

70. Mazenko, G.F., Response functions in phase-ordering kinetics, *Phys. Rev. E*, 69, 016114, 2004.

71. Henkel, M., Paessens, M., and Pleimling, M., Scaling of the linear response in simple aging systems without disorder, *Phys. Rev. E*, 69, 056109, 2004.

6 Kinetics of Dewetting

*Rajesh Khanna, Narendra Kumar Agnihotri,
and Ashutosh Sharma*

CONTENTS

6.1 INTRODUCTION

In this chapter, we will study the *kinetics of dewetting*, that is, the appearance and growth of dry (uncovered) portions on surfaces covered with liquids. This process is analogous to the process of phase separation, which has been discussed in previous chapters. The liquid cover breaks as a result of sustained local thinning due to the flow of liquid from thinner to thicker parts. This flow, which is the main operative mechanism for dewetting, can be achieved by a variety of mechanical means, for example, impact by a gas jet or dust particles. In some cases, this uphill liquid flow can also be spontaneous under the influence of chemical-potential gradients. In this case, dewetting occurs even if the system is free of any destabilizing mechanical input. This spontaneous dewetting is dependent on the thickness of the liquid cover. The thickness of the intervening film (see Figure 6.1) has to be small enough for the film surface to feel the attraction of molecules in the solid surface and get pulled (thin) toward the solid surface. Naturally, this thickness has to be less than the effective range of intermolecular interactions ($\sim$100 nm).

The occurrence of dewetting under favorable chemical-potential gradients gives rise to several scientific and technological challenges, as well as solutions. Dewetting is present in many situations in the industrial and scientific worlds. For example, the cleaning of household items such as dinner tables, mirrors, dishes, floors, and so on, requires the greasy oil films on the surface to break into drops (or the surface to dewet) so that their detachment can occur. In industry, dewetting is important in many processes such as flotation, coating and cleaning of surfaces, trickle-bed reactors, contact equipment for heat and mass transfer, and so on [1–17]. In flotation, the thinning and rupture of the liquid film between the particle and the bubble can be the rate-determining step of the process. The dewetting of coated surfaces where the coating layer breaks down is important in, for example, photographic films, paints, adhesives, dielectrics, and biocompatible and optical coatings [7–12]. Any breakup of these coatings will generally dilute, and in extreme cases nullify, their utility.

Dewetting is encountered in many situations of scientific interest also. Thin mucous and aqueous coatings of the cornea can become unstable in dry eyes [18–20]. The hydrophilic ingredients of dry-eye solutions restore the stability of the protective tear film [19] and prevent dewetting of the cornea. Furthermore, dewetting results in adhesion failure of fluid particles to surfaces [6,21]. It is also important in the formation of precursor films in wetting [22–27], heterogeneous nucleation, film boiling and condensation [28–30], morphology of nanostructures [10,31–35], multilayer adsorption [36–38], near-interface diffusion, and phase separation [39,40]. In another highly topical application, the instability of thin films (and associated dewetting) is now being used to make nano- and micro-patterns.

Now, why do some surfaces dewet whereas others do not? The answer lies in the amplification of any small perturbation on the thin film's surface under the attractive force due to the solid surface. The attractive pull, or the pressure experienced by the thin film's free surface, is characterized by the *disjoining pressure*. This is the extra pressure that has to be deducted from the interfacial pressure of the film-bounding fluid interface due to the presence of the nearby film-solid interface—its magnitude depends on the thickness of the film. If the film is thicker than the effective range

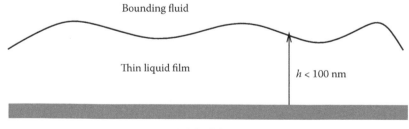

FIGURE 6.1 A solid surface covered by a liquid layer. The breakup of the liquid layer by local thinning leads to the dewetting of the solid.

of intermolecular forces (about 100 nm), then the disjoining pressure is zero as the film-solid interface is not seen by the film-bounding interface. If the variation of disjoining pressure [1–5] with thickness is such that thinner parts of the film are at higher pressure, then any fluctuation in the film's surface can amplify by the flow of liquid from thinner to thicker parts. The fluctuation of the molecules due to thermal energy, aided by any additional mechanical vibrations that may also be present, ensure that the film's surface is never completely flat. The initial fluctuations will amplify whenever the decrease in free energy of the system due to thinning of the film can compensate for the increase due to extra interface created during amplification. In these cases, the initial inhomogeneity triggers a series of morphological changes. These result in the transformation of a flat film into a high-curvature phase, consisting of an array of droplets surrounded by low-curvature (almost flat) and thinner film. This process is commonly known as *morphological phase separation* (MPS) [8,34,35,41,42].

In some cases, the changes lead to an assembly of circular dry spots that grow laterally. The neighboring dry spots coalesce, entrapping the remaining liquid as cylindrical ridges, which break into droplets due to the Rayleigh instability. The underlying solid becomes devoid of any liquid away from the droplets. This scenario is known as *true dewetting* (TD). In contrast, MPS is sometimes also referred to as *pseudo-dewetting* due to the absence of any true dewetted portion. In the case of MPS, the droplets coalesce though transport of liquid via the intervening liquid film, and ripening of the structure takes place. This occurs as all the droplets are not the same size and shape, leading to a gradient of Laplace pressure with bigger drops being at lower pressures due to lesser curvature. Thus, liquid is transferred from smaller to bigger drops, as discussed earlier in the context of phase separation (see Section 1.5). The coalescence of the droplets ultimately results in the thermo-dynamically stable situation of a single droplet bounded by a flat thin film. In TD, the absence of any intervening film prevents the coalescence of droplets via film fluid. However, the coalescence may still happen through transport mediated by the bounding fluid.

Both the dewetting processes, MPS and TD, can be understood in terms of the long-range and short-range components of the intermolecular force between the bounding fluid and the solid substrate. This is usually the case when more than one factor contributes to the intermolecular force field present across the film. For example, the long-range component can be provided by the *apolar van der Waals interaction* and the short-range component by the *polar hydrophobic interaction* [17]. Whereas the individual components show a monotonic decrease in strength with thickness, the net force may show a non-monotonic behavior. There can be four such combinations and corresponding thin-film systems:

- *Type I*: long-range attraction and short-range attraction.
- *Type II*: long-range attraction and short-range repulsion.
- *Type III*: long-range repulsion and short-range repulsion.
- *Type IV*: long-range repulsion and short-range attraction.

The typical force versus thickness plots for these four classes are shown in Figures 6.2 through 6.5. The figures show the individual components as well as the net force. For the initial fluctuations at the film surface to start amplifying, the net force between the bounding fluid and the solid should be attractive at the given film thickness. Understandably, Type III films are stable at all thicknesses, and Type I films are always unstable. Also, thicker and thinner films are unstable in the cases of Type II and Type IV, respectively. Type I films experience an increasing attraction as they thin, and the rate of thinning keeps increasing. This leads to explosive local thinning, breakup, and true dewetting. In contrast, Type II films experience short-range repulsion as they thin, so the rate of thinning decreases. This leads to a flattening of the thinner regions as surrounding regions catch up with the thinnest portion. The thicker portions become increasingly spherical under the influence of interfacial tension. This appearance of a distinct low-curvature flat film phase and a high-curvature droplet phase is characteristic of MPS [8,34,35,41,42]. Also, thinning stops when the film reaches some equilibrium thickness, determined by the balance of attractive and repulsive components of the forces, and true dewetting never happens. Type III films, in the range of unstable thicknesses, behave like Type I films as they also encounter increasing attraction while thinning. Thus, one has to investigate Type I and Type II systems if one wishes to study TD and MPS, respectively.

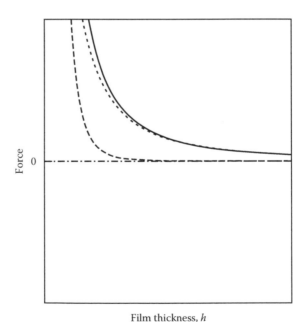

Film thickness, h

FIGURE 6.2 Force versus thickness diagram for *Type I* films. Positive values of the force refer to attraction, and negative values refer to repulsion. The dashed lines represent the components, and the solid line represents the combined effect.

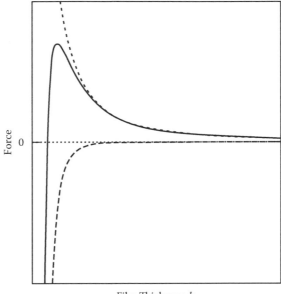

Film Thickness, h

FIGURE 6.3 Force versus thickness diagram for *Type II* films. The dashed lines represent the components, and the solid line represents the combined effect.

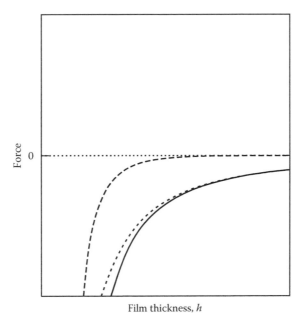

Film thickness, h

FIGURE 6.4 Force versus thickness diagram for *Type III* films. The dashed lines represent the components, and the solid line represents the combined effect.

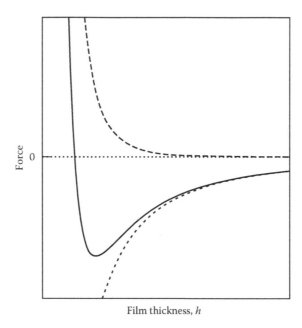

FIGURE 6.5 Force versus thickness diagram for *Type IV* films. The dashed lines represent the components, and the solid line represents the combined effect.

The evolution kinetics of the surface instability that leads to either TD or MPS is the subject matter of this chapter. The dewetting problem will be discussed in the following sequence of steps:

- Discussion of hydrodynamics of thin liquid films, and the forces that are operational in them.
- Discussion of linear stability analysis and methods for solving nonlinear governing equations.
- Discussion of the kinetics of TD and MPS processes.

More precisely, this chapter is organized as follows. In Section 6.2, we present the governing equation for thin-film flows and give a description of various relevant forces. The governing equation closely resembles the Cahn–Hilliard (CH) equation 1.177, which describes the dynamics of phase separation (see Section 1.5.1). In Section 6.3, we undertake a linear stability analysis of the governing equation. The linear stability criteria and the dominant length and time scales are presented in this section. In Section 6.4, we briefly discuss numerical techniques for solving the nonlinear governing equation. In Section 6.5, we discuss the initial stages of the evolution, focusing on the dynamics of reorganization of initial fluctuations in film thickness. This dynamic becomes universal in nature with the proper scaling of length and time scales. Section 6.6 discusses the dynamics of true dewetting, focusing on the timescales of onset of dewetting and subsequent growth of the dry spots. In Section 6.7, we study the ripening dynamics of droplets formed in MPS. The dynamics follow *Lifshitz–Slyozov (LS)* *kinetics $L(t) \sim t^{1/3}$* [43], which describes domain growth whenever material transport occurs via bulk diffusion under the gradient of interfacial pressure. In Section 6.8, we conclude this chapter with a summary and discussion.

6.2 CAHN–HILLIARD EQUATION FOR DEWETTING: HYDRODYNAMIC APPROACH

The process of dewetting is a consequence of liquid flow in thin films and can be studied via the equations governing such flows. Thin-film flows are adequately described by equations of motions simplified under the long-wave approximation [3,4,6,18,23,24,44–53] and including excess intermolecular forces [54,55] as body forces. These simplified equations have been derived for a variety of flow situations and boundary conditions [3,56,57], for example, two-dimensional (2-*d*) and three-dimensional (3-*d*) flows [58], evaporating films [59,60], films with surfactants [3,61,62], films on homogeneous and heterogeneous substrates [63–65], films on porous and non-porous substrates [4], and films that do not follow the no-slip condition at the solid surface [64–66]. Each variant uncovers a special feature of dewetting, and one is tempted to discuss the most general case here. There is a problem though—a unified picture of the dewetting process has not yet emerged and is still the subject of scientific discussion. Any attempt to describe the dewetting process in its entirety will lack the coherence and simplicity required for the general reader of this book. Fortunately, the basic kinetics of the dewetting process can be adequately captured by the least complicated of all the situations, that is, 2-*d* Newtonian flow on homogeneous substrates. Notice that kinetic studies based on 3-*d* simulations are almost nonexistent, even for the simplest case. Thus, we confine ourselves to 2-*d* Newtonian films to explain dewetting in this chapter. It is well understood that one is going to miss out on an important aspect of dewetting, the morphological pattern formation, unless one looks at complete 3-*d* simulations. But that is another story and is told elsewhere [67].

In Figure 6.6, we show a schematic of the system, which consists of a Newtonian liquid film (f) of mean thickness h_0 (<100 nm) on a solid substrate (s), bounded by a bulk inviscid fluid (b) from above. The liquid is assumed to be an incompressible

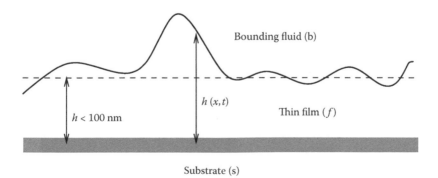

FIGURE 6.6 Schematic diagram of a thin liquid film.

viscous fluid with constant viscosity and interfacial tensions. The film is taken to be non-draining and laterally unbounded. Further, we neglect the effect of gravity and evaporation/condensation. The following long-wave equation [1–4] describes the evolution of the one-dimensional (1-d) film surface in time:

$$3\mu h_t + \left[h^3 (\gamma_{fb} h_{xx} - \phi)_x \right]_x = 0. \tag{6.1}$$

Here, μ is the viscosity of the film liquid, and γ_{fb} is the interfacial tension between the film and the bounding fluid. Further, t is time and x is the space coordinate along the substrate, h is the film thickness, and ϕ is the potential due to excess intermolecular forces. The above equation represents the sum of three forces: (i) unsteady/pseudo viscous force (term with μ), (ii) interfacial tension due to the local curvature of the free surface (term with γ_{fb}), and (iii) excess intermolecular forces (term with ϕ). The unsteady term, which includes viscous effects, merely retards or speeds up the amplification. The interfacial tension suppresses the formation of extra surface and plays a stabilizing role before the appearance of the equilibrium flat film (in MPS) or dry spots (in TD). The excess intermolecular forces are (de)stabilizing whenever the local film thickness is in the (attractive) repulsive range.

The above equation describes the evolution of the free surface of the film and is commonly known as the *thin-film equation*. It is analogous to the CH equation [68,69] (see Section 1.5.1) in film thickness h with an h-dependent mobility factor.

6.2.1 EXCESS INTERMOLECULAR FORCES

As such, intermolecular forces are completely represented in interfacial tensions for isolated interfaces. The situation is different for interacting or closely located interfaces. In the case of thicker films, the free surface of the film can be treated as an isolated interface. As the film thickness decreases below the effective range of intermolecular forces, the free surface starts interacting with the solid-film interface. We will now see how this interaction gives rise to a force that depends on the thickness of the film.

The force on any given molecule results from intermolecular forces between it and the neighboring molecules. Only those molecules that are closer than the effective range of intermolecular forces contribute to the resultant force. In the case of thick films, these consist of molecules in the b- and f-phases. The molecules in the s-phase do not contribute as they are too far away. This force, also represented as interfacial tension, remains constant as the film thickness decreases provided the molecules in the s-phase do not come close enough to contribute. A different situation arises when the film thickness is decreased so much that the molecules in the s-phase also start contributing. Now any change in the film thickness implies a corresponding change in the set of contributing molecules. Thus, the resultant force will now depend on the film thickness. This force, which is in excess to the intermolecular forces that contribute to the interfacial tension, is known as *excess intermolecular force*. It can be expressed in terms of the excess free energy per unit area, ΔG^E—the contribution to the potential ϕ in Equation 6.1 is the derivative of ΔG^E. It has dimensions of pressure and is also known as the *conjoining pressure*, which is the negative of the *disjoining pressure* [2–5,70].

The evaluation of these excess intermolecular forces requires knowledge about the various contributing forces that are operative at these ranges [71,72]. Two such forces, which are relatively better understood, are van der Waals forces and electrostatic double-layer forces. In addition, there are other less understood but important forces such as hydrophobic attraction, hydration repulsion, depletion attraction, and structural forces [71]. These contributing effects are assumed to be additive, so the total disjoining pressure is the sum total of the disjoining pressures due to various components.

6.2.1.1 Lifshitz–van der Waals Interactions

Among all the components of the disjoining pressure, the Lifshitz–van der Waals (LW) forces are most important as they are ubiquitous and universal. They arise due to the interactions of atomic and molecular electrical dipoles formed by the non-overlapping centers of charge of the electron cloud and the nuclear protons. These dipoles may be permanent or induced by already present neighboring dipoles. An instantaneous dipole exists in all atoms even if its average dipole moment is zero, and it induces a dipole in other neighboring atoms. The average interaction of these two dipoles is non-zero and results in an attraction.

The pair potential for LW forces is given as $V(d) = -ad^{-6}$, where a is a positive material-specific constant and d is the separation distance. The attractive nature of the pair potential requires that $a > 0$. We now need to scale up this pair potential to the case of two macroscopic interfaces separated by a thin film of thickness h. This is done by assuming pair-wise additivity of intermolecular potentials among the molecules of the three phases. The net result for the LW energy per unit area of the film is [71]

$$\Delta G^E = -\frac{A_{\text{eff}}}{12\pi h^2} + \text{constant}. \tag{6.2}$$

Here, A_{eff} is termed the *effective Hamaker constant* [73] and is defined by the pair-wise Hamaker constants for four types of binary interactions:

$$A_{\text{eff}} = A_{\text{ff}} + A_{\text{sb}} - A_{\text{bf}} - A_{\text{sf}}, \tag{6.3}$$

where A_{ij} is the Hamaker constant for interactions between phases i and j. If either i or j is a gas, then there is very little interaction and $A_{ij} = 0$. For example, if the bounding fluid is a gas, then $A_{\text{eff}} = A_{\text{ff}} - A_{\text{sf}}$. The pair-wise Hamaker constants, A_{ij}, can be estimated by the following relation:

$$A_{ij} = \pi^2 \rho_i \rho_j a_{ij}. \tag{6.4}$$

Here, a_{ij} is the coefficient for the LW pair potential between molecules of species i and j, and ρ_i and ρ_j are the molecular number densities of the i and j phases, respectively. Whereas the pair-wise Hamaker constants are always positive and denote attraction between the participating pairs, A_{eff} can take negative values. A positive (negative) value of A_{eff} indicates that there is an effective repulsion (attraction) between the bounding fluid and the solid substrate. Intuitively, repulsion indicates the stability of the intervening liquid film, and attraction means that the film is unstable and dewetting may occur. As we are interested in dewetting, we consider situations where attraction (negative value of A_{eff}) prevails.

The constant in Equation 6.2 is the value of ΔG^E as $h \to \infty$. The limit $h \to \infty$ means that the two interfaces are now isolated from each other, and the free energy is given by the sum of respective interfacial tensions, viz., $\gamma_{sf} + \gamma_{fb}$. This fixes the constant in Equation 6.2. Now, let the phases b and s come into contact by reducing h to some equilibrium cutoff distance d_0. The closest distance of approach is taken to be non-zero to prevent non-physical divergence of the force field as $h \to 0$. The film is absent now, and the energy per unit area, $\Delta G^E(d_0)$, is given by γ_{sb}. Now, what is the change in energy as the film goes from bulk ($h \to \infty$) to nothing ($h \to d_0$)? Macroscopically, this is ($\gamma_{sb} - \gamma_{sf} - \gamma_{fb}$), which is commonly known as the *spreading coefficient* (S) of the film fluid. Microscopically, this is $[\Delta G^E(d_0) - \Delta G^E(\infty)]$ and is equal to $-A_{eff}/(12\pi d_0^2)$. Thus, the relation between a macroscopic S and a microscopic A_{eff} is [17,74,75]

$$S = -\frac{A_{eff}}{12\pi d_0^2}. \tag{6.5}$$

The spreading coefficient S can be estimated from the equilibrium contact angle θ of the film liquid, provided θ is non-zero or $S < 0$. Figure 6.7 shows the following relationship between S and θ:

$$S = -\gamma_{fb}(1 - \cos \theta). \tag{6.6}$$

Once the value of S has been estimated, the value of A_{eff} can be found using Equation 6.5. Fortunately, d_0 is found to be almost constant ($0.158 \text{ nm} \pm 10\%$) for most materials [17]. The excess free energy per unit area, ΔG^E, can also be written

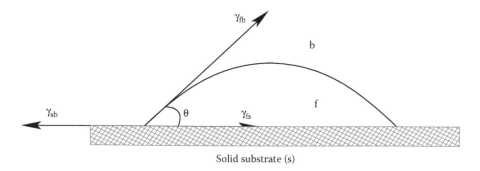

FIGURE 6.7 Equilibrium contact angle given by the resolution of interfacial tensions at the three-phase contact line.

in terms of the spreading coefficient as

$$\Delta G^E(h) = S \frac{d_0^2}{h^2}. \tag{6.7}$$

The required potential $\phi(h)$ is

$$\phi(h) = \frac{\partial(\Delta G^E)}{\partial h} = -2S \frac{d_0^2}{h^3}. \tag{6.8}$$

The spreading coefficient S has to be negative (that is, A_{eff} has to be positive) for the LW component of the disjoining pressure to be attractive and promote instability and dewetting. Here it is assumed that LW is the only operative force in the system. In the presence of other contributing forces, S can be written as a sum of spreading coefficients due to various components. In the case of purely LW interactions, S in Equations 6.5, 6.7, 6.8 will be replaced by S^{LW}, the spreading coefficient component due to LW forces. The contact angle θ would still be related to the overall S as one does not decompose θ into components due to various forces.

6.2.1.2 Short-Ranged Repulsion

In cases of instability, the film surface deforms continuously and the film thickness h reduces. As the film thins and $h \to 0$ then $\phi \to \infty$, which is of course not possible. One needs a short-range repulsion to arrest the thinning and provide a minimum cutoff distance. One source of this repulsion is ubiquitous and is present in the form of *Born repulsion*. This has its origin in the interaction between two unshielded nuclei, as well as unshielded electron clouds, as two atoms come very close to each other. Understandably, Born repulsion is extremely short-ranged with an effective range of only a fraction of a nanometer. Thus, the pair potential for Born repulsion is of the form $-bd^{-c}$, where b is a negative material-specific constant, c is a positive index, which is typically greater than 12 for steep rise, and d is the separation distance. For $c = 12$, the combined pair potential of LW and Born interactions is equivalent to the Lennard–Jones potential.

As in the case of LW interactions, this pair potential has to be scaled up for thin films. The excess free energy per unit area (ΔG^E) and per unit volume (ϕ) for the combination of LW and Born repulsion is

$$\Delta G^E(h) = S^{\text{LW}} \frac{d_0^2}{h^2} + \frac{B}{h^8}, \tag{6.9}$$

and

$$\phi(h) = -2S^{\text{LW}} \frac{d_0^2}{h^3} - \frac{8B}{h^9}. \tag{6.10}$$

Here, S^{LW} is the LW component of the spreading coefficient, and B is the coefficient for Born repulsion. The substrate is replaced by a combination of substrate with a pseudo-liquid film of extremely small thickness l_0 over it. The equilibrium cutoff thickness l_0 is related to d_0 by $l_0^2 = 3d_0^2/4$. This leads to two requirements: (a) the excess force becomes zero at l_0 and is negative (repulsive) at smaller distances, and (b) the spreading coefficient does not change. The above requirements are equivalent to

$$\phi(l_0) = 0, \tag{6.11}$$

and

$$\Delta G^E(l_0) = S^{LW}, \tag{6.12}$$

yielding

$$B = -\frac{27}{256} S^{LW} d_0^8. \tag{6.13}$$

With this value for B, one obtains

$$\Delta G^E(h) = S^{LW} \frac{d_0^2}{h^2} \left[1 - \frac{1}{4} \left(\frac{l_0}{h} \right)^6 \right], \tag{6.14}$$

and

$$\phi(h) = -2S^{LW} \frac{d_0^2}{h^3} \left[1 - \left(\frac{l_0}{h} \right)^6 \right]. \tag{6.15}$$

There are many other sources that can provide the required repulsion, chief among them being non-LW components of the disjoining pressure [4,70]. One simple way to achieve this while considering only LW interactions is to coat the original substrate with a material that repels the bounding fluid in the presence of the film liquid. The thickness of the coating (δ), as well as the relative magnitude of the substrate spreading coefficient (S_s^{LW}) and the coating spreading coefficient (S_c^{LW}), determine the range and strength of the repulsion. Intuitively, relatively thicker and repulsive coatings are likely to mask the attractive nature of the substrate more effectively. This will result in a stronger and longer-ranged repulsion. Mathematically, ΔG^E and ϕ for coated systems with only LW interactions are as follows:

$$\Delta G^E(h) = S_s^{LW} \frac{d_0^2}{h^2} \left[\frac{1 - R}{(1 + \delta/h)^2} + R \right], \tag{6.16}$$

$$\phi(h) = -2S_s^{LW} \frac{d_0^2}{h^3} \left[\frac{1 - R}{(1 + \delta/h)^3} + R \right]. \tag{6.17}$$

Here, $R = S_c^{LW}/S_s^{LW}$ and has a negative value when the coating is repulsive and the substrate is attractive.

Other non-LW sources, which can provide the required repulsive component, may be conveniently modeled by an exponential decay. In that case, ΔG^E and ϕ are given by

$$\Delta G^E(h) = S^{LW} \frac{d_0^2}{h^2} + S^{NLW} \exp\left(\frac{d_0 - h}{l_p}\right), \qquad (6.18)$$

$$\phi(h) = -2S^{LW} \frac{d_0^2}{h^3} - \frac{S^{NLW}}{l_p} \exp\left(\frac{d_0 - h}{l_p}\right), \qquad (6.19)$$

where S^{LW} and S^{NLW} are the LW and non-LW components of the spreading coefficient. The quantity l_p is the characteristic decay length, which has experimentally been found to vary from fractions of nanometers to tens of nanometers. A positive value of S^{NLW} signifies repulsion, whereas a negative value denotes attraction.

The above list is by no means complete, and there are many situation-specific forces. The reader is directed toward specialized texts [71] for an in-depth description of excess intermolecular forces.

6.2.1.3   Force Fields and Markers to Study Dewetting

Clearly, a general force field for studying the kinetics of dewetting has to be made up of antagonistic (attractive and repulsive) components with different rates of decay. In the absence of attractive forces near the free surface of the film, the surface instability cannot be initiated. In contrast, no study of subsequent dewetting is possible unless shorter-range repulsion is present near the solid surface. In this chapter, true dewetting is discussed in the context of Type I films realized by a purely attractive LW component and by a combination of LW and Born repulsion. These films are referred to as LW and LWBR films. On the other hand, pseudo-dewetting (or morphological phase separation) is discussed in the context of Type II films realized by a combination of an attractive substrate and a repulsive coating. These films are referred to as LWC films. The equilibrium film thickness provided by the balance of substrate and coating effects does not allow the film to thin sufficiently for Born repulsion to manifest itself in LWC films.

The kinetics of dewetting is usually studied by monitoring the following three characteristics:

- Time required for the onset of dewetting.
- Growth rate of the dewetted portion.
- Changes in the length scale of the film morphology with time.

The time of onset of dewetting is of prime concern in the case of TD as it marks the initial breach in the liquid coating on the substrate. As we will see later, the growth rate of the newly formed dry spot is extremely fast. Thus, the onset of dewetting becomes the deciding factor even in applications where a fair amount of dewetting is important for either success or failure of the application. In the case of MPS, one needs to focus on the third characteristic as the coarsening of droplets changes the length scale of the film's morphology until only one droplet remains. In the case of TD, the number of droplets remains approximately constant due to the absence of any connecting film between them. The change in length scale prior to (pseudo-) dewetting is important in both MPS and TD as it dictates the number of (droplets) dry spots formed. The linear analysis, which is considerably simpler than solving the nonlinear equations, provides answers for the change in the length scales before the dewetting process. However, it provides only upper bounds for the time required for onset of dewetting. The linear analysis is not of much use after dewetting, and a full nonlinear analysis is required.

6.3 LINEAR STABILITY ANALYSIS

The kinetics of dewetting can be studied by solving Equation 6.1 numerically. This is not easy as the equation is a fourth-order nonlinear partial differential equation. Can one obtain some insight into the solution, such as its length scales and timescales, without solving the equation? In this context, we undertake a linear stability analysis of the equation. In this analysis, a steady-state solution of the thin-film equation is found, and its stability is examined after linearizing the equation. Clearly, $h =$ constant is a steady-state solution of Equation 6.1 as all space-derivatives vanish. Let us consider the evolution of a small fluctuation about this solution, that is, we ask whether a disturbance in the free surface of the film will grow or decay to the steady-state solution. The advantage of considering a small fluctuation is that, as a first approximation, we can linearize all the terms in the equation by expanding them around the steady state. The resulting linear equation is much easier to solve than is the original nonlinear equation (6.1). Of course, the linear solution only remains valid for short times if the system strays away from the steady state, but it is enough to tell us the stability of the steady state. Let h_0 be the mean film thickness, which will be the steady-state value if the film surface is completely flat. We substitute $h(x,t) = h_0 + f(x,t)$ in Equation 6.1, where f denotes a small deviation from h_0, and neglect all nonlinear terms in f. Then, we obtain the following linear equation:

$$3\mu f_t + h_0^3 \left(\gamma_{fb} f_{xxxx} - \phi_{h_0} f_{xx} \right) = 0, \qquad (6.20)$$

where ϕ_{h_0} denotes ϕ_h (the partial derivative of ϕ with respect to h) evaluated at $h = h_0$.

It is simple to verify that any spatially periodic function of the form $f = \epsilon \sin(kx) \exp(\omega t)$ or $f = \epsilon \cos(kx) \exp(\omega t)$ satisfies Equation 6.20. Here, ω represents the growth rate. If $\omega > 0$, the film is unstable, but for $\omega < 0$, the film is stable—more precisely, the steady state $h = h_0$ is stable as $f \to 0$ when $t \to \infty$. Here, k is the wave number of the instability. We have $k = 2\pi/\lambda$, where λ is a characteristic length scale or wavelength of surface deformation. Further, ϵ is the initial amplitude $(t = 0)$ of the surface deformation. If we substitute the above trial solutions in the linear equation (6.20), we obtain a *dispersion relation* or a *characteristic equation* that relates ω to k:

$$\omega(k) = -\frac{h_0^3}{3\mu} k^2 \left(\phi_{h_0} + \gamma_{fb} k^2 \right). \qquad (6.21)$$

Therefore, the film can be stable ($\omega < 0$) or unstable ($\omega > 0$), depending on the wavelength of the surface deformation. The first term of Equation 6.21 is positive (and destabilizing) whenever $\phi_{h_0} < 0$. Negative values of ϕ_{h_0} mean that the pressure due to excess intermolecular forces increases as the film thickness decreases locally. Clearly, such a pressure distribution where lower thicknesses are at higher pressure is destabilizing as it encourages flow from thinner portions to thicker portions. This causes local breakup of the film and dewetting of the underlying solid. The second term of Equation 6.21 is always negative and represents the stabilizing influence of interfacial tension (γ_{fb})—any deformation of the planar surface increases the interfacial area and results in an energy penalty. The sum of the first and second terms decides whether ω is positive (unstable film, dewetting) or negative (stable film, no dewetting). The film viscosity, μ, does not influence the stability of the film. It merely changes the rate at which the film's surface evolves. Less viscous films evolve faster, because $|\omega|$ is higher. The wavenumbers for which the film becomes unstable lie between 0 and k_c, where k_c is given by

$$k_c^2 = -\frac{\phi_{h_0}}{\gamma_{fb}}. \tag{6.22}$$

The corresponding wavelength l_c, also known as the *neutral wavelength*, is given by

$$l_c^2 = \frac{4\pi^2}{k_c^2} = -\frac{4\pi^2 \gamma_{fb}}{\phi_{h_0}}. \tag{6.23}$$

In the interval $[0, k_c]$, $\omega(k)$ is maximum at a wave number k_m determined by $d\omega/dk = 0$, or

$$k_m^2 = -\frac{\phi_{h_0}}{2\gamma_{fb}}. \tag{6.24}$$

The corresponding *dominant wavelength* l_m is

$$l_m^2 = \frac{4\pi^2}{k_m^2} = -\frac{8\pi^2 \gamma_{fb}}{\phi_{h_0}}. \tag{6.25}$$

Only the long-range deformations of the film surface, satisfying $0 < k < k_c$ or $\infty > \lambda > l_c$, are unstable. Shorter waves, or disturbances with $\lambda < l_c$, disappear in time because the corresponding ω's are negative. The driving force for moving the film liquid increases as the waves grow progressively longer, and the rate of fluid transport increases. On the other hand, longer waves require film liquid to be transported across larger distances. This results in the intermediate value k_m, for which the growth of instability is fastest. One can estimate the rate of thinning of the film's surface by following the linear evolution of the wavenumber k_m. This gives the first estimate (linear) of the time of appearance of dry spots or onset of dewetting (t_{lm}) as t corresponding with $h \rightarrow 0$:

$$t_{lm} = \frac{12\mu\gamma_{fb}}{h_0^3 \phi_{h_0}^2} \ln\left(\frac{h_0}{\epsilon}\right). \tag{6.26}$$

Now, any real disturbance of the film surface contains all the different wavelengths, called Fourier modes. The unstable modes ($l_c < \lambda < \infty$) all grow, but those modes with wavelengths near l_m grow at the fastest rate. Thus, in practice, one sees the instability of the film's surface manifesting itself on a length scale close to l_m. The above result for the length scale is true for any force field, provided ϕ_{h_0} is known for the field. In the case of LWBR films, ϕ_{h_0} is given by

$$\phi_{h_0} = 6S^{\text{LW}}\frac{d_0^2}{h_0^4}\left(1 - \frac{3l_0^6}{h_0^6}\right). \tag{6.27}$$

For $h_0 \gg l_0$, which is generally the case,

$$\phi_{h_0} \simeq 6S^{\text{LW}}\frac{d_0^2}{h_0^4}. \tag{6.28}$$

The linear length and time scales are then given by

$$l_m^2 = -\frac{4\pi^2\gamma_{\text{fb}}h_0^4}{3S^{\text{LW}}d_0^2}, \tag{6.29}$$

$$t_{lm} = \frac{\mu\gamma_{\text{fb}}h_0^5}{3S^{\text{LW}2}d_0^4}\ln\left(\frac{h_0}{\epsilon}\right). \tag{6.30}$$

As $S^{\text{LW}} = -\gamma_{\text{fb}}(1 - \cos\theta)$, where θ is the equilibrium contact angle, one gets $S^{\text{LW}} \simeq -\gamma_{\text{fb}}\theta^2/2$ for small values of θ. In terms of the equilibrium contact angle,

$$l_m^2 = \frac{8\pi^2 h_0^4}{3\theta^2 d_0^2}, \tag{6.31}$$

and

$$t_{lm} = \frac{4\mu h_0^5}{3\gamma_{\text{fb}}\theta^4 d_0^4}\ln\left(\frac{h_0}{\epsilon}\right). \tag{6.32}$$

The most useful measure of the length scale for dewetting studies is the number density of dry spots formed due to breakup of the film. The linear estimate of this number density, n_{ld}, is the inverse of the dominant wavelength l_m, and is given by

$$n_{ld}^2 = -\frac{3S^{\text{LW}}d_0^2}{4\pi^2\gamma_{\text{fb}}h_0^4}. \tag{6.33}$$

The linear theory tells us that the onset of dewetting will be delayed in thicker and more non-wettable films. Also, the dewetting will be initiated at larger separations of dry spots (lower number density) for these films.

6.4 NONLINEAR ANALYSIS

6.4.1 INITIAL AND BOUNDARY CONDITIONS

In general, the nonlinear equation (6.1) has to be integrated numerically in time and space. The initial conditions, that is, the local film thickness at the start, are often modeled as random fluctuations from a flat film of constant thickness. (Similar initial conditions were used to obtain the evolution pictures in Figures 1.4, 1.10 and 1.11.) This mimics the thermal fluctuations at the film surface. Sometimes, harmonic (in space) perturbations are also used to have better control over the amplitude of the wave corresponding with the dominant wavelength. In the case of random fluctuations, one would need a Fourier decomposition of the initial fluctuations to estimate this amplitude.

The equation is solved in a fixed-length domain (l) with periodic boundary conditions. These are justified as linear analysis predicts that the evolution of the film surface is characterized by a dominant wavelength l_m. Also, nonlinear simulations over domains much larger than l_m show the appearance of a periodic film surface with a well-defined wavelength. Interestingly, this nonlinear dominant wavelength (l_{nm}) is independent of l and comes out to be close to l_m. The solution domain, l, is taken to be an integral multiple of l_m.

6.4.2 NON-DIMENSIONALIZATION

The governing equation is non-dimensionalized before numerical solution to reduce the number of parameters and for compact presentation of results. The characteristic length and time scales, x^* and t^*, are obtained from Equations 6.29 and 6.30 as follows:

$$x^* = h_0 \left(\frac{h_0}{d_0} \right) \left(\frac{\gamma_{fb}}{6|S^{LW}|} \right)^{1/2}, \tag{6.34}$$

and

$$t^* = h_0 \left(\frac{h_0}{d_0} \right)^4 \left(\frac{\mu \gamma_{fb}}{12\, S^{LW2}} \right). \tag{6.35}$$

The characteristic height is taken to be the mean thickness of the film, h_0. In order to reduce the number of parameters further, a characteristic conjoining pressure, ϕ^*, is defined as

$$\phi^* = 6h_0 \left(\frac{d_0}{h_0} \right)^2 |S^{LW}|. \tag{6.36}$$

Then, the non-dimensional variables are defined as

$$T = \frac{t}{t^*},$$

$$X = \frac{x}{x^*},$$

$$H = \frac{h}{h_0},$$

$$\Phi = \frac{\phi}{\phi^*}. \tag{6.37}$$

The non-dimensional governing equation in terms of the above variables is

$$H_T + \left[H^3(H_{XX} - \Phi)_X\right]_X = 0. \tag{6.38}$$

It is interesting to note that the dependence of the linear estimates of breakup time and number density of dry spots is inherent in the nonlinear scalings, leading to the above non-dimensional governing equation. Let us do a thought experiment in which a large-scale non-dimensional simulation produces N dry spots per unit length. The unit non-dimensional length is actually proportional to $h_0^2(\gamma_{fb}/|S^{LW}|)^{1/2}$, as given by the above scaling. Thus, the actual number density is proportional to $h_0^2(\gamma_{fb}/|S^{LW}|)^{1/2}$. Similarly, the time of onset of dewetting will come out to be proportional to $\mu\gamma_{fb}h_0^5/S^{LW^2}$.

6.4.3 NUMERICAL SCHEME

Many numerical schemes can now be used to solve Equation 6.38 with periodic boundary conditions in space [76–88]. One of the schemes that has been quite successful is the *method of lines*. The above equation is solved in its conservative form without expanding the successive derivatives. The space domain $L = l/x^*$ is divided into grids with uniform spacing. The space domain is taken to be an integral multiple of the non-dimensional dominant wavelength, $L_M = l_m/x^*$. Then, the partial differential equation is converted to a set of ordinary differential equations of the form:

$$\frac{DH_i}{DT} = -\left[H^3(H_{XX} - \Phi)_X\right]_X\bigg|_i, \tag{6.39}$$

where H_i is the local film thickness at the i-th grid location. The expression on the right-hand side is also evaluated at the i-th grid location. This evaluation is done by a successive central-differencing scheme employing half-nodes to prevent decoupling of the odd and even grid locations. Any suitable interpolation formula can be used to evaluate the half-node values from full-node values. The resultant set of coupled ordinary differential equations is simultaneously integrated in time using *Gear's algorithm*, which is specially suited for stiff equations. Computer codes based on these algorithms are readily available, and NAG Library's subroutine D02EJF has been found to be adequate to solve these equations.

6.5 EMERGENCE OF A DOMINANT WAVELENGTH: A NEW UNIVERSAL DYNAMICS

The process of (pseudo-) dewetting begins by rearrangement of surface disturbances in the free surface of the overlying thin film. These disturbances quickly organize themselves on a distinct length scale. As discussed in Section 6.3, the amplitudes of the Fourier components of the initial surface fluctuations having wavelength shorter than a critical wavelength (l_c) decay whereas others amplify. The component having the dominant wavelength grows at the fastest rate and dominates the evolution. The dominant nonlinear wavelength comes out to be very similar to the linear one in most cases as it emerges quite early in the evolution, when the film thickness is close to the initial value. For such small deviations from the initial values, the linearized equations remain a good approximation to the nonlinear ones.

This reorganization step is best studied by the simplest system of LW films with attractive van der Waals forces as one is interested only in the early stages of evolution. The reorganization is complete, and the dominant wavelength emerges much before the repulsion (as provided in more complicated LWBR and LWC films) becomes relevant. Figure 6.8 shows the morphological changes during reorganization of initial random fluctuations over the length scale of the dominant wavelength. The initial inhomogeneities, shown as pluses in the figure, quickly lose their randomness and start showing some structure (curve 2). Shortly afterwards, a distinct wave with a wavelength much smaller than L_M appears (curve 2). As time proceeds, the dominant wave on the length scale of L_M emerges (curve 3). Further evolution until the breakup of the film takes place at the length scale of the dominant wave, as shown in Figure 6.9. The rapidity of the reorganization with respect to the dewetting process is highlighted when one compares the time of reorganization (~ 1, curve 3 in Figure 6.8) and the time of onset of dewetting (~ 22, curve 5 in Figure 6.9).

Figure 6.10 shows the early stages of evolution of morphological features for an LW film in domain size L_M. Apart from the surface roughness, all features show an

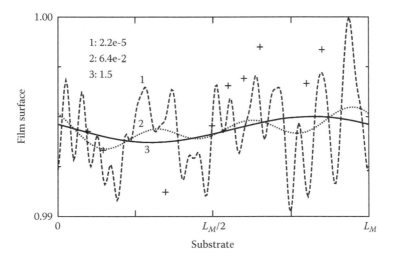

FIGURE 6.8 Initial reorganization of a random surface perturbation of the free surface of a thin film in the domain size L_M. Different curves show the surface at the non-dimensional times mentioned against the corresponding number. The pluses denote the initial random fluctuations. (From Khanna, R., *Wave Dynamics and Stability of Thin Film Flow Systems*, Usha, R., Sharma, A., and Dandapet, B.S., (Eds), Narosa Publishing House, New Delhi, Figure 12, p. 486, 2006. With permission.)

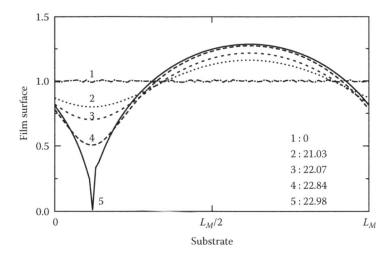

FIGURE 6.9 Evolution of the free surface of a typical LW film in the domain size L_M. The different curves show the surface at the non-dimensional times mentioned against the corresponding number.

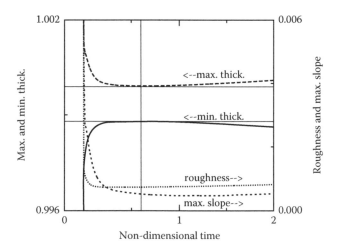

FIGURE 6.10 Early stages of the evolution of the maximum thickness, minimum thickness, roughness, and maximum slope of an LW film in a domain of size L_M. (From Khanna, R., *Wave Dynamics and Stability of Thin Film Flow Systems*, Usha, R., Sharma, A., and Dandapet, B.S., (Eds), Narosa Publishing House, New Delhi, Figure 12, p. 486, 2006. With permission.)

unusual behavior during the time of reorganization. The maximum thickness first decreases and then starts increasing. The minimum thickness first increases and then starts decreasing on similar time scales. The maximum slope also dips first and then increases. The surface roughness, which is the root-mean-square deviation of the film thickness, decreases continuously. Again, these anomalous behaviors can be explained

by the variable growth rate of the different Fourier components of the surface disturbance. The increase in minimum thickness and decrease in maximum thickness indicate that the film is leveling off rather than breaking in this period. This corresponds
with the decay of the stable components before the unstable ones grow. The maximum
slope is estimated from a second-order finite-difference analog of the first derivative
and can have a very high value for the random initial fluctuations. It dips to a small
value as the initial smoothening takes place. Notice that it starts increasing again after
the dip. Interestingly, this takes place on a much shorter timescale than that of the variation of the thickness. The surface roughness is a characteristic of the whole surface
and not just the local film, as is the case with the other three markers. The random initial
fluctuations give a value that is close to the mean amplitude of the fluctuation. As the
film smoothens and coarsens, the roughness value first dips and then starts increasing.

The increase in length scale (or the *coarsening* of the film surface) during this
reorganization can be tracked by counting the number of local maxima (hills) with
time. Simulations of coarsening have to be done in a larger domain so as to average
over a larger number of drops and over longer length scales. Figure 6.11 shows the
time-dependence of the number of hills (on a log-log scale) for the LW film in domain
sizes of $8L_M$ to $1024L_M$. The regularity of vertical spacing between consecutive curves
suggests that, with suitable scaling, these curves can be merged into a single curve.
This is achieved by scaling the number of hills with N_L, the domain size in units
of the dominant wavelength. Figure 6.12 plots the scaled number of hills vs. non-
dimensional time. Indeed, the curves collapse onto a single curve. Three important
conclusions can be drawn from this result:

- The coarsening dynamics in the reorganization stage follows a power-
  law dynamics of the type $N \propto t^{-\alpha}$, where N is the number of hills. As

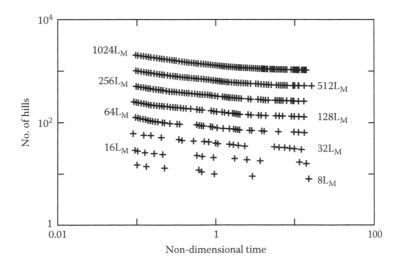

FIGURE 6.11 Variation of the number of hills with non-dimensional time in LW films in
domain sizes of $8L_M$ to $1024L_M$.

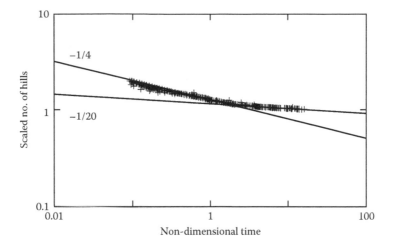

FIGURE 6.12 Variation of the scaled number of hills with non-dimensional time in LW films in domain sizes of $8L_M$ to $1024L_M$.

we have seen in previous chapters, this power-law behavior is a common characteristic of systems undergoing domain growth.
- The rate of coarsening is inversely proportional to the mean spacing between the hills, if one is looking at the same number of hills in different domains.
- The power-law exponent $\alpha \simeq 1/4$ in the initial regime, and after the number of hills reaches N_L, $\alpha \simeq 0$.

Let us next examine whether we can include other force fields in the same results. Figure 6.13 presents results for some other attractive potentials with different decay characteristics such as $1/H^4, 1/H^5$, etc. Although the growth exponents follow the same trend as earlier, the curves are separated from each other. A scaling plot of the data in Figure 6.13 can be obtained as follows. The representative timescale for each force field is the corresponding time of onset of dewetting or film-rupture T_R. This is similar to the time of the last curve in Figure 6.9. As the film thickness never reaches zero, rupture is deemed to have occurred when its minimum value reaches a low value of 0.01 (in dimensionless units). Any comparable choice for cutoff thickness will have very little effect on the estimate of T_R as the film thickness falls sharply near the rupture. Thus, T/T_R can be thought of as an evolution coordinate similar to reaction coordinates in the study of chemical reactions. Figure 6.14 replots the data of Figure 6.13 in terms of the evolution coordinate and scaled number of hills. The curves collapse onto a single curve. Thus, one can say that

- The time of rupture, T_R, is the representative timescale for different films with respect to their coarsening dynamics in the initial stages.

This is an important conclusion as it shows the universality of the coarsening process in the reorganization stage, that is, it would apply to any kind of film irrespective

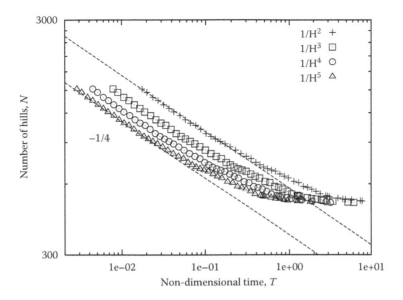

FIGURE 6.13 Variation of the number of hills with non-dimensional time for films in a domain size of $512\,L_M$ under various attractive potentials with different decay characteristics.

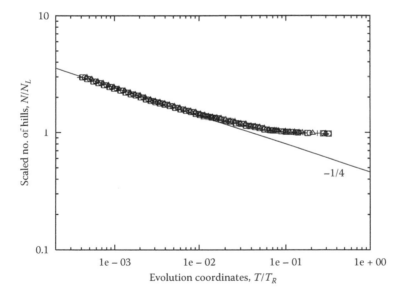

FIGURE 6.14 Variation of the number of hills with the evolution coordinate T/T_R for films in a domain size of $512\,L_M$ under various attractive potentials with different decay characteristics.

of the force fields, domain size, thickness, or any other conceivable parameter. Interestingly, universal dynamics is often encountered in other phase transformations, as discussed in detail in Chapter 1 [89,90].

6.6   DYNAMICS OF TRUE DEWETTING

Figures 6.15 and 6.16 show a typical dewetting process starting from random surface fluctuations. The renormalized length and time coordinates are described later in this section. The initial random disturbances are quickly reorganized on the length scale of the linear dominant wavelength, and further evolution continues on this scale until the film ruptures and holes are formed. The holes, formed at the instant of film rupture, expand until their growth is stopped by the growth of neighboring holes, leading to the formation of cylindrical threads between them. The $d = 2$ simulations cannot track any further evolution of the instability beyond this time, when drainage (normal to the plane of the paper) occurs, resulting in hole coalescence. The long and thin liquid cylinders then break into smaller droplets due to the Rayleigh instability. These droplets become increasingly spherical under the action of interfacial tension.

The simplest potential that results in true dewetting is the van der Waals attraction. The major problem with it is that the thin-film analysis breaks down at the instant of film breakup. The non-physical divergence of the hydrodynamic model at the three-phase contact line ($h \to 0$) is removed by inclusion of a short-range Born repulsion. This repulsion provides an equilibrium cutoff distance and thereby prevents the (non-physical) penetration of the liquid into solid at the point of rupture. Thus, the process of true dewetting can be best demonstrated by LWBR films. We discuss the kinetics of true dewetting in two distinct stages, separated by the breakup of the liquid film.

6.6.1   BEFORE FILM BREAKUP

The evolution from a random initial fluctuation up to the emergence of a dominant wavelength has been discussed in detail in the previous section. We now describe the further evolution until the film breaks. The main features of this instability are shown in Figures 6.17 through 6.20, which present results for the growth of the instability starting with a random surface disturbance ($\epsilon = 0.01$) in a domain of size $2L_M$.

Figure 6.17 shows the morphological changes during the evolution of the instability. The initial random disturbance (profile 1) is reorganized on the length scale (L_M) of the dominant wavelength, and further evolution continues on this scale until film rupture (profile 7). The film thickness away from the hole remains largely undisturbed during the later phase of growth of the instability ($H_{\min} < 0.6$).

Figure 6.18 shows the variation of the minimum and maximum thicknesses ($H_{\min}$ and $H_{\max}$, respectively) as the instability proceeds. Figure 6.19 shows the variation of the surface roughness (S_R) during the evolution of the instability. Figure 6.20 shows the corresponding variation of the maximum slope (H_{XM}) with time T. These figures indicate that the growth of the instability until film rupture takes place in three distinct phases. During the first phase, $H_{\max}$, S_R, and H_{XM} decrease whereas $H_{\min}$ increases. During the second phase, there is a slight but concurrent increase in $H_{\max}$, S_R, and H_{XM}, together with a corresponding decrease in $H_{\min}$. During the third phase, $H_{\max}$, S_R, and H_{XM} increase rapidly, and $H_{\min}$ shows a rapid decline. Three different timescales, corresponding with the three different phases of the growth of instability, can be identified from the figures, viz.,

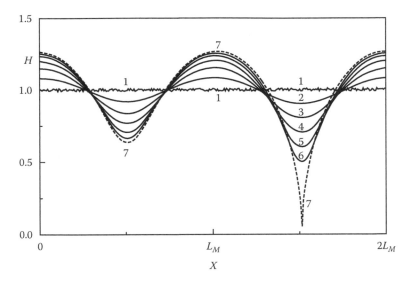

FIGURE 6.15 Evolution of a random initial surface perturbation ($\epsilon = 0.01$) over a large domain ($2L_M$) into a pair of holes in an LWBR film. Profiles 1–7 correspond with renormalized times ($t_R \times 10^{-6}$) = 0, 8.22, 9.15, 9.56, 9.76, 9.85, and 9.91, respectively. The mean thickness of the film (h_0) is 3 nm.

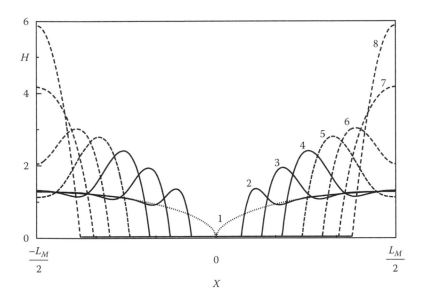

FIGURE 6.16 Various stages of hole growth in a bounded LWBR film of size L_M. Profiles 1–7 correspond with renormalized times ($t_R \times 10^{-3}$) = 0, 1.6, 3.5, 5.6, 7.7, 10.0, 11.7, and 100.0, respectively. The time t_R is reset to 0 at the time of first appearance of the hole. The mean thickness of the film (h_0) is 3 nm.

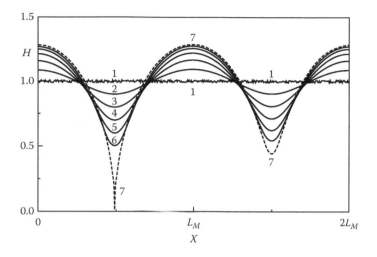

FIGURE 6.17 Morphological changes during the evolution of random surface disturbances ($\epsilon = 0.01$) in a non-slipping LW film over a domain of size $2L_M$. Profiles 1–7 correspond with non-dimensional times (T) equal to 0, 20.3, 22.6, 23.6, 24.2, 24.4, and 24.6, respectively.

1. A short-time regime of initial rearrangement.
2. A larger-time regime of mostly linear evolution of the instability.
3. An extremely short-time regime with explosive growth of instability close to the breakup of the film.

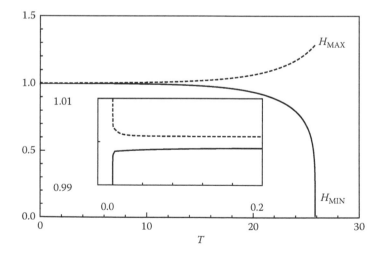

FIGURE 6.18 Variation of non-dimensional minimum and maximum thicknesses ($H_{\min}$ and $H_{\max}$, respectively) with non-dimensional time (T) during the evolution of random surface disturbances ($\epsilon = 0.01$) in a non-slipping LW film over a domain of size $2L_M$.

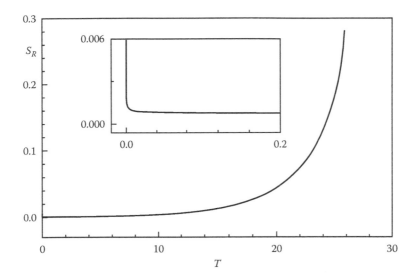

FIGURE 6.19 Variation of non-dimensional surface roughness (S_R) with non-dimensional time (T) during the evolution of random surface disturbances ($\epsilon = 0.01$) in a non-slipping LW film over a domain of size $2L_M$.

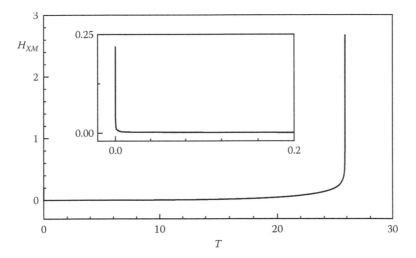

FIGURE 6.20 Variation of non-dimensional maximum slope (H_{XM}) with non-dimensional time (T) during the evolution of random surface disturbances ($\epsilon = 0.01$) in a non-slipping LW film over a domain of size $2L_M$.

It may be noted that Figure 6.18 shows an explosive rate of thinning when the minimum film thickness is below 0.6, which justifies the cutoff choice of $H_{min} = 0.01$ for film breakup. A comparison of the three timescales clearly shows that the second phase is the rate-determining step for film breakup. The nonlinear time of rupture T_{NM} is always less than the linear time of rupture T_{LM}. Nonlinearities accelerate film

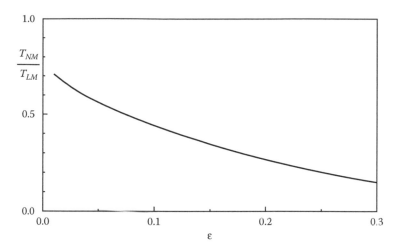

FIGURE 6.21 Variation of the ratio of nonlinear and linear minimum times of rupture (T_{NM}/T_{LM}) with the amplitude (ϵ) of the initial cosine disturbance in a non-slipping LW film.

breakup because the destabilizing term (H^{-1}) grows stronger at locally thin spots, whereas the stabilizing term due to surface tension (H^3) becomes weaker as $H \rightarrow 0$. Thus, in linear theory, the film thins at a rate given by the strength of the force field (ϕ_h) at the mean film thickness, whereas in reality the rate of thinning increases as the film thins. The effect of the initial amplitude becomes more pronounced in the nonlinear analysis compared with the linear analysis. Figure 6.21 clearly shows that the ratio T_{NM}/T_{LM} decreases as ϵ increases.

6.6.2 HOLE GROWTH

Figure 6.16 shows a typical hole growth process starting from the formation of the hole. The expansion of the hole takes place in three phases. In the first *transient phase* of hole growth immediately after its appearance, the velocity of hole growth is maximum and declines rapidly with time, whereas the maximum slope shows a rapid increase. This phase only lasts for a very short time. In the second phase, which may be termed as the *quasi-steady expansion phase*, the maximum slope increases very slowly and the hole velocity also declines very slowly. The quasi-steady phase lasts until the neighboring rims begin to overlap. In the third phase, the maximum slope first decreases slightly and then climbs rapidly to its equilibrium value. The hole velocity first increases slightly and then decreases rapidly, bringing an abrupt end to hole growth. This final phase of hole growth may be termed as *hindered expansion*. The hole growth stops when neighboring rims merge to produce a cylindrical liquid thread.

As discussed earlier, LW films can only be used to study the evolution prior to the dewetting. LWBR films are best suited to study the further process of hole growth due to the presence of strong short-range Born repulsion. This has a serious repercussion on the numerical simulation of the thin-film equation as the resultant equations become very stiff. Further, the non-dimensional evolution equation now has a parameter $L_0 = l_0/h_0$. The evolution of films of different thicknesses will be obtained by putting the corresponding value of L_0 in the non-dimensional equation, and each equation will be different. Thus, any comparison between different films based on non-dimensional results will not be straightforward, as it was in the case of parameterless LW films. Thus, the data for hole growth is best analyzed by defining renormalized scales that remove the effect of the film thickness but retain the effects of viscosity and surface properties. The relations between the renormalized quantities (denoted by the subscript R) and the previously defined non-dimensional and dimensional variables are as follows:

$$x_R = Xh_0 \left(\frac{h_0}{d_0}\right) = x \left(\frac{-6S^{LW}}{\gamma_{fb}}\right)^{1/2}, \tag{6.40}$$

$$t_R = Th_0 \left(\frac{h_0}{d_0}\right)^4 = t \left(\frac{12S^{LW^2}}{\mu \gamma_{fb}}\right). \tag{6.41}$$

Furthermore,

$$\frac{\partial h}{\partial x_R} = (h_x)_R = H_X \left(\frac{d_0}{h_0}\right) = h_x \left(\frac{-\gamma_{fb}}{6S^{LW}}\right)^{1/2}, \tag{6.42}$$

$$r_R = Rh_0 \left(\frac{h_0}{d_0}\right) = r \left(\frac{-6S^{LW}}{\gamma_{fb}}\right)^{1/2}, \tag{6.43}$$

$$(\tan \theta)_R = (h_x)_{0R} = (H_X)_0 \left(\frac{d_0}{h_0}\right) = (h_x)_0 \left(\frac{-\gamma_{fb}}{6S^{LW}}\right)^{1/2}, \tag{6.44}$$

$$u_{hR} = \frac{dr_R}{dt_R} = U_H \left(\frac{d_0}{h_0}\right)^3 = u_h \left(\frac{\mu}{12S^{LW}}\right) \left(\frac{-6\gamma_{fb}}{S^{LW}}\right)^{1/2}. \tag{6.45}$$

Here, h_0 and d_0 are in nm; r and R denote the dimensional and non-dimensional (in X coordinate) hole radii, respectively; u_h and U_H are the dimensional (dr/dt) and non-dimensional (dR/dT) velocities of hole growth, respectively; and $(h_x)_0$ and $(H_X)_0$ refer to the dimensional and non-dimensional slopes at the three-phase contact line. It should be noted that the renormalized variables may have dimensions, but they do not always match with the dimensions of the original variables. For example, $x_R, t_R,$ and r_R have dimensions of length (nm), whereas $(h_x)_R$ and u_{hR} are non-dimensional.

Figure 6.16 shows the various stages of hole growth until the rim reaches the cell corner and merges with the rim of the neighboring hole to produce a cylindrical liquid thread. The newly formed hole is shown by dotted lines and corresponds with $t_R = 0$ in simulations. The quasi-steady phase is shown by solid lines, and the hindered phase is shown by dashed lines. Figure 6.22 shows the unhindered growth of a hole and its rim in an unbounded flat film. A comparison of Figures 6.16 and 6.22 shows the same general features of hole growth, both for the *free expansion* and the *hindered expansion*, until the far side of the rim is close to the boundary of the unit cell.

Three important morphological features emerge from the simulations: (a) the liquid rim surrounding the hole is non-circular and asymmetric, with higher slopes near the

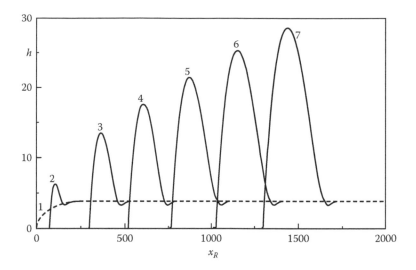

FIGURE 6.22 Unhindered growth of a hole in an unbounded LWBR flat film. Profiles 1–7 correspond with renormalized times $(t_R \times 10^{-4}) = 0, 0.4, 2.3, 4.5, 7.0, 9.8,$ and 12.8, respectively. The time t_R is reset to 0 at the time of first appearance of the hole. The mean thickness of the film (h_0) is 3 nm.

contact line, (b) the far side of the rim does not merge with the film monotonically, but a slight depression is created ahead of the moving rim, and (c) a portion of the film beyond the far end of the moving rim remains undisturbed.

Figure 6.23 is a schematic representation of the hole-rim geometry (solid line) and the initial shape of the film (broken line) at the instant of the first appearance of the

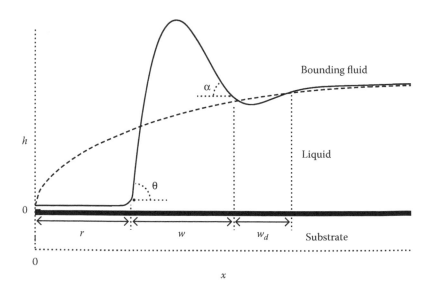

FIGURE 6.23 Schematic diagram of a growing hole in an LWBR film.

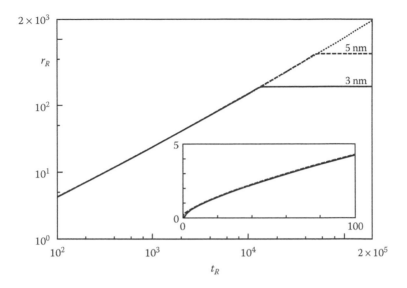

FIGURE 6.24 Variation of renormalized hole radius (r_R) with renormalized time (t_R) in non-slipping LWBR films.

hole (and the three-phase contact line). A pseudo-film of thickness close to l_0 is left behind during the hole expansion. The radius of the hole is defined as the distance from the hole center to the point where the thickness is $1.05l_0$. Any other reasonable cutoff (say $1.1l_0$) produces an identical value of the radius due to the steepness of the liquid wedge close to the contact line. The maximum slope near the retracting contact line is defined as the dynamic contact angle, θ, and the maximum angle on the descending branch of the rim is labeled α. The rim width w extends to the point where the descending branch of the rim intersects the initial profile of the film at the instant of its rupture. The width of the depression ahead of the elevated portion of the rim is denoted by w_d.

Figure 6.24 shows the variation of the renormalized hole radius with renormalized time for 3-nm-thick and 5-nm-thick films. Figures 6.25 and 6.26 show the variation of the renormalized dynamic contact angle and the renormalized velocity of hole growth, respectively. In these figures, t_R is reset to zero at the instant of appearance of the hole. Numerical values of t_R are reported in units of nm. The solid lines correspond with holes in 3- and 5-nm-thick films growing in their respective unit cells of size λ_R. The broken lines correspond with the free expansion of a hole in a 3-nm-thick film. The three phases—transient, quasi-steady, hindered—of hole growth are apparent from these figures. The size of an average unit cell grows as $l_R \sim h_0^2$, because of which the merger of neighboring rims occurs at increasing times (and hole radii) for thicker films. However, the quasi-steady regime continues in the case of free expansion of a lone hole unconstrained by its neighbors. Conceptually, the free expansion is nothing but expansion in a unit cell of very large dimensions. Thus, differences in the dynamics of the free expansion and expansion in a unit cell of finite size arise only during the hindered phase of growth in the latter case.

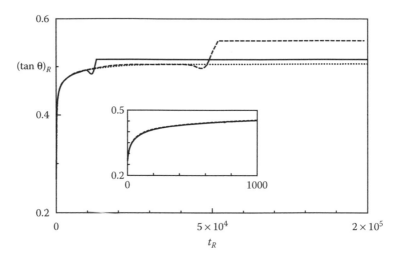

FIGURE 6.25 Variation of renormalized maximum slope [$(\tan\theta)_R$] with renormalized time (t_R) in non-slipping LWBR films. The solid and dashed lines correspond with hindered expansion in 3- and 5-nm-thick films, respectively. The dotted line represents results for unhindered expansion ($h_0 = 3$ nm).

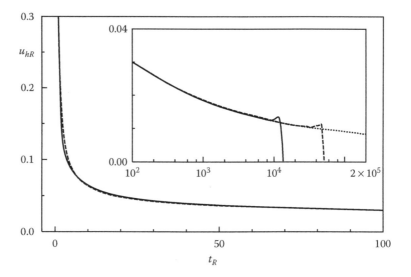

FIGURE 6.26 Variation of renormalized velocity of hole growth (u_{hR}) with renormalized time (t_R) in non-slipping LWBR films. The solid and dashed lines correspond with hindered expansion in 3- and 5-nm-thick films, respectively. The dotted line represents results for unhindered expansion ($h_0 = 3$ nm).

During the unhindered quasi-steady growth of holes away from the cell corners, there is no significant influence of the film thickness on the hole radius or the velocity. This point is illustrated by Figure 6.27, where the best-fit values of the apparent exponent q_a in films of different thicknesses are shown. The data on hole radii for

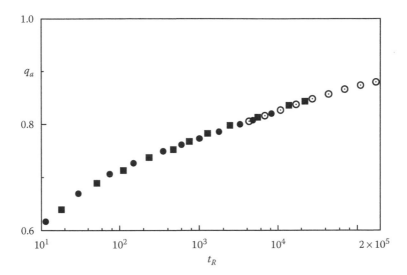

FIGURE 6.27 Variation of apparent exponent (q_a) of hole growth with renormalized time (t_R) in non-slipping LWBR films. The filled circles and squares correspond with hindered expansion for 3- and 5-nm-thick films, respectively. The open circles represent results for unhindered expansion ($h_0 = 3$ nm).

3- and 5-nm films are combined for the analysis of the quasi-steady phase. The data of Figure 6.24 (excluding the hindered phase) were fitted to the form $\log r_R \propto q_a \log(t_R)$ over small time-windows. The best local fits show that q_a increases from an initially low value of about 0.6 to a maximum of about 0.87 at the end of the simulation. After a short transient phase (not shown), the variation in q_a becomes fairly regular in time, viz., $q_a \sim \log(t_R)$. Extrapolating this dependence yields an estimate of the minimum time at which the apparent exponent becomes 1. Physically, q_a can never be greater than 1 as this will imply increase in the velocity with time. The extremely slow variation of q_a with time is evident from the fact that a fit with a single mean exponent $q_m = 0.84$ can adequately describe the quasi-steady data of Figure 6.24 spread over three orders of magnitude in time. This issue is especially significant in the interpretation of experimental data acquired over a limited range of time. Such data may suggest a fixed value of the apparent exponent, as measured from the local slopes on a log-log plot of r versus t.

Figure 6.28 shows the variation of the renormalized velocity of hole expansion u_{hR} with the renormalized maximum slope $(\tan\theta)_R$ in the quasi-steady phase. The figure shows an almost linear decline of velocity with increase in slope. A linear fit to u_{hR} vs. $(\tan\theta)_R$ gives the slope as -0.16 ± 0.0008 and intercept as 0.0904 ± 0.0004. In the hindered phase, when the maximum slope decreases for a while, there is a corresponding increase in the velocity (see Figure 6.26). The linear variation of u_{hR} with $(\tan\theta)_R$ translates into an approximate form of Tanner's law [91] as follows. From Equations 6.44 and 6.45, the dimensional slope and hole velocity are given as

$$\tan\theta = \sqrt{6}\,(\tan\theta)_R\,(1 - \cos\theta_0)^{1/2}, \qquad (6.46)$$

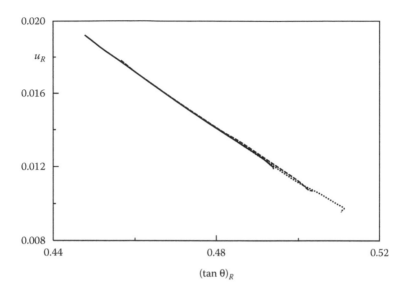

FIGURE 6.28 Variation of renormalized velocity of hole growth (u_{hR}) with renormalized maximum slope [$(\tan \theta)_R$] in non-slipping LWBR films. The solid and dashed lines correspond with hindered expansion in 3- and 5-nm-thick films, respectively. The dotted line represents results for unhindered expansion ($h_0 = 3.77$ nm).

$$u_h = \frac{12\gamma_{fb} u_{hR}}{\sqrt{6}\mu} (1 - \cos \theta_0)^{3/2}, \tag{6.47}$$

where $\cos \theta_0$ is the equilibrium contact angle and is given by the Young–Dupré relation:

$$\cos \theta_0 = 1 + \frac{S^{LW}}{\gamma_{fb}}. \tag{6.48}$$

For small values of θ_0, the linear fit can be written in terms of dimensional variables as

$$u_h = 0.16 \frac{\gamma_{fb}}{\mu} \theta_0^2 (\theta_0 - \theta). \tag{6.49}$$

6.7 DYNAMICS OF MORPHOLOGICAL PHASE SEPARATION

LWC films with a combination of long-range attraction provided by the substrate and short-range repulsion provided by the coating can be used to study MPS (see Section 6.2.1). Figure 6.29 shows the morphological evolution of a typical LWC film in a domain size L_M. The parameters are $R = -1$ and $D = 0.1$. These correspond with a coating that is as repulsive as the substrate is attractive and is of 1/10 thickness of the overlying film. In Figure 6.29, curve 1 is the initial random fluctuation and it

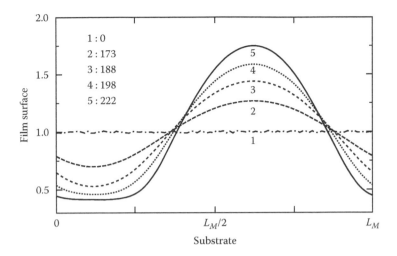

FIGURE 6.29 Evolution of the film's surface for an LWC film in a domain of size L_M. The different curves show the surface at non-dimensional times mentioned against the corresponding number. The parameters are $R = -1$ and $D = 0.1$.

reorganizes over a length scale L_M, as shown in curve 2. The wave then amplifies over a length scale L_M, with the lower portion becoming increasingly flat as it approaches the equilibrium film thickness ($h \simeq 0.45$ in this case) due to the short-range repulsion provided by the coating. The upper portion, in contrast, does not face any such repulsion and becomes more circular under the increasing influence of interfacial tension. The evolution reaches an equilibrium stage (curve 5), where the central *high-curvature drop phase* is in equilibrium with the surrounding *almost flat film phase*. Thus, the film coarsens from a molecularly fine initial condition (curve 1) to the equilibrium length scale of L_M. Figure 6.30 shows the evolution of the maximum and minimum thickness, maximum slope, and surface roughness of the same film. Both the thicknesses change only slightly in the beginning—they start to evolve only for $T > 100$ and gradually reach their equilibrium value. The maximum slope also shows a gradual increase to its equilibrium value ($\simeq 0.5$ in this case). The roughness first dips, remains constant for a long time, and then increases gradually to its equilibrium value ($\simeq 0.4$ in this case).

Figures 6.31 and 6.32 show results for the same film in a bigger domain size of $2L_M$. The random initial fluctuations (not shown) again reorganize to form a wave of wavelength L_M (curve 1). This wave transforms into two separate drops connected by an almost flat film of near-equilibrium thickness (curve 2). The drops appear to be of similar size, but there is a slight mismatch in size. This provides a gradient of Laplace pressure, whereby the bigger drop is at lower pressure due to lower curvature. This results in coarsening with the bigger drops growing by feeding on the smaller drop (curves 3 and 4). The smaller drop eventually disappears (curve 5), leading to coarsening of the film's surface from L_M to $2L_M$ (one drop in $2L_M$). The minimum thickness does not evolve any further once it has reached the equilibrium value, as shown in the magnified view in Figure 6.33.

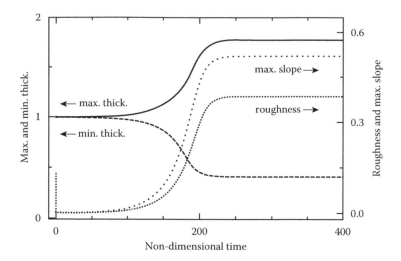

FIGURE 6.30 Evolution of the maximum thickness, minimum thickness, maximum slope, and surface roughness of an LWC film in a domain of size L_M. The parameters are $R = -1$ and $D = 0.1$.

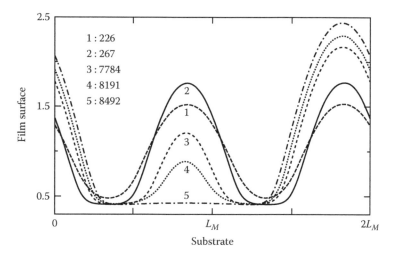

FIGURE 6.31 Evolution of the film's surface for an LWC film in a domain of size $2L_M$. The different curves show the surface at non-dimensional times mentioned against the corresponding number. The parameters are $R = -1$ and $D = 0.1$.

The kinetics of coarsening of the thin film is of interest in MPS, as true dewetting and hole growth do not occur. The results presented next refer to a system where the ratio of the substrate and coating spreading coefficients is taken to be -0.1. Physically, this means that the repulsive effect of the coating is 10 times larger than the attractive effect of the substrate. The thickness of the coating is taken to be half the film thickness, that is, $D = 0.5$. Once again, simulations have to be done in large domains so as to

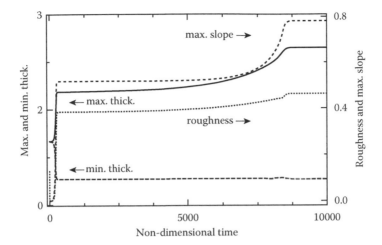

FIGURE 6.32 Evolution of the maximum thickness, minimum thickness, maximum slope, and surface roughness of an LWC film in a domain of size $2L_M$. The parameters are $R = -1$ and $D = 0.1$.

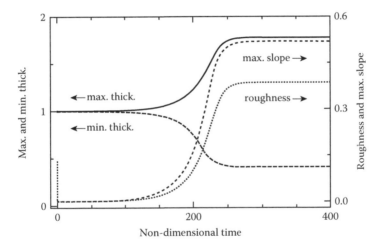

FIGURE 6.33 Early stages of evolution of the maximum thickness, minimum thickness, maximum slope, and surface roughness of an LWC film in a domain of size $2L_M$. The parameters are $R = -1$ and $D = 0.1$.

observe a large number of droplets. Figure 6.34 shows the time dependence of the number of hills for a typical LWC film. Once again, the nearly parallel nature of the curves suggests scaling the number of hills by N_L, the size of the simulation domain in units of L_M. The scaled results are presented in Figure 6.35. As expected, the data sets for different domain sizes collapse onto a single curve. This indicates that the rate of coarsening is inversely proportional to the mean distance between the hills not only during the initial stages of the dewetting process but also during the late stages in MPS. The later part of the curve fits to a $-1/3$ exponent, as suggested by the LS law for

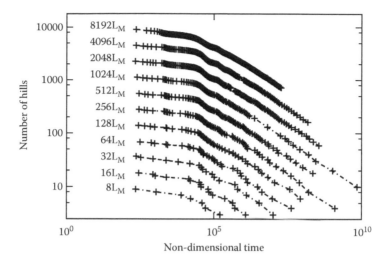

FIGURE 6.34 Variation of the number of hills with non-dimensional time in an LWC film in the domain sizes of 8, 16, 32, 64, 128, 256, 512, 1024, 2048, 4096, and 8192 L_M. The parameters are $R = -0.1, D = 0.5$.

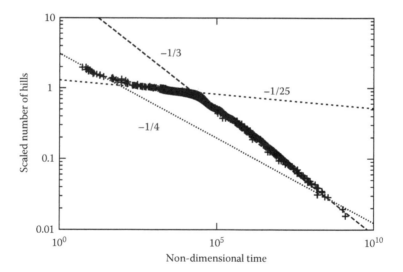

FIGURE 6.35 Variation of the scaled number of hills with non-dimensional time in an LWC film in the domain sizes of 32, 64, 128, 256, 512, 1024, 2048, 4096, and 8192 L_M. The parameters are $R = -0.1, D = 0.5$.

conserved kinetics [43]. This result may not be so important for the 2-d films studied here but has important implications for 3-d studies. The results also show the utility of tracking the number of hills to study the coarsening process. It should be noted that each simulation has started from a different initial condition and even the direct morphological evolution does not follow the same timescales. But when the number of hills is plotted with respect to time, then all data collapse onto a single curve.

6.8 SUMMARY AND CONCLUSIONS

The dewetting of solid surfaces can be studied by following the evolution of the free surface of the liquid film covering the solid. This evolution can be studied by solving the appropriate thin-film equation, which is derived from the fluid equations of motion under the lubrication approximation and includes excess intermolecular forces, for example, Lifshitz–van der Waals forces. The governing equation is a high-order nonlinear partial differential equation. It decomposes into a set of coupled, nonlinear, and stiff ordinary differential equations when the space-derivatives are approximated by their finite-difference analogs.

The irregular free surface of the film deforms spontaneously under the influence of excess intermolecular forces whenever the disjoining pressure reduces with decrease in film thickness. The initial surface fluctuations quickly rearrange themselves into a wavy pattern over a dominant length scale. This length scale is accurately predicted by a linear stability analysis of the thin-film equation and is related to the disjoining pressure evaluated at the mean thickness of the film. Depending on the nature of the disjoining pressure isotherm, the further evolution of the instability leads to one of the following scenarios.

- *True dewetting* (TD) of the substrate, whereby the film breaks up into dry spots or holes. These holes grow laterally and the liquid between them breaks and rearranges to form isolated droplets.
- *Morphological phase separation* (MPS), whereby the film separates into distinct high-curvature drops connected by low-curvature flat film. These drops coarsen by the transport of liquid through the film, and eventually only a single drop surrounded by a flat film remains.

TD occurs whenever the disjoining pressure is devoid of a repulsive component, and MPS occurs in the presence of a short-range repulsive component. In both processes, the morphological pattern starts from the molecular length scale of random thermal noise. During the common initial reorganization stage, the pattern quickly coarsens to the length scale of the dominant wavelength with the number of hills decreasing as $t^{-1/4}$. In TD, the film breaks into circular holes with the linear theory providing an upper bound on the breakup time. The growth of these dry spots follows a power law of the form $r \sim t^{\alpha}$, where r is the radius of the hole. The exponent α starts from a low value of ~ 0.6 and increases to 1 asymptotically. The velocity of hole growth follows an approximate form of Tanner's law [91], which relates the velocity to the dynamic contact angle. In MPS, the coarsening kinetics beyond the dominant wavelength follows the Lifshitz–Slyozov (LS) law [43], whereby the number of droplets decreases as $t^{-1/3}$.

The above description applies to an idealized system of a Newtonian film on a homogeneous and impermeable solid. A brief description of the deviations from this idealized behavior due to various non-idealities or non-classic effects is presented here to put the ideal case in the proper perspective. Detailed discussions are available elsewhere in the literature [77,78,87,92].

- *3-d versus 2-d systems*: Simulations for 3-*d* flows are costly in terms of computing resources and, as a result, not many exist in the literature. In general, as is the case with other instabilities, one would expect the flow to be faster in 3-*d* systems. The spatial gradients in the extra direction are likely to contribute toward the driving force for fluid movement. For 2-*d* simulations, symmetry is assumed in this direction.
- *Non-Newtonian versus Newtonian liquids*: Typically, the characteristic length scales are unchanged for power-law and Maxwell liquids, as viscosity and high-frequency elasticity do not participate in the balance between the destabilizing disjoining pressure and stabilizing interfacial tension. Timescales, on the other hand, are profoundly reduced.
- *No-slip versus slip at the solid surface*: Many applications of dewetting involve long-chain polymers that are known to disobey the no-slip condition at the solid interface. Some results for slipping films exist in the literature, and all of them indicate that slippage leads to increase in length scales and reduction in timescales. The rimmed morphology of the growing holes becomes increasingly flatter as slippage increases. Also, the hole grows with the power-law $r \sim t^{2/3}$ rather than $r \sim t$.
- *Homogeneous versus inhomogeneous solids*: Any inhomogeneity on the solid surface provides an extra gradient in the driving force and thus facilitates evolution. In most cases, a critical size of the heterogeneity is required to destabilize an otherwise stable system. In these systems, the length scales depend on the distribution of heterogeneities and time-scales are extremely fast. In otherwise unstable systems, the interplay of homogeneous and heterogeneous parts, as well as the extreme gradients at their interface, result in many possibilities for morphological patterns, length scales, and timescales.

ACKNOWLEDGMENTS

This chapter is based on A.S. and R.K.'s work on the dewetting of solid surfaces over the years. The description of the coarsening dynamics forms part of the doctoral thesis of N.A. Anigmanshu Ghatak contributed to the work on hole growth. R.K. acknowledges financial support from the Department of Science and Technology, India. A.S. acknowledges support from the Department of Science and Technology Nanoscience Unit at IIT Kanpur.

REFERENCES

1. Sheludko, A., Thin liquid films, *Adv. Colloid Interface Sci.*, 1, 391, 1967.
2. Vrij, A. and Overbeek, J.Th.G., Rupture of thin liquid films due to spontaneous fluctuations in thickness, *J. Am. Chem. Soc.*, 90, 3074, 1968.
3. Ruckenstein, E. and Jain, R.K., Spontaneous rupture of thin liquid films, *J. Chem. Soc. Faraday Trans. II*, 70, 132, 1974.

4. Williams, M.B. and Davis, S.H., Nonlinear theory of film rupture, *J. Colloid Interface Sci.*, 90, 220, 1982.
5. Sharma, A., Equilibrium contact angles and film thicknesses in the apolar and polar systems: Role of intermolecular interactions in coexistence of drops with thin films, *Langmuir*, 9, 3580, 1993.
6. Dimitrov, D.S., Dynamic interactions between approaching surfaces of biological interest, *Prog. Surf. Sci.*, 14, 295, 1983.
7. Kheshgi, H.S. and Scriven, L.E., Dewetting: Nucleation and growth of dry regions, *Chem. Eng. Sci.*, 46, 519, 1991.
8. Reiter, G., Dewetting of thin polymer films, *Phys. Rev. Lett.*, 68, 75, 1992.
9. Reiter, G., Unstable thin polymer films: Rupture and dewetting processes, *Langmuir*, 9, 1344, 1993.
10. Zhao, W., Rafailovich, M.H., Sokolov, J., Fetters L.J., Plano, R., Sanyal, M.K., Sinha, S.K., and Sauer, B.B., Wetting properties of thin liquid polyethylene propylene films, *Phys. Rev. Lett.*, 70, 1453, 1993.
11. Extrand, C.W., Continuity of very thin polymer films, *Langmuir*, 9, 475, 1993.
12. Jain, R.K, Ivanov, I.B., Maldarelli, C., and Ruckenstein, E., in *Dynamics and Instability of Fluid Interfaces*, Sorensen, T.S., Ed., Springer-Verlag, Berlin, 1979, 140.
13. Maldarelli, C., Jain, R.K., Ivanov, I.B., and Ruckenstein, E., Stability of symmetric and unsymmetric, thin liquid films to short and long wavelength perturbations, *J. Colloid Interface Sci.*, 78, 118, 1980.
14. Maldarelli, C. and Jain, R.K., The linear, hydrodynamic stability of an interfacially perturbed, transversely isotropic, thin, planar viscoelastic film. I. General formulation and a derivation of the dispersion equation, *J. Colloid Interface Sci.*, 90, 233, 1982.
15. Gallez, D. and Coakley, W.T., Interfacial instability at cell membranes, *Prog. Biophys. Mol. Biol.*, 48, 155, 1986.
16. Ivanov, I.B. and Kralchevsky, P.A., *Thin Liquid Films—Fundamentals and Applications*, Marcell Dekker Inc., New York, 1988.
17. Van Oss, C.J., Acid-base interfacial interactions in aqueous media, *Colloids Surf. A*, 78A, 1, 1993.
18. Sharma, A. and Ruckenstein, E., Mechanism of tear film rupture and formation of dry spots on cornea, *J. Colloid Interface Sci.*, 106, 12, 1985.
19. Sharma, A., Energetics of corneal epithelial cell-ocular mucus-tear film interactions: Some surface chemical pathways of corneal defense, *Biophys. Chem.*, 47, 87, 1993.
20. Ruckenstein, E. and Sharma, A., A surface chemical explanation of tear film break-up and its implications, in *The Preocular Tear Film: In Health, Disease and Contact Lens Wear*, Holly, F.J., Ed., Dry Eye Institute, Lubbock, Texas, 1986, 697.
21. Chen, J.-D., Effects of London–van der Waals and electric double layer forces on the thinning of a dimpled film between a small drop or bubble and a horizontal solid plane, *J. Colloid Interface Sci.*, 98, 329, 1984.
22. Miller, C.A. and Ruckenstein, E., The origin of flow during wetting of solids, *J. Colloid Interface Sci.*, 48, 368, 1974.
23. de Gennes, P.G., Wetting: Statics and dynamics, *Rev. Mod. Phys.*, 57, 827, 1985.
24. Teletzke, G.F., Davis, H.T., and Scriven, L.E., Wetting hydrodynamics, *Rev. Phys. Appl.*, 23, 989, 1988.
25. Huh, C. and Scriven, L.E., Hydrodynamic model of steady movement of a solid/liquid/fluid contact line, *J. Colloid Interface Sci.*, 35, 85, 1971.
26. Lopez, J., Miller, C.A., and Ruckenstein, E., Spreading kinetics of liquid drops on solids, *J. Colloid Interface Sci.*, 56, 460, 1976.

27. Neogi, P. and Miller, C.A., in *Thin Liquid Film Phenomena*, AICHE Symposium Series, Krantz, W.B., Wasan, D.T., and Jain, R.K., Eds., AICHE, New York, 1986, 145.

28. Neogi, P. and Berryman, J.P., Stability of thin liquid films evaporating into saturated vapor, *J. Colloid Interface Sci.*, 88, 100, 1982.

29. Bankoff, S.G., Dynamics and stability of thin heated liquid films, *J. Heat Transfer*, 112, 538, 1990.

30. Wayner, P.C., Spreading of a liquid film with a finite contact angle by the evaporation/condensation process, *Langmuir*, 9, 294, 1993.

31. Guerra, J.M., Srinivasrao, M., and Stein R.S., Photon tunneling microscopy of polymeric surfaces, *Science*, 262, 1395, 1993.

32. De Coninck, J., Fraysse, N., Valignat, M.P., and Cazabat, A.M., A microscopic simulation of the spreading of layered droplets, *Langmuir*, 9, 1906, 1993.

33. Forcada, M.L. and Mate, C.M., Molecular layering during evaporation of ultrathin liquid films, *Nature*, 363, 527, 1993.

34. Jameel, A.T. and Sharma, A., Morphological phase separation in thin liquid films: II. Equilibrium contact angles of nanodrops coexisting with thin films, *J. Colloid Interface Sci.*, 164, 416, 1994.

35. Sharma, A. and Jameel, A.T., Nonlinear stability, rupture, and morphological phase separation of thin fluid films on apolar and polar substrates, *J. Colloid Interface Sci.*, 161, 190, 1993.

36. Philip, J.R., Unitary approach to capillary condensation and adsorption, *J. Chem. Phys.*, 66, 5069, 1977.

37. Hu, P. and Adamson, A.W., Adsorption and contact angle studies: II. Water and organic substances on polished polytetrafluoroethylene, *J. Colloid Interface Sci.*, 59, 605, 1977.

38. Fowkes, F.M., McCarthy, D.C., and Mostafa, M.A., Contact angles and the equilibrium spreading pressures of liquids on hydrophobic solids, *J. Colloid Interface Sci.*, 78, 200, 1980.

39. Reiter, G., Dewetting as a probe of polymer mobility in thin films, *Macromolecules*, 27, 3046, 1994.

40. Reiter, G., Mobility of polymers in films thinner than their unperturbed size, *Europhys. Lett.*, 23, 579, 1993.

41. Reiter, G., Sharma, A., Casoli, A., David, M.O., Khanna, R., and Auroy, P., Thin film instability induced by long-range forces, *Langmuir*, 15, 2551, 1999.

42. Reiter, G., Auroy, P., and Auvray, L., Instabilities of thin polymer films on layers of chemically identical grafted molecules, *Macromolecules*, 29, 2150, 1996.

43. Lifshitz, I.M. and Slyozov, V.V., The kinetics of precipitation from supersaturated solid solutions, *J. Phys. Chem. Solids*, 19, 35, 1961.

44. Glasner, K.B. and Witelski, T.P., Collision versus collapse of droplets in coarsening of dewetting thin films, *Physica D*, 209, 80, 2005.

45. Glasner, K.B. and Witelski, T.P., Coarsening dynamics of dewetting films, *Phys. Rev. E*, 67, 016302, 2003.

46. Benney, D.J., Long waves on liquid films, *J. Math. Phys.*, 45, 150, 1966.

47. Atherton, R.W. and Homsy, G.M., On the derivation of evolution equations for interfacial waves, *Chem. Eng. Commun.*, 2, 57, 1976.

48. Teletzke, G.F., Davis, H.T., and Scriven, L.E., How liquids spread on solids, *Chem. Eng. Commun.*, 55, 41, 1987.

49. Sharma, A and Ruckenstein, E., An analytical nonlinear theory of thin film rupture and its application to wetting films, *J. Colloid Interface Sci.*, 113, 456, 1986.

50. Borgas, M. and Grotberg, J., Monolayer flow on a thin film, *J. Fluid Mech.*, 193, 151, 1988.
51. Thiele, U., Brusch, L., Bestehorn, M., and Bar, M., Modelling thin-film dewetting on structured substrates and templates: Bifurcation analysis and numerical simulations, *Eur. Phys. J. E*, 11, 255, 2003.
52. Stockelhuber, K.W., Stability and rupture of aqueous wetting films, *Eur. Phys. J. E*, 12, 431, 2003.
53. Shull, K.R. and Karis, T.E., Dewetting dynamics for large equilibrium contact angles, *Langmuir*, 10, 334, 1994.
54. Lifshitz, E.M., The theory of molecular attractive forces between solids, *J. Exp. Theor. Phys. USSR*, 29, 94, 1955.
55. Dzyaloshinskii, J.E, Lifshitz, E.M., and Pitaevskii, L.P., The general theory of van der Waals forces, *Adv. Phys.*, 10, 165, 1961.
56. Sharma, A., Khanna, R., and Reiter, G., Adhesion failure and dewetting of thin liquid films, in *Mittal Festschrift on Adhesion Science and Technology*, van Ooij, W.J. and Anderson, H.R., Eds., VSP Publishers, Utrecht, 1998, 199.
57. Mitlin, V.S., Dewetting of solid surface: Analogy with spinodal decomposition, *J. Colloid Interface Sci.*, 156, 491, 1993.
58. Sharma, A. and Khanna, R., Pattern formation in unstable thin liquid films under the influence of antagonistic short- and long-range forces, *J. Chem. Phys.*, 110, 4929, 1999.
59. Oron, A. and Bankoff, S.G., Dewetting of a heated surface by an evaporating liquid film under conjoining/disjoining pressures, *J. Colloid Interface Sci.*, 218, 152, 1999.
60. Elbaum, M., Lipson, S.G., and Wettlaufer, J.F., Evaporation preempts complete wetting, *Europhys. Lett.*, 29, 457, 1995.
61. Matar, O.K. and Troian, S.M., The development of transient fingering patterns during the spreading of surfactant coated films, *Phys. Fluids*, 11, 3232, 1999.
62. Matar, O.K., Dynamics, stability and pattern formation in surfactant driven thin film flows, in *Wave Dynamics and Stability of Thin Film Flow Systems*, Narosa, Delhi, 2006, 334.
63. Jacobs, K., Herminghaus, S., and Mecke, K.R., Thin liquid polymer films rupture via defects, *Langmuir*, 14, 965, 1998.
64. Kargupta, K. and Sharma, A., Templating of thin films induced by dewetting on patterned surfaces, *Phys. Rev. Lett.*, 86, 4536, 2001.
65. Kargupta, K., Konnur, R., and Sharma, A., Instability and pattern formation in thin liquid films on chemically heterogeneous substrates, *Langmuir*, 16, 10243, 2000.
66. Kargupta, K. and Sharma, A., Creation of ordered patterns by dewetting of thin films on homogeneous and heterogeneous substrates, *J. Colloid Interface Sci.*, 245, 99, 2002.
67. Sharma, A., Self-organized structures in soft confined thin films, *Pramana*, 65, 601, 2005.
68. Cahn, J.W. and Hilliard, J.E., Free energy of a nonuniform system. I. Interfacial free energy, *J. Chem. Phys.*, 28, 258, 1958.
69. Pego, R.L., Front migration in the nonlinear Cahn–Hilliard equation, *Proc. R. Soc. London A*, 422, 261, 1989.
70. Derjaguin, B.V and Churaev, N.V., The current state of the theory of long-range surface forces, *Colloids Surf.*, 41, 223, 1989.
71. Israelachvili, J.N., *Intermolecular and Surface Forces*, Second Edition, Academic Press, New York, 1985.
72. Israelachvili, J.N. and Adams, G.E., Measurement of forces between two mica surfaces in aqueous electrolyte solutions in the range 0–100 nm, *J. Chem. Soc. Faraday Trans.*, 74, 975, 1978.

73. Hamaker, H.C., The London–van der Waals attraction between spherical particles, *Physica*, 4, 1058, 1937.
74. Van Oss, C.J., Chaudhury, M.K., and Good, R.J., Monopolar surfaces, *Adv. Colloid Interface Sci.*, 28, 35, 1987.
75. Van Oss, C.J., Interaction forces between biological and other polar entities in water: How many different primary forces are there?, *J. Dispersion Sci. Technol.*, 12, 201, 1991.
76. Sharma, A. and Khanna, R., Pattern formation in unstable thin liquid films, *Phys. Rev. Lett.*, 81, 3463, 1998.
77. Konnur, R., Kargupta, K., and Sharma, A., Instability and morphology of thin liquid films on chemically heterogeneous substrates, *Phys. Rev. Lett.*, 84, 931, 2000.
78. Zope, M., Kargupta, K., and Sharma, A., Self-organized structures in thin liquid films on chemically heterogeneous substrates: Effect of antagonistic short and long range interactions, *J. Chem. Phys.*, 114, 7211, 2001.
79. van Gemmert, S., Barkema, G.T., and Puri, S., Phase separation driven by surface diffusion: A Monte Carlo study, *Phys. Rev. E*, 72, 046131, 2005.
80. Bergeron, V. and Radke, C.J., Equilibrium measurements of oscillatory disjoining pressures in aqueous foam films, *Langmuir*, 8, 3020, 1992.
81. Bergeron, V., Fagan, M.E., and Radke, C.J., Generalized entering coefficients: a criterion for foam stability against oil in porous media, *Langmuir*, 9, 1704, 1993.
82. Redon, C., Brzoska, J.B., and Brochard-Wyart, F., Dewetting and slippage of microscopic polymer films, *Macromolecules*, 27, 468, 1994.
83. Redon, C., Brochard-Wyart, F., and Rondelez, F., Dynamics of dewetting, *Phys. Rev. Lett.*, 66, 715, 1991.
84. Khanna, R., *Ph.D. Thesis*, IIT Kanpur, Kanpur, 1998.
85. Thiele, U., Open questions and promising new fields in dewetting, *Eur. Phys. J. E*, 12, 409, 2003.
86. Thiele, U., Velarde, M.G., Neuffer, K., and Pomeau, Y., Film rupture in the diffuse interface model coupled to hydrodynamics, *Phys. Rev. E*, 64, 031602, 2001.
87. Thiele, U., Velarde, M.G., and Neuffer, K., Dewetting: Film rupture by nucleation in the spinodal regime, *Phys. Rev. Lett.*, 87, 016104, 2001.
88. Bankoff, S.G., Significant questions in thin film heat transfer, *J. Heat Transfer*, 116, 10, 1994.
89. Puri, S., Bray, A.J., and Lebowitz, J.L., Segregation dynamics in systems with order-parameter dependent mobilities, *Phys. Rev. E*, 56, 758, 1997.
90. Mocuta, C., Reichert, H., Mecke, K., Dosch, H., and Drakopoulos, M., Scaling in the time domain: Universal dynamics of order fluctuations in Fe_3Al, *Science*, 308 1287, 2005.
91. Tanner, L.H., The spreading of silicone oil drops on horizontal surfaces, *J. Phys. D*, 12, 1473, 1979.
92. Sharma, A., Many paths to dewetting of thin films: Anatomy and physiology of surface instability, *Eur. Phys. J. E*, 12, 397, 2003.
93. Khanna, R., *Wave Dynamics and Stability of Thin Film Flow Systems*, Usha, R., Sharma, A., and Dandapet, B.S., (Eds), Narosa Publishing House, New Delhi, 486, 2006.

7 Morphological Transitions in Microphase-Separated Diblock Copolymers

Takao Ohta

CONTENTS

7.1   INTRODUCTION

A binary (AB) polymer mixture undergoes macrophase separation at low temperatures if there is a short-range repulsive interaction between the A and B monomers. The kinetics of phase separation has been discussed extensively in Chapter 1 and subsequent chapters. Let us now consider a situation where the different polymer chains are connected by a covalent bond at the chain ends, as shown in Figure 7.1. In this case, macrophase separation is impossible—what happens instead is microphase separation, where A-rich and B-rich domains are arrayed in a lamellar structure with a typical spatial period of the order 50 nm, which is comparable with the size of each polymer chain [1]. (Lamellar ordering was briefly discussed

FIGURE 7.1 AB diblock copolymers, where one end of the block (with black monomers) is connected to the other end of the block (with gray monomers) by a covalent bond.

in Section 4.4.1.) If the ratio of the molecular weights of the A and B blocks is different, other morphologies appear such as gyroids, hexagonal structures of cylindrical domains, and body-centered-cubic (BCC) structures of spherical domains. These four fundamental structures are displayed in Figures 7.2 and 7.3.

We mention here that there are two different kinds of gyroid structures—one is called a *single gyroid*, and the other is called a *double gyroid*. The single gyroid is an interconnected cylindrical domain structure such that three domains meet at each vertex. The surface of the domains is approximately represented by a minimal surface on which the mean curvature is zero. A double gyroid consists of two such interconnected cylindrical domains, which are untouched by each other. An example of the double gyroid is shown in Figure 7.2b, and its Bragg spots in wave-number space are shown in Figure 7.4. Because only the double gyroid appears in diblock copolymer melts, we shall refer to the "double gyroid" as the "gyroid" throughout this chapter.

These mesoscopic structures are common in soft-matter systems, which include polymers, liquid crystals, amphiphilic molecules, colloids, and so on [2]. For example, similar structures have been observed in water-surfactant mixtures [3]. It should be mentioned that a greater variety of three-dimensional periodic network structures (apart from the gyroid) appear in lyotropic liquid crystals and are called *lipidic cubic* structures [4].

(a)   (b)

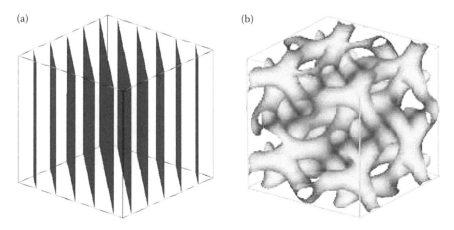

FIGURE 7.2 (a) Lamellar structure and (b) gyroid structure.

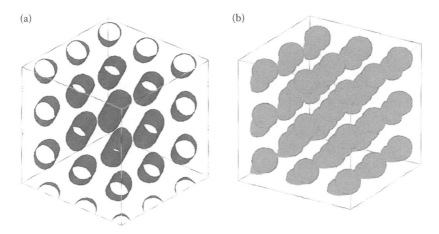

FIGURE 7.3   (a) Hexagonal structure of cylindrical domains and (b) BCC structure of spherical domains.

Although we shall consider regular periodic structures in this chapter, there are also non-periodic meso-structures that occur in ternary mixtures such as microemulsions and copolymer-homopolymer mixtures [5]. Some typical examples are micelles, sponge phases, and vesicles. A *micelle* is a closed monolayer of surfactant molecules formed between oil–water interfaces. When the volume fraction of oil is comparable with that of water, a bicontinuous structure appears, which consists of interconnected monolayers of amphiphilic molecules separating oil-rich and water-rich domains. The *sponge phase* is a random domain structure formed by a bilayer of amphiphilic

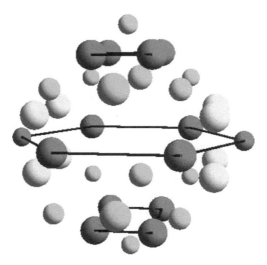

FIGURE 7.4   **(See color insert following page 180.)** Bragg spots of the gyroid structure. The size of the Bragg spot is proportional to the intensity.

molecules in water. Finally, a *vesicle* is a closed bilayer membrane of diblock copolymers in selective homopolymers (or of surfactants in water).

The existence of these curious mesoscopic structures provides us with a number of interesting problems. In contrast with macrophase separation, there are many equilibrium structures in the microphase-separated state. Therefore, the first problem is to predict and identify the stable structures and obtain a phase diagram. Recalling that similar mesophases have been observed in various soft-matter systems, it is important to elucidate the universal mechanism of mesophase formation. It should also be stressed that increasing the internal degrees of freedom of molecules (such as triblocks, tetrablocks, and star polymers) and changing the rigidity of the blocks produces a diversity of mesoscopic structures [6–10]. Furthermore, mixtures of different soft matter, for example, liquid crystals and colloids, also provide us with spectacular structures having unique properties [11].

The second problem is to investigate the ordering (or coarsening) process of these mesoscopic structures starting from the high-temperature disordered phase [12]. There are two main issues specific to the kinetics of microphase separation. The first issue is to investigate how an interconnected structure like the gyroid is self-organized. As discussed in Section 4.4.1, lattice Boltzmann simulations have been carried out for ternary amphiphilic fluids to clarify the self-assembly of the gyroid structure [13]. (See also Refs. [14] and [15], which are related papers dealing with gyroids.) The second issue concerns the slow dynamics of domain growth. A spontaneously formed structure is only ordered locally. Globally, the structure is disordered, having many defects and grain boundaries. The slow evolution of these grains has been studied both theoretically and numerically [16–18]. The alignment of these structures by external flows and fields has also been investigated theoretically and experimentally [19–25].

The third problem is to investigate the flow properties of mesoscopic structures formed in fluids. The rheology of soft matter has now been studied for many years [26–28] but our understanding remains incomplete. For instance, the de Gennes–Doi–Edwards theory for viscoelasticity of a uniform polymeric system, which deals with chain dynamics constrained by the surrounding chains, should be extended to inhomogeneous systems where mesoscopic structures are self-organized. A large-scale approach for modeling the deformation and viscoelasticity of mesoscopic structures has also been introduced for numerical simulations [29–33].

The fourth problem is the kinetics of the structural transitions between different mesophases. These transitions are achieved, for example, by changing the temperature. There are many studies of structural transitions in metallurgy [34]. However, it should be emphasized that structural transitions in soft matter have some unique features. Recall that the mesophases consist of domains where many molecules are self-assembled. Hence, shape change and the reconnection of domains are essential processes for structural transitions in soft matter. This should be contrasted with structural transitions in metals, where the exchange or displacement of atoms at lattice points is the origin of the transitions. Recently, experimental results for morphological transitions of mesophases have been accumulating [35–46]. However, there are relatively few theoretical investigations of the kinetics of structural transitions [47].

From the academic perspective, these are the reasons for our interest in soft matter. We stress that there are also other good reasons for studying soft matter, for example, technological applications. For example, liquid crystal displays have become ubiquitous due to lower consumption of energy. Furthermore, attempts have been made to produce mesoscopic porous carbon networks by using mesoscopic structures as a template [48]. It should also be emphasized that soft matter is a constituent of most biological systems. For example, lipid molecules (which are typical amphiphilic molecules) self-assemble to constitute biomembranes. Furthermore, micelles, vesicles, and various interconnected structures are contained or produced in living cells to maintain biological functionality [49–51]. In the context of the above-mentioned applications, it is clear that a systematic theory is needed for the structural kinetics of soft matter and morphological transitions under various external fields and non-equilibrium conditions.

This chapter is organized as follows. In Section 7.2, we briefly review the coarse-grained approach necessary for understanding the morphology and dynamics of mesoscopic structures. In the preceding chapters, an order-parameter—based description has frequently been used to study the kinetics of phase transitions in different contexts. A Ginzburg–Landau (GL) theory for microphase separation is described in Section 7.3, and we demonstrate its utility in extracting the universal features of the system. In Section 7.4, an interface version of the GL theory is shown to apply to the strong-segregation regime. The term *weak segregation* refers to a phase-separated state near the phase-separation temperature, where the concentration profile is sinusoidal. On the other hand, the term *strong segregation* refers to a strongly phase-separated state with a sharp interface between two different domains (see Section 1.4.1). In Section 7.5, we discuss the application of *self-consistent mean-field theory* to this problem. The time-dependent Ginzburg–Landau (TDGL) equation of motion is introduced in Section 7.6. The so-called *mode-expansion method* is described in Section 7.7. We shall apply it to study the formation of mesophases in Section 7.8, and the kinetics of morphological transitions in Section 7.9. A comparison with recent experiments will also be made. In Section 7.10, we present results from direct numerical simulations of the evolution equation. An analytical theory of linear elasticity is presented in Section 7.11. We conclude this chapter with a summary and discussion in Section 7.12.

7.2   COARSE-GRAINED APPROACH

Despite the fact that there are diverse soft-matter systems (for example, liquid crystals, copolymers, and surfactants), it appears that ordered structures have some common properties, independent of the microscopic details of the materials. This is partly because the structure is not at the atomic scale but at a mesoscopic scale. This scale difference causes, on one hand, a technical difficulty of dealing with the structures by taking full account of the microscopic degrees of freedom. On the other hand, it justifies the coarse-grained approach where molecular-scale degrees of freedom are averaged out. More importantly, the coarse-grained approach is most convenient to understand the universal features of mesoscopic structures.

The coarse-grained approach has a long history in polymer physics. In early work, Edwards introduced a Hamiltonian for flexible polymer chains [52]

$$H = \frac{1}{2} \sum_{\alpha} \sum_{i=1}^{n_\alpha} \int_0^{N_\alpha} \left(\frac{\partial \vec{c}_i^\alpha(\tau)}{\partial \tau} \right)^2 d\tau$$

$$+ \frac{1}{2} \sum_{\alpha} \sum_{\beta} \sum_{i=1}^{n_\alpha} \sum_{j=1}^{n_\beta} v_{\alpha\beta} \int_0^{N_\alpha} d\tau \int_0^{N_\beta} d\tau' \delta \left[\vec{c}_i^\alpha(\tau) - \vec{c}_j^\beta(\tau') \right], \quad (7.1)$$

where $\vec{c}_i^\alpha(\tau)$ represents the conformation of the ith chain of the αth species, N_α is the polymerization index of the αth species, and n_α is the corresponding number of molecules. The coefficient $v_{\alpha\beta}$ is the parameter for the short-range monomer-monomer interactions.

The Hamiltonian in Equation 7.1 ignores the microscopic chemical details of monomers by regarding each monomer as a bead, and then using the continuum representation of the molecular conformation as well as a delta-function for the monomer interaction. It is well known that this type of coarse-grained model, together with renormalization-group theory, was useful to develop the statistical theory of dilute and semi-dilute polymer solutions [53].

In order to apply the Edwards Hamiltonian to the formation of mesoscopic structures, and to develop a dynamical theory, one needs further reduction of the degrees of freedom. This is because Equation 7.1 still contains a lot of conformational information about each chain. In the next two sections, we shall describe two typical coarse-grained theories.

7.3   GINZBURG–LANDAU THEORY

In this section, we formulate a coarse-grained theory in terms of local order parameters, such as the GL free-energy functional for macrophase separation in Equation 1.63 [54] (or the Landau–de Gennes expansion of the free energy for liquid crystals). It will be shown below that this approach indeed makes it possible to represent both macrophase separation and microphase separation in a unified way.

The reduction of the degrees of freedom from the chain conformation to the local concentration can be done as follows. First, we define a partition function that has a constraint such that

$$Z = \int d\{\vec{c}_i^\alpha\} \Pi_{\vec{r},\alpha} \delta \left[\phi_\alpha(\vec{r}) - \hat{\phi}_\alpha(\vec{r}) \right] \exp(-H). \quad (7.2)$$

(Throughout this chapter, we use the unit of energy $k_B T = 1$.) In Equation 7.2, $\hat{\phi}_\alpha$ is the local concentration of the α-th monomer, defined by

$$\hat{\phi}_\alpha(\vec{r}) = \sum_i \int_0^{N_\alpha} d\tau \, \delta \left[\vec{r} - \vec{c}_i^\alpha(\tau) \right], \quad (7.3)$$

whereas $\phi_\alpha(\vec{r})$ is an arbitrary smooth function. The "hat" indicates that this quantity is a functional of the stochastic variable $\{\hat{c}_i^\alpha\}$. Thus, if we evaluate

$$F\{\phi\} = -\ln Z, \tag{7.4}$$

the free energy can be obtained as a functional of the concentration profile $\phi_\alpha(\vec{r})$. Unfortunately, the integral in Equation 7.2 cannot be carried out exactly and we have to often employ the *mean-field approximation*. It should be noted that the mean-field approximation is justified for concentrated polymer solutions or polymer melts, where the concentration fluctuations are not strongly correlated.

Using the above arguments, Leibler derived the free-energy functional for diblock copolymers as [55]

$$F\{\phi\} = \int d\vec{r} \left[\frac{K}{2}(\nabla\phi)^2 + \frac{1}{2}(\nabla^2\phi)^2 + \frac{g_2}{2}\phi^2 + \frac{g_3}{3}\phi^3 + \frac{g_4}{4}\phi^4 \right], \tag{7.5}$$

where $\phi = \phi_A - \phi_B$ and K, g_2, g_3, and $g_4 > 0$ are parameters (cf. Equation 4.66).

One of the characteristic features of AB diblock copolymer melts is *osmotic incompressibility*:

$$\lim_{\vec{q}\to 0} \langle \phi_{\vec{q}}\,\phi_{-\vec{q}} \rangle = 0, \tag{7.6}$$

where $\phi_{\vec{q}}$ is the Fourier transform of ϕ:

$$\phi_{\vec{q}} = \int d\vec{r}\, e^{i\vec{q}\cdot\vec{r}}\phi(\vec{r}). \tag{7.7}$$

The free-energy functional that satisfies the osmotic-incompressibility condition in Equation 7.6 within the mean-field approximation is given by [56,57]

$$F\{\phi\} = \int d\vec{r} \left[\frac{K}{2}(\nabla\phi)^2 + f(\phi) \right]$$
$$+ \frac{\alpha}{2} \int d\vec{r} \int d\vec{r'} G(\vec{r},\vec{r'}) [\phi(\vec{r}) - \bar{\phi}] [\phi(\vec{r'}) - \bar{\phi}], \tag{7.8}$$

where $\bar{\phi}$ denotes the spatial average of ϕ, and α is a positive constant. The Green's function $G(\vec{r},\vec{r'})$ is defined through

$$-\nabla^2 G(\vec{r} - \vec{r'}) = \delta(\vec{r} - \vec{r'}). \tag{7.9}$$

The local part $f(\phi)$ is an algebraic function of ϕ. One example is the usual ϕ^4-energy (cf. Equation 1.63):

$$f(\phi) = -\frac{\tau}{2}\phi^2 + \frac{g}{4}\phi^4, \tag{7.10}$$

where $\tau, g > 0$ in the microphase-separated state. These parameters are assumed to be independent of the block ratio. This requirement comes from the fact that the interfacial tension should be independent of the block ratio at low temperatures. Another choice of $f(\phi)$ is the Flory–Huggins free energy:

$$f(\phi) = \chi\phi(1 - \phi) + \phi\ln\phi + (1 - \phi)\ln(1 - \phi), \tag{7.11}$$

where χ is the Flory–Huggins parameter. For the purpose of exploring universal properties, we are not concerned with the specific form of f [58]. It is sufficient that it guarantees a saturation of the concentration ϕ at some appropriate value in each state.

The molecular-weight dependence of the constant α can be determined by a mean-field calculation. However, it is easier to determine it using dimensional analysis. Notice that the Green's function $G(\vec{r}, \vec{r}')$ defined by Equation 7.9 has a dimension of L^{2-d}. By comparing the long-range term with the gradient term in Equation 7.8, one sees that the constant α should have a dimension L^{-4}. Because the radius of gyration of a chain $R_G \sim N^{1/2}$ is the only relevant material parameter, one may conclude that

$$\alpha \propto R_G^{-4} \propto N^{-2}. \tag{7.12}$$

The last term of Equation 7.8, which has a Coulomb-type long-range interaction, arises from the osmotic-incompressibility condition in Equation 7.6. If this term is absent, Equation 7.8 is the usual GL free energy for macrophase separation. The competition of the long-range repulsive interaction and the short-range attractive interaction (the first term of Equation 7.8) gives rise to mesoscopic periodic structures.

One may argue that a similar long-range interaction (or q^{-2}-divergence in Fourier space) appears in the higher-order vertices in the expansion of the free energy in powers of ϕ. However, the long-range interaction in Equation 7.8 is crucial because it (together with the short-range interaction) gives rise to an optimal wave-number. Therefore, one may employ a local approximation for the higher vertices by equating all the wave-numbers to the optimal one.

As another example of competition between short-range and long-range interactions, we describe the free energy in the nematic-smectic transition of liquid crystals. In this context, an entropic non-local interaction is found to appear as a restriction of rotational degrees of freedom due to the excluded volume effect of rod molecules. Assuming that the molecular density ρ is periodic in the z-direction,

$$\rho(z) = \rho_0 \left[1 + 2\sigma_q \cos(qz)\right], \tag{7.13}$$

the non-local part of the free energy assumes the following form [59]:

$$F_{\text{nonlocal}} = \int dq \left[(r_0 q)^2 + C\frac{\sin(qL)}{qL}\right]\sigma_q^2. \tag{7.14}$$

Here, r_0 is the characteristic length of the short-range attractive interaction, C is a positive constant, and L denotes the length of the rod molecules. A local part is added to Equation 7.14, which comes from the translational entropy of the molecules. It is possible that Equation 7.14 has a minimum at a finite value of the wave-number, and hence a layered structure can be stable in equilibrium.

A similar mechanism governs domain formation in thin magnetic films. It is well known that a stripe structure, or a hexagonal domain structure, can be observed in magnetic thin films because the diamagnetic field caused by the local magnetization produces a long-range interaction that competes with the short-range ferromagnetic interaction [60].

We emphasize that the free energies in Equation 7.8 and Equation 7.5 are universal in the sense that they are independent of the microscopic details of the system. In particular, the free energy in Equation 7.8 is very useful because it predicts (without being confined to soft matter) that a stable periodic structure emerges when there is a competition between *short-range attractive* and *long-range repulsive* interactions. This issue will be discussed further in Section 7.12.

7.4   FREE ENERGY IN TERMS OF THE INTERFACIAL TENSION

In the original form of the free energy in Equation 7.8, the coefficient K depends on the block ratio f [56], which is the ratio of the molecular weight of one of the blocks to the total molecular weight of a chain. This makes the interfacial tension dependent on the block ratio. This is clearly invalid in the strong-segregation regime because the interfacial width is much smaller than the spatial period of the mesoscopic structures, and hence the interfacial energy should be independent of the molecular weight and the block ratio. This is the reason why the free energy in Equation 7.8 is restricted to the weak-segregation regime near the order-disorder transition temperature.

However, it should be pointed out that the free energy in Equation 7.8 is meaningful even for the strong-segregation limit if we write it as follows:

$$F_{\mathrm{I}} = \sigma \int dA + \frac{\alpha}{2} \int d\vec{r} \int d\vec{r'} G\left(\vec{r}, \vec{r'}\right) \left[\phi(\vec{r}) - \bar{\phi}\right]\left[\phi(\vec{r'}) - \bar{\phi}\right], \qquad (7.15)$$

where σ is the interfacial tension, and the integration runs over all interfaces. The free energy in Equation 7.15 is obtained from Equation 7.8 by noting that the short-range interaction term and $f(\phi)$ determine the interfacial energy and the bulk constant energy inside domains in the strong-segregation limit. Thus, if we use the correct interfacial energy, Equation 7.15 can be applied to the strong-segregation case.

As an example of this, let us derive an expression for the spatial period of an equilibrium lamellar structure. Consider a lamellar structure with period ℓ in a system with linear dimension L. There are L/ℓ flat interfaces so the first term in Equation 7.15 yields

$$\sigma L^2 \frac{L}{\ell} = \sigma \frac{L^3}{\ell}. \qquad (7.16)$$

The second term in Equation 7.15 can be estimated by dimensional considerations as

$$\frac{\alpha}{2\ell} \ell^3 L^3 = \frac{\alpha}{2} \ell^2 L^3, \qquad (7.17)$$

where we have used $G \propto \ell^{-1}$ in $d = 3$. The factors ℓ^3 and L^3 come from the two spatial integrals, with the L^3-factor ensuring the extensivity of the free energy. We combine Equations 7.16 and 7.17, and minimize the expression with respect to ℓ to obtain [56]

$$\ell \propto N^{2/3}, \tag{7.18}$$

where we have used Equation 7.12. The prediction in Equation 7.18 is consistent with experiments [61].

Next, we provide an alternative argument for the derivation of Equation 7.18. This will further clarify the meaning of the long-range interaction. In the strongly segregated microphase-separated lamellar state, each chain is elongated normal to the interface to make the density uniform everywhere. This elongation causes a decrease of conformational entropy. For a Gaussian chain, the probability to find the end-point of a chain at $\vec{r}$ when the other end is fixed at $\vec{r} = 0$ is

$$Q(\vec{r}, N) = \exp(-S) \propto \exp\left(-\frac{3r^2}{2N}\right). \tag{7.19}$$

Assuming that one end is at the interface, the entropy decrease per chain is estimated for $N \gg 1$ as

$$S = -\frac{3\ell^2}{8N}. \tag{7.20}$$

From Equations 7.16 and 7.20, we have the free energy per chain

$$F = \frac{\sigma N}{\rho_0 \ell} + \frac{3\ell^2}{8N}, \tag{7.21}$$

where ρ_0 is the monomer number density. We can minimize this expression with respect to ℓ to obtain essentially the same result as Equation 7.18. (See also Ref. [62].)

Comparing Equations 7.17 and 7.20, we note that the long-range interaction has an entropic origin. This is an important fact in soft-matter systems, where each molecule has large internal degrees of freedom. Then, the entropic force often dominates the primitive interaction (for example, van der Waals interaction) between monomers and plays a decisive role in the formation of mesoscopic structures. Because the entropic force or potential is insensitive to the atomic-scale details of the system, we expect a corresponding universality of the statistical properties of mesophases.

7.5   SELF-CONSISTENT MEAN-FIELD THEORY

In Section 7.3, we have described a coarse-graining procedure in terms of the local concentration field. In the current section, a more refined theory is presented, which is termed the *self-consistent mean-field theory*. In this theory, information about the chain conformation is retained to some extent by keeping the spirit of the Edwards

Hamiltonian in Equation 7.1. The effect of the surrounding chains is treated in a mean-field approximation. The self-consistent mean-field theory was introduced by Helfand [63] for diblock copolymer melts. After that, the theory has been developed systematically to enable large-scale numerical simulations [64–66]. It should also be mentioned that the theory has been extended to investigate the dynamical properties of polymeric systems [66–68].

The self-consistent mean-field theory for AB diblock copolymer melts is formulated as follows [69]. Assume that each chain has the polymerization index N, and the block for $0 < \tau < Nf$ consists of the A monomers, whereas the block for $Nf < \tau < N$ consists of the B monomers. The Edwards Hamiltonian for a diblock chain can be written as

$$H_{DB}(0, N) = \int_0^N \left[\frac{1}{2} \left(\frac{\partial \vec{c}_i(\tau)}{\partial \tau} \right)^2 + \omega(\vec{c}_i, \tau) \right] d\tau, \qquad (7.22)$$

where the mean field is defined by $\omega(\vec{c}_i, \tau) = \omega_A(\vec{r})$ $(0 < \tau < Nf)$ and $\omega(\vec{c}_i, \tau) = \omega_B(\vec{r})$ $(Nf < \tau < N)$. We introduce the partition functions $q(\vec{r}, \tau)$ and $q^\dagger(\vec{r}, \tau)$ under the constraint that the conformation $\vec{c}_i(\tau)$ (the monomer at the contour position τ) of the ith chain is located at $\vec{r}$:

$$q(\vec{r}, \tau) = \int d\{\vec{c}_i\} \delta \left[\vec{r} - \vec{c}_i(\tau) \right] \exp[-H_{DB}(0, \tau)], \qquad (7.23)$$

$$q^\dagger(\vec{r}, \tau) = \int d\{\vec{c}_i\} \delta \left[\vec{r} - \vec{c}_i(\tau) \right] \exp[-H_{DB}(\tau, N)]. \qquad (7.24)$$

It is readily proved that $q(\vec{r}, \tau)$ satisfies

$$\frac{\partial q(\vec{r}, \tau)}{\partial \tau} = \frac{1}{2} \nabla^2 q(\vec{r}, \tau) - \omega(\vec{r}, \tau) q(\vec{r}, \tau), \qquad (7.25)$$

with the "initial" condition $q(\vec{r}, 0) = 1$. Similarly, $q^\dagger(\vec{r}, \tau)$ satisfies Equation 7.25 with a minus sign on the right-hand side and should be solved with the condition $q^\dagger(\vec{r}, N) = 1$. The local concentration of A monomers is given by

$$\phi_A(\vec{r}) = \frac{n}{Q} \int_0^{fN} d\tau q(\vec{r}, \tau) q^\dagger(\vec{r}, \tau). \qquad (7.26)$$

Here, n is the total number of chains, and Q is the partition function without any constraint:

$$Q = \int d\vec{r} q(\vec{r}, \tau) q^\dagger(\vec{r}, \tau). \qquad (7.27)$$

The local concentration $\phi_B(\vec{r})$ of B monomers can be defined similarly to Equation 7.26 with the integral domain being $fN < \tau < N$. In terms of the local concentration, the mean field can be represented by

$$\omega_A(\vec{r}) = \chi \phi_B(\vec{r}) + P(\vec{r}), \qquad (7.28)$$

and similarly for ω_B. The constant P is determined by the requirement that the incompressibility condition $\phi_A(\vec{r}) + \phi_B(\vec{r}) = 1$ is satisfied.

We start the iterative procedure with an arbitrary function for the mean field in Equation 7.22. Then, a new mean field is generated via the process given by Equations 7.26 and 7.28. By substituting the obtained mean field into Equation 7.22, the procedure is repeated until the mean field converges to a fixed function. By using the asymptotically converged functions, the equilibrium free energy per polymer chain is obtained as

$$F = -n \ln \frac{Q}{n} + \int d\vec{r} \left[\chi \phi_A(\vec{r}) \phi_B(\vec{r}) - \omega_A(\vec{r}) \phi_A(\vec{r}) - \omega_B(\vec{r}) \phi_B(\vec{r}) \right]. \quad (7.29)$$

Now we compare the self-consistent mean-field theory and the GL theory in terms of the local concentration in Section 7.3. The self-consistent theory can be applied to both strong-segregation and weak-segregation regimes. However, the free energies in Equations 7.8 and 7.5 are not quantitatively accurate in the case of strong segregation. (In Section 7.4, we showed that Equation 7.8 can be applied to the strong-segregation regime if it consists of the interfacial energy and the long-range interaction [56].)

The disadvantageous aspect of self-consistent field theory is that it is not easy to directly extract the universal properties of the mesoscopic structures because they involve the microscopic polymer conformations. Furthermore, a completely analytical theory is not possible because the iteration process from Equation 7.22 to Equation 7.28 relies heavily on numerical simulations. On the other hand, the GL theory for copolymer melts should be extended to copolymer-homopolymer mixtures and copolymer solutions. This aspect of the GL theory is currently incomplete.

7.6 TIME-EVOLUTION EQUATION

Let us now derive the evolution equation by using the free energy in Equation 7.8 to study the dynamics of mesoscopic structures. We will follow the derivation that resulted in the Cahn–Hilliard (CH) equation in Section 1.5.1. Because ϕ is a conserved quantity, it must satisfy the continuity equation

$$\frac{\partial}{\partial t} \phi(\vec{r}, t) = -\vec{\nabla} \cdot \vec{J}(\vec{r}, t). \quad (7.30)$$

The current vector $\vec{J}$ should be determined such that Equation 7.30 is consistent with thermodynamic principles, that is, the free energy should be a decreasing function of time in a closed system in contact with a heat bath:

$$\begin{aligned}
0 > \frac{dF}{dt} &= \int d\vec{r} \frac{\partial \phi}{\partial t} \frac{\delta F}{\delta \phi} \\
&= -\int d\vec{r} \left(\vec{\nabla} \cdot \vec{J} \right) \frac{\delta F}{\delta \phi} \\
&= \int d\vec{r} \vec{J} \cdot \vec{\nabla} \left(\frac{\delta F}{\delta \phi} \right).
\end{aligned} \quad (7.31)$$

In the last step, we have used the fact that there is no current at the system boundary. In order to satisfy the inequality in Equation 7.31, we may choose

$$\vec{J}(\vec{r}) = - \int d\vec{r}' L(\vec{r}, \vec{r}') \vec{\nabla}' \left(\frac{\delta F}{\delta \phi(\vec{r}')} \right). \tag{7.32}$$

The positive function $L(\vec{r}, \vec{r}')$ corresponds with the transport coefficient, which was taken to be a delta-function in Equation 1.75. Equation 7.32 accounts for the fact that the relation between the thermodynamic force $\delta F/\delta \phi$ and the current is generally non-local because a polymer chain is extended in a finite volume. Substituting Equation 7.32 into the continuity equation 7.30, we obtain

$$\frac{\partial}{\partial t} \phi(\vec{r}, t) = \vec{\nabla} \int d\vec{r}' L(\vec{r}, \vec{r}') \vec{\nabla}' \left(\frac{\delta F}{\delta \phi(\vec{r}')} \right). \tag{7.33}$$

If we ignore the non-locality and put $L(\vec{r}, \vec{r}') = \delta(\vec{r} - \vec{r}')$, we obtain by using the free energy in Equation 7.8

$$\frac{\partial}{\partial t} \phi(\vec{r}, t) = \nabla^2 \left(\frac{\delta F}{\delta \phi} \right)$$
$$= \nabla^2 \left[\left(-\nabla^2 \phi - \tau \phi + g \phi^3 \right) - \alpha \left(\phi - \bar{\phi} \right) \right]. \tag{7.34}$$

Here, we have set $K = 1$, ignoring the block-ratio dependence in the free energy in Equation 7.8. (Notice that the most relevant dependence has been taken into account through $\bar{\phi}$.) Equation 7.34 is perhaps the simplest equation that produces a mesoscopic structure. It should be emphasized that the long-range interaction in the free energy in Equation 7.8 has disappeared in Equation 7.34, but gives rise to the *local* linear damping term. Furthermore, Equation 7.34 enables us to study both macrophase separation and microphase separation in a unified way by putting either $\alpha = 0$ or $\alpha \neq 0$. In Ref. [71], Equation 7.34 was introduced via general arguments for microphase separation, without using Equation 7.8.

The expression for $L(\vec{r}, \vec{r}')$ should be fixed to make the dynamic theory complete. There are several studies that derive the non-local mobility $L(\vec{r}, \vec{r}')$ starting from the motion of polymer chains [72,73]. However, the mobility is expected to be a functional of the local concentration in the following sense. The evolution of domains is due to the motion of the interacting polymer chains, but the chain motion itself is constrained by the potential due to the domain structure and hence influenced by the domain deformation. This implies that there is a dynamic correlation between the microscopic chain motion and the mesoscopic domain dynamics. This remains an unsolved problem inherent to soft matter, where hierarchical non-uniform structures are self-organized from the molecular scale to the mesoscopic scale.

Finally, we discuss the effects of hydrodynamic flows on the ordering process (see Section 1.5.2). It has been shown by simulations that hydrodynamic effects are highly relevant for the kinetics of morphological ordering, for example, flows can prevent trapping in metastable states during the coarsening process [74].

We follow the mode-coupling theory for critical fluids [75], as discussed in Section 1.5.2. The local velocity of monomers obeys [cf. Equation 1.187]

$$\eta \nabla^2 \vec{v} - \vec{\nabla} p - \phi \vec{\nabla} \left(\frac{\delta F}{\delta \phi} \right) = 0, \tag{7.35}$$

with the incompressibility condition $\vec{\nabla} \cdot \vec{v} = 0$. The pressure p is a Lagrange multiplier to be determined by this condition. In Equation 7.35, we have ignored the time-derivative and the nonlinear term in the velocity, assuming that the relaxation of the velocity field is much faster than that of the concentration field. Equation 7.34 should also be modified as [cf. Equation 1.185]

$$\frac{\partial \phi}{\partial t} + \vec{\nabla} \cdot (\vec{v}\phi) = D \left[\nabla^2 \left(-\nabla^2 \phi - \tau\phi + g\phi^3 \right) - \alpha \left(\phi - \phi^- \right) \right], \tag{7.36}$$

where we have explicitly introduced the constant mobility D for later convenience. The local velocity can be obtained from Equation 7.35 as

$$v_\alpha(\vec{r}, t) = \frac{1}{\eta} \int d\vec{r}' \sum_\beta T_{\alpha\beta}(\vec{r}, \vec{r}')(\nabla_\beta \phi) \frac{\delta F}{\delta \phi \left(\vec{r}' \right)}, \tag{7.37}$$

where $T_{\alpha\beta}(\vec{r}, \vec{r}')$ is the Oseen tensor, whose Fourier transform is given by

$$T_{\alpha\beta}(\vec{q}) = \frac{1}{q^2} \left(\delta_{\alpha\beta} - \frac{q_\alpha q_\beta}{q^2} \right). \tag{7.38}$$

It is readily shown that the local part $f(\phi)$ of the free energy does not contribute to Equation 7.37. Substituting Equation 7.37 into Equation 7.36, we obtain a closed evolution equation for ϕ.

Next, let us estimate the order of the relevant terms in Equation 7.36. For simplicity, we consider a lamellar structure with $\bar{\phi} = 0$, and assume that the system is near the order-disorder transition temperature. The relaxation rate of ϕ arising from the hydrodynamic interaction is estimated as

$$\Omega_h \simeq \frac{\langle \phi^2 \rangle}{\eta \ell^2}, \tag{7.39}$$

where ℓ is the characteristic length scale (equilibrium spatial period). On the other hand, the linear part of the right-hand side is estimated as [76,77]

$$\Omega_d \simeq \frac{D}{\ell^4}. \tag{7.40}$$

It is well known that

$$\eta D \simeq R_G^2 \frac{N}{N_e}, \tag{7.41}$$

where N is the molecular weight of each polymer chain, and N_e is the molecular weight between entanglement points. From Equations 7.39 through 7.41, we obtain

$$\frac{\Omega_h}{\Omega_d} \simeq \frac{N_e \langle \phi^2 \rangle \ell^2}{N R_G^2}. \tag{7.42}$$

In the limiting case of Rouse dynamics, the factor N_e/N is absent. If we employ $\langle \phi^2 \rangle \simeq \ell^{-2}$, which is obtained from the free energy in Equation 7.8, we note that $\Omega_h/\Omega_d \simeq O(1)$. Therefore, both hydrodynamic effects and ordinary diffusion are equally important in this case. In the case of macrophase separation, we have a different situation, that is, $\langle \phi^2 \rangle \simeq \xi_b^{-2}$ where ξ_b is the correlation length of concentration fluctuations, and the domain size ℓ increases indefinitely. Then, Equation 7.42 implies that $\Omega_h/\Omega_d \gg 1$, and hence the hydrodynamic effects are dominant as the domains become larger and larger (see Section 1.5.2).

7.7 MODE EXPANSION

Systematic computer studies of Equation 7.34 have not been carried out for structural transitions of 3-d mesophases, including the gyroid. The main reason is that the spatial period of a gyroid is slightly different (several percent) from the period of other fundamental structures, for example, a lamellar structure. In simulations, it is necessary to provide a system size commensurate with the spatial period to avoid a finite-size effect in a system with periodic boundary conditions. Thus, if the spatial period of the initial structure is different from that of the final structure, one needs to vary the system size during the evolution. A possible method of avoiding this technical difficulty is to carry out simulations for a sufficiently large system compared with the period. As far as the formation of a gyroid starting from the high-temperature disordered state is concerned, simulations with a fairly large system size have been done recently [13]. However, no comparable results are available for the kinetics of morphological transitions involving interconnected structures.

In this chapter, in order to avoid finite-size effects and boundary effects, we employ the mode-expansion method for the local concentration field $\phi(\vec{r}, t)$ [47,78,79]:

$$\phi(\vec{r}, t) = \bar{\phi} + \left[\sum_{n=1}^{N} a_n(t) e^{i\vec{q}_n \cdot \vec{r}} + \text{c.c.} \right], \tag{7.43}$$

where the vectors $\vec{q}_n$ are the fundamental reciprocal lattice vectors and some higher combinations thereof. The amplitudes a_n are approximated to be real. Their evolution equations can be obtained in a closed form by substituting Equation 7.43 into Equation 7.34 and ignoring the higher harmonics. We do not present this set of equations here because they are very lengthy. These are available in Ref. [79]. As mentioned above, the wave-number $Q = |\vec{q}_n|$ generally changes during the structural transitions. Its time-dependence is given by Equation 7.56 below.

The mode-expansion method is restricted to the weak-segregation regime because only a finite number of modes are considered in the Fourier series expansion. It should be noted, however, that the mode-expansion method is realistic because a gyroid is formed by a temperature quench only in the weak-segregation regime.

Here, we present a simple example of how to solve the amplitude equations. In the case of a lamellar structure with $\bar{\phi} = 0$, the expansion in Equation 7.43 takes the form

$$\phi(x, t) = A(t) \cos(qx). \tag{7.44}$$

Substituting this into the right-hand side of Equation 7.34, we obtain

$$\frac{dA}{dt} = (-q^4 + \tau q^2 - \alpha)A + 3gq^2 A^3. \tag{7.45}$$

The equilibrium solution is given by

$$-q^2[\Gamma(q) + 3gA^2]A = 0, \tag{7.46}$$

where

$$\Gamma(q) = q^2 - \tau + \frac{\alpha}{q^2}. \tag{7.47}$$

The magnitude of q is determined by the minimization of the free energy, which is equivalent to

$$\frac{\partial \Gamma(q)}{\partial q} = 0, \tag{7.48}$$

so that

$$q^2 = \sqrt{\alpha}. \tag{7.49}$$

For this value of q,

$$\Gamma(q) = 2\sqrt{\alpha} - \tau. \tag{7.50}$$

Therefore, microphase separation takes place for $\tau > 2\sqrt{\alpha} = \tau_c$. The spatial period of the lamellar structure is approximately given by $\ell = 2\pi/q$ for $\tau \simeq \tau_c$. The amplitude A of the concentration profile is obtained from Equation 7.46 as

$$A^2 = -\frac{\Gamma(q)}{3g}. \tag{7.51}$$

Apart from the truncation mentioned above, the expansion in Equation 7.43 employs the following further approximations. First, the spatial dependence of the amplitudes is not considered. Therefore, the current theory is unable to study an interface separating two different structures. We shall return to this problem in Section 7.10, where direct numerical simulations of Equation 7.34 will be described. Second, we omit the phase degree of freedom in a periodic structure because the amplitudes are assumed to be real. This means that weak deformations around equilibrium structures are not considered. The elasticity of mesoscopic structures for long-wavelength deformations will be studied in Section 7.11.

It is known that a double gyroid is approximately represented as [80]

$$0 = 8(1 - s)\big[\sin(2x)\sin z \cos y + \sin(2y)\sin x \cos z + \sin(2z)\sin y \cos x\big]$$
$$- 4s\big[\cos(2x)\cos(2y) + \cos(2y)\cos(2z) + \cos(2z)\cos 2x\big] - u, \quad (7.52)$$

where s and u are free parameters. Noting that Equation 7.52 has 12 wave-vectors, the expansion in Equation 7.43 takes the following form

$$\phi(\vec{r}) = \bar{\phi} + \left[-\phi_a \sum_{j=1,4,5,8,9,12} e^{i\vec{q}_j \cdot \vec{r}} + \phi_a \sum_{j=2,3,6,7,10,11} e^{i\vec{q}_j \cdot \vec{r}} + \phi_b \sum_{j=13}^{18} e^{i\vec{q}_j \cdot \vec{r}} + \text{c.c.} \right].$$

$$(7.53)$$

The constants $\bar{\phi}$ and ϕ_a are determined as a linear combination of s and u in Equation 7.52. The modes with the amplitudes ϕ_b are necessary to take account of the BCC structure. The reciprocal vectors $\vec{q}_j$ in Equation 7.53 are given by

$$\vec{q}_1 = (2, -1, 1), \quad \vec{q}_2 = (-2, 1, 1), \quad \vec{q}_3 = (-2, -1, 1),$$
$$\vec{q}_4 = (2, 1, 1), \quad \vec{q}_5 = (-1, -2, 1), \quad \vec{q}_6 = (1, -2, 1),$$
$$\vec{q}_7 = (-1, 2, 1), \quad \vec{q}_8 = (1, 2, 1), \quad \vec{q}_9 = (1, -1, -2), \quad (7.54)$$
$$\vec{q}_{10} = (1, 1, -2), \quad \vec{q}_{11} = (-1, 1, -2), \quad \vec{q}_{12} = (-1, -1, -2),$$
$$\vec{q}_{13} = (2, 2, 0), \quad \vec{q}_{14} = (2, -2, 0), \quad \vec{q}_{15} = (0, 2, 2),$$
$$\vec{q}_{16} = (0, -2, 2), \quad \vec{q}_{17} = (2, 0, 2), \quad \vec{q}_{18} = (-2, 0, 2).$$

For convenience, the actual wave-vectors are defined by multiplying the factor $q_c/\sqrt{6}$ with $q_c^2 = \sqrt{\alpha}$ so that the magnitudes are normalized as $|\vec{q}_j| = q_c$ ($j = 1, \ldots, 12$), and $|\vec{q}_j| = 2q_c/\sqrt{3}$ ($j = 13, \ldots, 18$). In terms of these 18 reciprocal lattice vectors, lamellar, hexagonal, BCC, Fddd, and perforated lamellar structures (the latter two will be explained later) are represented. The Bragg spots for the gyroid in Figure 7.4 are given by $\pm\vec{q}_i$ ($i = 1, \ldots, 12$).

The free energy in Equation 7.8 can also be written in terms of the amplitudes as

$$F = F(\{a_n\}, Q). \quad (7.55)$$

Substituting the equilibrium values of the amplitudes into Equation 7.55, one can compare the free energies of various structures to obtain the phase diagram, which will be shown in the next section.

As mentioned above, the evolution of the period or the wave-vectors should be considered during the structural transitions. Here we simply impose the phenomenological evolution [79]:

$$\frac{dQ}{dt} = -\frac{\partial F(\{a_n\}, Q)}{\partial Q}. \tag{7.56}$$

An alternative way to determine the evolution of the period is to evaluate the first moment $\int dq\, q S_{\bar{q}}(t) / \int dq\, S_{\bar{q}}(t)$, where $S_{\bar{q}}(t) = \langle \phi_{\bar{q}}(t)\, \phi_{-\bar{q}}(t) \rangle$.

7.8 FORMATION OF EQUILIBRIUM STRUCTURES

We can study the formation of mesoscopic structures by solving the coupled set of amplitude equations. The initial condition is the high-temperature disordered state with $a_n = 0 +$ small random numbers for all n. Because all the transitions here are first order, we have performed simulations by adding random noises to the amplitude equations. The parameters are fixed as $\alpha = g = 1$ in Equation 7.34, unless stated otherwise.

An example of mesophase formation is shown in Figure 7.5. This corresponds with the formation of a hexagonal structure. It is evident that the system reaches a hexagonal structure asymptotically but is trapped transiently in a metastable BCC structure [79]. Figure 7.6 displays another example where a lamellar structure is formed. It is seen that a perforated lamellar structure appears as a metastable state before the system reaches the final lamellar structure [81]. The perforated lamellar structure and its Bragg spots are shown in Figure 7.7. These results imply that one of the characteristic features of morphological transitions in mesoscopic structures is the appearance of intermediate structures in the kinetics of these transitions.

The equilibrium solutions of the coupled amplitude equations, and the evaluation of the free energy for each solution, give us the phase diagram. In order to obtain a precise result, we have increased the number of modes in the mode expansion. Actually, 33 amplitude equations are considered in deriving the phase diagram. By comparing the equilibrium free energies for stable time-independent solutions of the amplitude

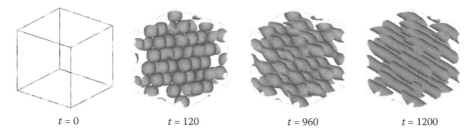

$t = 0$ $t = 120$ $t = 960$ $t = 1200$

FIGURE 7.5 Formation of a hexagonal structure via a BCC structure starting from a uniform state. The equi-surface with $\phi = -0.05$ is drawn and, therefore, no domains are represented at $t = 0$. The parameters are chosen as $\bar{\phi} = -0.15$ and $\tau = 2.2$. (From Yamada, K. et al., *Macromolecules*, 37, 5762, 2004. With permission.)

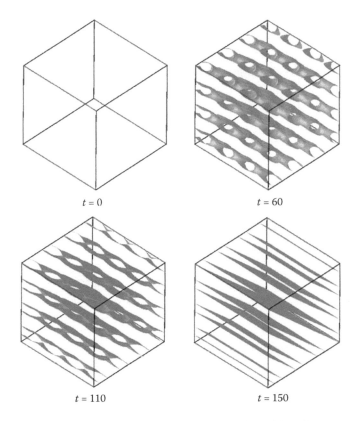

FIGURE 7.6   Formation of a lamellar structure via a perforated lamellar structure starting from a uniform state. The equi-surface with $\phi = -0.15$ is drawn. The parameters are chosen as $\bar{\phi} = 0.1$ and $\tau = 2.3$.

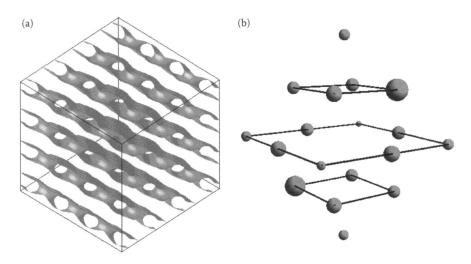

FIGURE 7.7   (a) Perforated lamellar structure and (b) Bragg spots.

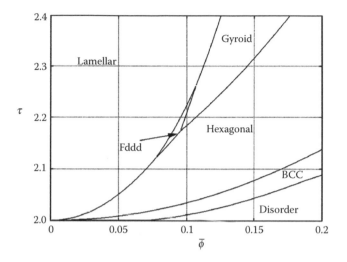

FIGURE 7.8 Phase diagram in the $(\tau, \bar{\phi})$-plane. "Disorder" indicates the high-temperature uniform phase. (From Yamada, K. et al., *J. Phys.: Condens. Matter*, 18, L421, 2006. With permission.)

equations, the phase diagram displayed in Figure 7.8 has been obtained [81]. It should be noted that there is a parameter regime where the gyroid has the lowest free energy. This is consistent with results obtained by earlier studies based on the Brazovskii free energy in Equation 7.5 [82] or by the self-consistent mean-field theory [64]. A more remarkable property in Figure 7.8 is that the Fddd structure exists as an equilibrium structure in AB diblock copolymers. This is consistent with a result obtained recently using self-consistent mean-field theory [83]. The 3-*d* picture and the Bragg spots of the Fddd structure are shown in Figure 7.9. Note that the Fddd structure has a uniaxial symmetry, which is in contrast with the gyroid structure with a cubic symmetry. We

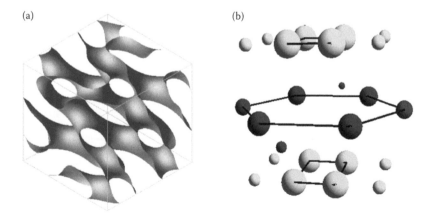

FIGURE 7.9 (a) Fddd structure and (b) Bragg spots. (From Yamada, K. et al., *J. Phys.: Condens. Matter*, 18, L421, 2006. With permission.)

FIGURE 7.10 (See color insert following page 180.) Three successive layers of the perforated lamellar structure. The location of holes is indicated by the dark areas.

shall discuss the Fddd structure again in the next section on the kinetics of structural transitions.

Notice that the perforated lamellar structure (displayed in Figure 7.7) does not appear in the phase diagram in Figure 7.8 and is a metastable solution. This is at variance with an earlier conclusion that it is an unstable solution of the amplitude equations [47,79]. Figure 7.10 shows the location of holes in three successive lamellar layers—we see that the holes have the same configuration every two layers (ABAB stacking). Furthermore, the symmetry of the holes is not perfectly hexagonal but distorted. The origin of this distorted structure might be the mode-expansion approximation because, in order to precisely represent a hexagonally perforated lamella, an infinite series of the mode expansion is required. Experimentally, a hexagonally perforated lamellar structure in diblock copolymers was found by Hashimoto et al. [84]. It is believed to be metastable by analyzing the experimental data [85]. Recent scattering experiments on diblock copolymers indicate an ABCABC stacking [86] of a perforated lamellar structure. Morphological transitions involving perforated lamellar structures have also been reported [87,88].

7.9 KINETICS OF MORPHOLOGICAL TRANSITIONS

The kinetics of morphological transitions can be investigated by solving the set of amplitude equations [79]. (We shall again add random forces to avoid freezing in a metastable state.) An example of the transition from a gyroid structure to a lamellar structure is shown in Figure 7.11. The gyroid structure arises for parameter values $\tau = 2.2$ and $\bar{\phi} = -0.1$, and then τ is changed abruptly to $\tau = 2.5$ at $t = 0$. It is seen that the Fddd structure appears at $t \simeq 14,400$ as an intermediate structure. The transition from a lamellar structure for $\tau = 2.5$ and $\bar{\phi} = -0.13$ to a gyroid structure for $\tau = 2.4$ and $\bar{\phi} = -0.13$ is shown in Figure 7.12. In this reverse transition, a hexagonal structure appears shortly after the temperature change and it is followed by an Fddd structure. As another example of a morphological transition, Figure 7.13 displays the domain kinetics in the transition from a hexagonal structure for $\tau = 2.2$ and $\bar{\phi} = -0.16$ to a gyroid for $\tau = 2.5$ and $\bar{\phi} = -0.16$. We can see that the Fddd structure appears again as an intermediate structure.

In these simulations, we see that the Fddd structure is rather persistent as a metastable state during structural transitions. Recall that it exists only in a narrow region in the equilibrium phase diagram in Figure 7.8. Let us briefly discuss this unique

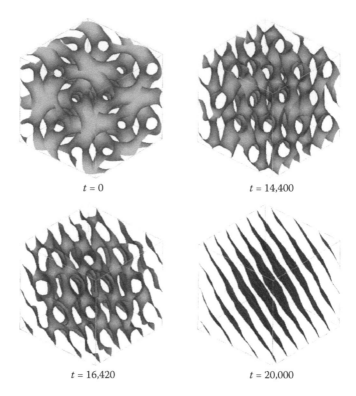

$t = 0$ $t = 14,400$

$t = 16,420$ $t = 20,000$

FIGURE 7.11 Time evolution from the gyroid structure to the lamellar structure via the intermediate Fddd structure. (From Yamada, K. et al., *Macromolecules*, 37, 5762, 2004. With permission.)

structure. The Fddd structure was first obtained by direct numerical simulations of Equation 7.34 [89]. In thermal equilibrium, its existence was suggested in experiments on triblock copolymers [90]. Theoretically, the Fddd structure was shown to exist in the equilibrium phase diagrams for both triblock and diblock copolymers [83]. (This was obtained from a self-consistent mean-field theory.) As described in Section 7.8, we have also shown using GL theory that the Fddd structure can be a stable equilibrium structure in diblock copolymers [81]. Finally, we mention that recent experiments have also observed the Fddd structure. Takenaka et al. [91] have performed small-angle X-ray scattering experiments on poly(styrene-b-isoprene) diblock copolymers. The temperature-dependence of the scattering intensity is shown in Figure 7.14. From 120°C to 130°C, the scattering intensity indicates a lamellar structure. A Fddd structure has been observed from 131°C to 136°C, and at higher temperatures scattering with a gyroid symmetry is obtained. Takenaka et al. have also observed that the Fddd structure appears in the transition from the lamellar structure to the gyroid structure [92]. This is consistent with our numerical results in Figures 7.11 and 7.12.

Recall that our simulations show that the Fddd structure appears in the transitions from gyroid to hexagonal and lamellar structures, and in the reverse transitions. It does not appear in the transition between gyroid and BCC structures as shown in

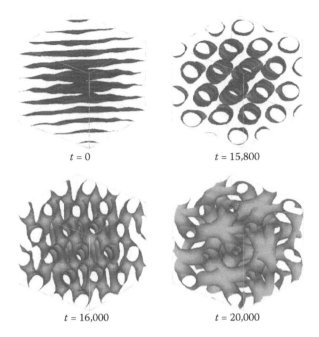

$t = 0$ $t = 15,800$

$t = 16,000$ $t = 20,000$

FIGURE 7.12 Time evolution from the lamellar structure to the gyroid structure via the intermediate Fddd structure. (From Yamada, K. et al., *Macromolecules*, 37, 5762, 2004. With permission.)

Figure 7.15, where the initial gyroid arises for $\tau = 2.5$ and $\bar{\phi} = -0.16$, and the final BCC structure arises for $\tau = 2.07$ and $\bar{\phi} = -0.16$. This fact can be understood from symmetry considerations, as summarized in Figure 7.16. A gyroid structure has a cubic symmetry, whereas lamellar and hexagonal structures are uniaxial in the sense that the *lamellar normal* and the *cylindrical axis* are particular directions for these structures. Therefore, it is natural that the Fddd structure, which is an interconnected structure but with a uniaxial symmetry, appears as an intermediate structure in the symmetry-breaking transitions.

7.10 DIRECT SIMULATIONS OF THE EVOLUTION EQUATION

In Section 7.7, we mentioned that direct simulations of Equation 7.34 are not easy in $d = 3$ because the period of a gyroid is not the same as that of other structures. However, such a difficulty does not exist in $d = 2$ because only lamellar and hexagonal structures exist. Thus, there are many simulations of Equation 7.34 in $d = 2$. For example, the formation of microphase-separated structures starting from the disordered state has been studied by Bahiana and Oono [71]. Furthermore, the kinetics of the transition between the hexagonal and lamellar structures has also been studied [89]. In this case, the spatial period is unchanged during the transition. Figure 7.17 shows a hexagonal-to-lamellar transition. Notice that the lamellar structure tends to nucleate at the grain boundaries. The final lamellar structure is locally deformed and is influenced

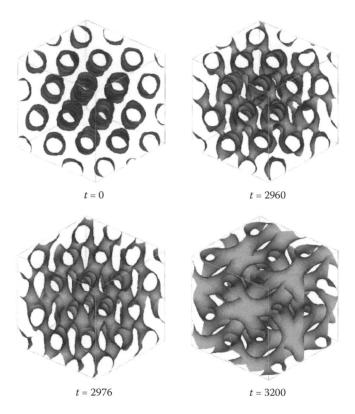

$t = 0$ $t = 2960$

$t = 2976$ $t = 3200$

FIGURE 7.13 Hexagonal-to-gyroid transition. It is seen that the Fddd structure appears at $t \simeq 2976$. (From Yamada, K. et al., *Macromolecules*, 37, 5762, 2004. With permission.)

by the initial grains of the hexagonal structure. Figure 7.18 shows a nucleation-driven growth process in a hexagonal-to-lamellar transition. The dislocation in the initial hexagonal structure, located at the center of the system, acts as a nucleation center after the temperature quench, and the lamellar domain develops around that dislocation. There are two different types of interfaces between the lamellar domain and the surrounding hexagonal matrix, depending on the relative primitive vectors of the two structures.

A similar problem also exists in $d = 3$. A systematic numerical study of Equation 7.34 is required to analyze the motion of interfaces with two different structures. In Figure 7.19, we show a gyroid structure invading a lamellar structure under the condition that the lamellar structure is less stable. We see that the lamellar structure changes to the gyroid structure in two steps. First, the lamellar structure is deformed to a perforated lamellar structure, and then it evolves to a hexagonal structure. A reconnection of the cylindrical domains starts near the interfaces and propagates to the entire region. We remark that the interface kinetics is not completely symmetric. We have also performed simulations where the gyroid is less stable than the lamellar structure. In this case, the interface separating the two structures does not always move but is pinned occasionally.

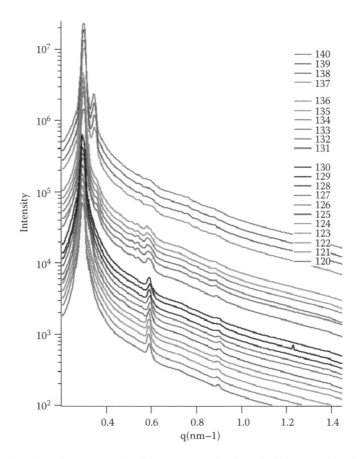

FIGURE 7.14   (**See color insert following page 180.**) Small-angle X-ray scattering intensities for different temperatures [91]. The material is poly(styrene-b-isoprene) with PS/PI = 36.2/63.8 vol/vol, $M_n = 2.64 \times 10^4$, and $M_w/M_n = 1.02$.

Before concluding this section, we show an alternative analysis to confirm the presence of the Fddd structure in the equilibrium phase diagram in Figure 7.8. In $d = 3$, it is difficult to perform numerical simulations of the transition kinetics due to the reasons mentioned above. However, in order to verify the equilibrium phase diagram, we are free from these difficulties. In fact, direct simulations of Equation 7.34 (with adjustment of the system size) are more accurate because it is not necessary to introduce any approximations, for example, truncation of the number of modes in the mode-expansion method.

Starting from the high-temperature uniform state, we solved the evolution equation to obtain the asymptotic stationary solutions. We repeat this by changing the system size and calculate the free energy as a function of the system size [81]. The results for $\bar{\phi} = 0.097$ are plotted in Figure 7.20. It should be stressed that the scale of the vertical axis is extremely small and hence the difference of the free energy is very small. However, it is clear that a hexagonal structure is most stable for $\tau = 2.15$, an

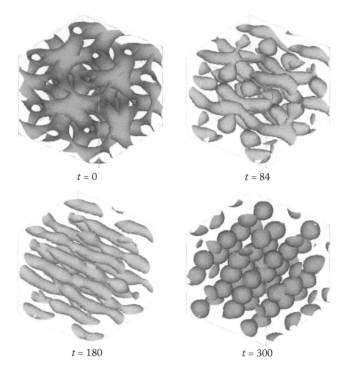

$t = 0$ $t = 84$

$t = 180$ $t = 300$

FIGURE 7.15   Gyroid-to-BCC transition. It is seen that a hexagonal structure appears as an intermediate structure, but the Fddd structure does not appear.

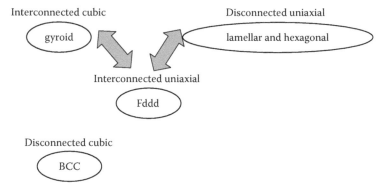

FIGURE 7.16   Schematic to demonstrate why the Fddd structure appears in the lamellar-gyroid and hexagonal-gyroid transitions but does not appear in the gyroid-BCC transition.

Fddd structure for $\tau = 2.19$, and a lamellar structure for $\tau = 2.2$—as is consistent with Figure 7.8. A small discrepancy of the parameter values where the Fddd phase appears in Figure 7.8 and in Figure 7.20 is due to the truncation approximation used to obtain Figure 7.8.

FIGURE 7.17 Transition from the hexagonal structure with grains for $\bar{\phi} = 0.04$ and $\hat{\tau} = 0.0117$ to the lamellar structure for $\bar{\phi} = 0.04$ and $\hat{\tau} = 0.045$ with $\hat{\tau} = (\tau - 2\sqrt{\alpha})/g$. The other parameters are $\alpha = 0.9$ and $g = 4.5$. The evolution times are $t = 0, 2 \times 10^4 \Delta t, 4 \times 10^4 \Delta t, 6 \times 10^4 \Delta t, 8 \times 10^4 \Delta t$, and $10 \times 10^4 \Delta t$ with $\Delta t = 1/45$ from (a) to (f). (From Nonomura, M. and Ohta, T., *J. Phys.: Condens. Matter*, 13, 9089, 2001. With permission.)

7.11   ELASTIC THEORY FOR MESOSCOPIC STRUCTURES

As was mentioned in Section 7.1, rheology in soft matter is one of the most important problems. Despite a number of experimental studies, theoretical investigations (in particular, an analytical theory of viscoelasticity) are very limited. This is due to the mathematical difficulty of describing deformations and reconnections of interconnected domains such as gyroids. However, if one focuses on linear elastic properties, there are several studies of the elasticity of mesoscopic periodic structures [93,94]. In fact, the expression for the elastic free energy of a lamellar structure is essentially the same as that for the smectic phase of liquid crystals. Recently, analytical elastic theory for more complicated interconnected structures has also been developed, as we will show below [96].

We introduce the set of wave-vectors $\vec{k}_1, \vec{k}_2, \ldots, \vec{k}_n$, which denote the fundamental reciprocal lattice vectors of the relevant structure. The local concentration $\phi(\vec{r})$ for the periodic structure can be written in terms of these vectors as

$$\phi(\vec{r}) = \Phi\left(\vec{k}_1 \cdot \vec{r}, \vec{k}_2 \cdot \vec{r}, \ldots, \vec{k}_n \cdot \vec{r}\right). \tag{7.57}$$

FIGURE 7.18 Transition from the hexagonal structure with a dislocation for $\bar{\phi} = 0.04$ and $\hat{\tau} = 0.00117$ to the lamellar structure for $\bar{\phi} = 0.04$ and $\hat{\tau} = 0.045$. The other parameters are $\alpha = 0.9$ and $g = 4.5$. The evolution times are $t = 0, 1.5 \times 10^4 \Delta t, 3 \times 10^4 \Delta t, 4.5 \times 10^4 \Delta t, 6 \times 10^4 \Delta t$, and $7.5 \times 10^4 \Delta t$ with $\Delta t = 1/45$ from (a) to (f). (From Nonomura, M. and Ohta, T., *J. Phys.: Condens. Matter*, 13, 9089, 2002. With permission.)

Long-wavelength gentle deformations of the structure can be represented by introducing phase variables, that is, we define the deformation field $\vec{u}(\vec{r})$ through

$$\theta_j(\vec{r}) = \vec{k}_j \cdot \left[\vec{r} - \vec{u}(\vec{r})\right], \quad j = 1, 2, \ldots, n. \tag{7.58}$$

The concentration profile of the deformed structure is given in terms of the phase variables by

$$\phi(\vec{r}) = \Phi\left[\theta_1(\vec{r}), \theta_2(\vec{r}), \ldots, \theta_n(\vec{r})\right]. \tag{7.59}$$

This is justified for $|\nabla \vec{u}| \ll 1$.

Because no free-energy increase due to phase deformation occurs from the local part of the free energy in Equation 7.8, the elastic energy comes from the non-local part. Substituting Equation 7.59 into Equation 7.8, and retaining only terms up to the second spatial derivative, we obtain

$$F_{\text{deform}} = \frac{1}{4} \int d\vec{r} \left[D_{st}^{\beta\gamma} \nabla_\beta u_s(\vec{r}) \nabla_\gamma u_t(\vec{r})\right], \tag{7.60}$$

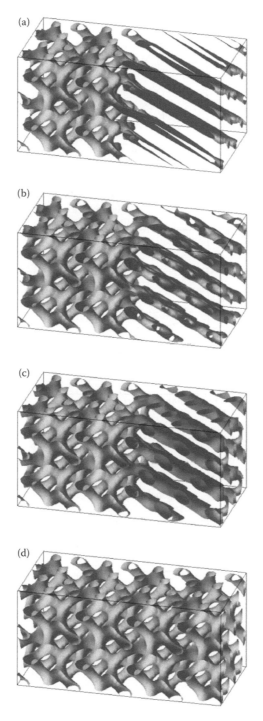

FIGURE 7.19   Evolution of the interface between the gyroid and lamellar structures for $\tau =$ 2.25, $\bar{\phi} = 0.12$, and $\alpha = g = 1$. (Reprinted from Yamada, K. and Ohta, T., *J. Phys. Soc. Jpn.*, 76, 084801–084805, 2007. With permission.)

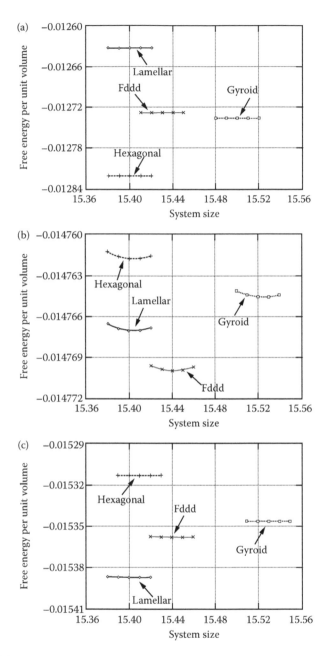

FIGURE 7.20 Comparison of the free energies for the lamellar, hexagonal, and Fddd structures, (a) $\tau = 2.15$ (b) $\tau = 2.19$, and (c) $\tau = 2.2$, and $\bar{\phi} = 0.097$, as a function of the system size. (From Yamada, K. et al., *J. Phys.: Condens. Matter*, 18, L421, 2006. With permission.)

where the repeated indices imply summation. The coefficient $D_{st}^{\beta\gamma}$ is given by

$$D_{st}^{\beta\gamma} = 2 \sum_{j=1}^{n} \left[|\phi_j|^2 k_{js} k_{jt} \left(\frac{\partial^2 v_q}{\partial q_\beta \partial q_\gamma} \right)_{\vec{q}=\vec{k}_j} \right], \tag{7.61}$$

where v_q is given in the free energy in Equation 7.8 by

$$v_q = q^2 + \frac{\alpha}{q^2}. \tag{7.62}$$

The Fourier series expansion is defined by

$$\phi(\vec{r}) = \sum_j \phi_j e^{i\vec{k}_j \cdot \vec{r}}. \tag{7.63}$$

Equations 7.60 and 7.61 can be applied to both strong segregation and weak segregation [93]. Here, we present results for the case of weak segregation.

The elastic free energy of a lamellar structure is calculated as

$$F_{\text{lam}} = \frac{1}{2} \int d\vec{r} K_{11}^L u_{xx}^2, \tag{7.64}$$

where the stress tensor has been introduced as

$$u_{\alpha\beta} \equiv \frac{1}{2} \left(\nabla_\beta u_\alpha + \nabla_\alpha u_\beta \right). \tag{7.65}$$

The elastic modulus is given by

$$K_{11}^L = 8\phi_L^2 k_c^2, \tag{7.66}$$

with ϕ_L and k_c given by

$$\phi_L^2 = \frac{1}{3gk_c^2} \left(-k_c^4 + \tau k_c^2 - \alpha - 3g\bar{\phi}^2 k_c^2 \right), \tag{7.67}$$

$$k_c^2 = \sqrt{\alpha}. \tag{7.68}$$

Similarly, the elastic free energy for a hexagonal structure takes the form

$$F_{\text{hex}} = \frac{1}{2} \int d\vec{r} \left[K_{11}^H \left(u_{xx}^2 + u_{yy}^2 \right) + 2K_{12}^H u_{xx} u_{yy} + 2 \left(K_{11}^H - K_{12}^H \right) u_{xy}^2 \right], \tag{7.69}$$

with

$$K_{11}^H = 6\phi_H^2 k_c^2, \tag{7.70}$$

$$K_{12}^H = 3\phi_H^2 k_c^2. \tag{7.71}$$

Here, the amplitude ϕ_H is given by

$$\phi_H = \frac{1}{15gk_c^2} \left\{ 3g |\bar{\phi}| k_c^2 + \left[9g^2\bar{\phi}^2 k_c^4 - 15gk_c^2(k_c^4 - \tau k_c^2 + \alpha + 3g\bar{\phi}^2 k_c^2) \right]^{1/2} \right\}.$$

(7.72)

The elastic free energy for the BCC structure is given by

$$F_{BCC} = \frac{1}{2} \int d\bar{r} \left[K_{11}^S \left(u_{xx}^2 + u_{yy}^2 + u_{zz}^2 \right) + 2K_{12}^S \left(u_{xx}u_{yy} + u_{yy}u_{zz} + u_{zz}u_{xx} \right) \right.$$
$$\left. + 4K_{44}^S \left(u_{xy}^2 + u_{yz}^2 + u_{zx}^2 \right) \right],$$

(7.73)

where

$$K_{11}^S = 8\phi s^2 k_c^2,$$

(7.74)

$$K_{12}^S = 4\phi s^2 k_c^2,$$

(7.75)

$$K_{44}^S = 4\phi s^2 k_c^2.$$

(7.76)

Here,

$$\phi_S = \frac{1}{45gk_c^2} \left\{ -6g |\bar{\phi}| k_c^2 - \left[36g^2\bar{\phi}^2 k_c^4 - 45gk_c^2(k_c^4 - \tau k_c^2 + \alpha + 3g\bar{\phi}^2 k_c^2) \right]^{1/2} \right\}.$$

(7.77)

The derivation of these results is presented in Ref. [93].

The elastic free energy of a gyroid was not considered in Ref. [93] because the gyroid was not known to exist in block copolymers at that time. Here, we show that Equations 7.60 and 7.61 are also applicable to the interconnected structure [96]. Because the gyroid has a cubic symmetry, the free energy takes the same form as for the BCC structure in Equation 7.73. However, the coefficients are different. The modulus K_{11}^S should be replaced by

$$K_{11}^G = 16k_c^2 \left(\phi_a^2 + \frac{2}{3}\phi_b^2 \right),$$

(7.78)

where the amplitudes ϕ_a and ϕ_b have been defined in Equation 7.53 and are the equilibrium solutions of the amplitude equations. The remaining two elastic constants are found to be related to K_{11}^G as

$$K_{12}^G = K_{44}^G = \frac{1}{2}K_{11}^G.$$

(7.79)

In these derivations, we have used the relation obtained from $\partial F/\partial k_c^2 = 0$:

$$2\left(1 - \frac{\alpha}{k_c^4}\right)\phi_a^2 + \left(\frac{4}{3} - \frac{3\alpha}{4k_c^4}\right)\phi_b^2 = 0.$$

(7.80)

Equation 7.79 shows that the three elastic constants are not independent in the weak-segregation regime. We note that this relation also holds for the BCC structure.

Before closing this section, we mention some recent related studies. Tyler and Morse have evaluated the elastic constants for a gyroid numerically via self-consistent field theory [97]. (A similar theory for the lamellar structure is available in Ref. [98].) It is found that the relations in Equation 7.79 are satisfied within the numerical uncertainty.

7.12   SUMMARY AND DISCUSSION

In this chapter, we have studied the morphology and dynamics of mesoscopic structures in diblock copolymers, emphasizing both their universality and diversity. In metallurgy, there is a long history of studying structural transitions. However, the essential difference in soft matter is that the structures are meso-scale, and domains (rather than atoms) are the fundamental objects of the periodic structures. Therefore, one of the characteristic features in soft matter is that various interconnected structures exist. Furthermore, the connectivity of domains plays an important role in the kinetics of structural transitions. This is also related to the fact that several intermediate metastable structures are often seen in the kinetics of these transitions.

The fact that the structures are meso-scale requires the formulation of a suitable theoretical approach. In the current chapter, we have focused on Ginzburg–Landau (GL) theory, where the local concentration is the relevant variable. In particular, the free-energy functional in Equation 7.8 has a simple structure, with the long-range interaction originating from the osmotic incompressibility of diblock copolymer melts. It is also remarkable that this non-local term only gives rise to a simple linear damping term in the evolution equation 7.34. This model describes both macro-phase and micro-phase separation in a unified way. It should be emphasized that the free energy and the evolution equation, despite their simplicity, predict complicated interconnected structures such as the gyroid and Fddd structures.

Because of the simplicity of the current coarse-grained theory, one can recognize relationships with entirely different systems or phenomenology. For example, the formation of various periodic structures called *pasta phases* in nuclear matter [99,100] is essentially modeled by the free-energy functional in Equation 7.8. In this context, the Coulomb interaction of protons and the interfacial energy between nuclear matter and the electron-rich phase are responsible for the structure formation.

There are also several other numerical methods for studying mesoscopic structures in soft matter. The dynamical *self-consistent mean-field theory* is useful for the quantitative analysis of block copolymers with specific molecular architecture although an analytical treatment is less probable. Kawakatsu and his co-workers have used this approach to study morphological changes in gyroids on the application of external fields like shear flows and electric fields [101,102]. The development of new algorithms that enable us to simulate larger systems in $d = 3$ is also required. Here, it is mentioned that the lattice Boltzmann method (see Chapter 4) has been applied to study the formation of gyroids [13,15], and their viscoelastic properties [33].

Another important problem is to extend the current theory to other soft-matter systems. At present, the free energy in Equation 7.8 has only been applied to block copolymer melts. However, a much greater variety of mesophases is possible in ternary mixtures or in copolymers with solvents. For example, the formation of vesicles is observed in computer simulations in $d = 3$ [103,104]. Apart from the gyroid, other cubic structures have also been observed in lyotropic liquid crystals and biomaterials [4] and their transitions have been investigated. (See Ref. [105] and the references cited therein.) At present, the Helfrich membrane free energy [106] is employed to analytically study equilibrium mesoscale structures in these systems [107]. To study the dynamics, one has to derive a suitable free energy and kinetic equation. There are also several other important outstanding problems. For example, a systematic study taking account of hydrodynamic effects will deepen our understanding of the mesophases. Furthermore, the effects of charges are also of considerable importance.

ACKNOWLEDGMENTS

We are grateful to M. Nonomura and K. Yamada for their collaboration in the current work. This study was supported by a scientific grant of the Ministry of Education, Science and Culture, Japan.

REFERENCES

1. Matsen, M.W., Self-consistent field theory and its applications, in *Soft Matter Vol. 1: Polymer Melts and Mixtures*, Gompper, G. and Schick, M., Eds., Wiley-VCH, Weinheim, 2005, chap. 2.
2. Hamley, I.W., *Introduction to Soft Matter*, John Wiley, Chichester, 2000.
3. Luzzati, V. and Spegt, P.A., Polymorphism of lipids, *Nature*, 215, 701, 1967.
4. *Bicontinuous Liquid Crystals*, Lynch, M.L. and Spicer, P.T., Eds., CRC Press, Boca Raton, 2005.
5. Gommper, G. and Schick, M., Self-assembling amphiphilic systems, in *Phase Transitions and Critical Phenomena*, Vol. 16, Domb, C. and Lebowitz, J.L., Academic Press, London, 1994.
6. Stadler, R., Auschra, C., Beckmann, J., Krappe, U., Voight-Martin, I., and Leibler, L., Morphology and thermodynamics of symmetric poly(A-block-B-block-C) triblock copolymers, *Macromolecules*, 28, 3080, 1995.
7. Breiner, U., Krappe, U., Abetz, V., Stadler, R., Cylindrical morphologies in asymmetric ABC triblock copolymers, *Macromolecular Chemistry and Physics*, 198, 1051, 1997.
8. Takano, A., Soga, K., Suzuki, J., and Matsushita, Y., Noncentrosymmetric structure from a tetrablock quarterpolymer of the ABCA type, *Macromolecules*, 36, 9288, 2003.
9. Stupp, S.I., LeBonheur, V., Walker, K., Li, L.S., Huggins, K.E., Keser, M., and Amstutz, A., Supramolecular materials: Self-organized nanostructures, *Science*, 276, 384, 1997.
10. Tschierske, C., Micro-segregation, molecular shape and molecular topology—partners for the design of liquid crystalline materials with complex mesophase morphologies, *J. Mater. Chem.*, 11, 2647, 2001.
11. Meeker, S.P., Poon, W.C.K., Crain, J., and Terentjev, E.M., Colloid-liquid-crystal composites: An unusual soft solid, *Phys. Rev. E*, 61, R6083, 2000.

12. Chastek, T.Q. and Lodge, T.P., Measurement of gyroid single grain growth rates in block copolymer solutions, *Macromolecules*, 36, 7672, 2003.

13. Gonzalez-Segredo, N. and Coveney, P.V., Coarsening dynamics of ternary amphiphilic fluids and the self-assembly of the gyroid and sponge mesophases: Lattice-Boltzmann simulations, *Phys. Rev. E*, 69, 061501, 2004.

14. Sun, P., Yin, Y., Li, B., Chen, T., Jin, Q., Ding, D., and Shi, A.-C., Simulated annealing study of gyroid formation in diblock copolymer solutions, *Phys. Rev. E*, 72, 061408, 2005.

15. Martines-Veracoechea, F.J. and Escobedo, F.A., Lattice Monte Carlo simulations of the gyroid phase in monodisperse and bidisperse block copolymer systems, *Macromolecules*, 38, 8522, 2005.

16. Boyer, D. and Vinals, J., Domain coarsening of stripe patterns close to onset, *Phys. Rev. E*, 64, 050101, 2001.

17. Harrison, C., Cheng, Z., Sethuraman, S., Huse, D.A., Chaikin, P.M., Vega, D.A., Sebastian, J.M., Register, R.A., and Adamson, D.H., Dynamics of pattern coarsening in a two-dimensional smectic system, *Phys. Rev. E*, 66, 011706, 2002.

18. Kyurylyuk, A., and Fraaije, J.G.E.M., Three-dimensional structure and motion of twist grain boundaries in block copolymer melts, *Macromolecules*, 38, 8546, 2005.

19. Diat, O., Roux, D., and Nallet, F., Effect of shear on a lyotropic lamellar phase, *J. Phys. II France*, 3, 1427, 1993.

20. Buchanan, M., Egelhaaf, S.U., and Cates, M.E., Dynamics of interface instabilities in nonionic lamellar phases, *Langmuir*, 16, 3718, 2000.

21. Cochran, E.W. and Bates, F.S., Shear-induced network-to-network transition in a block copolymer melt, *Phys. Rev. Lett.*, 93, 087802, 2004.

22. Tsori, Y. and Andelman, D., Thin film diblock copolymers in electric field: Transition from perpendicular to parallel lamellae, *Macromolecules*, 35, 5161, 2002.

23. Matsen, M.W., Stability of a block-copolymer lamella in a strong electric field, *Phys. Rev. Lett.*, 95, 258302, 2005.

24. Boker, A., Elbs, H., Hansel, H., Knoll, A., Ludwigs, S., Zettl, H., Urban, V., Abetz, V., Muller, A.H.E., and Krausch, G., Microscopic mechanisms of electric-field-induced alignment of block copolymer microdomains, *Phys. Rev. Lett.*, 89, 135502, 2002.

25. Osuji, C., Ferreira, P.J., Mao, G., Ober, C.K., Van der Sande, J.B., and Thomas, E.L., Alignment of self-assembled hierarchical microstructure in liquid crystalline diblock copolymers using high magnetic fields, *Macromolecules*, 37, 9903, 2004.

26. Jones, R.A.L., *Soft Condensed Matter*, Oxford University Press, Oxford, 2003.

27. Yamamoto, J. and Tanaka, H., Shear-induced sponge-to-lamellar transition in a hyperswollen lyotropic system, *Phys. Rev. Lett.*, 77, 4390, 1996.

28. Porcar, L., Hamilton, W.A., Butler, P.D., and Warr, G.G., Topological relaxation of a shear-induced lamellar phase to sponge equilibrium and the energetics of membrane fusion, *Phys. Rev. Lett.*, 93, 198301, 2004.

29. Ohta, T., Enomoto, Y., Harden, J.L., and Doi, M., Anomalous rheological behavior of ordered phases of block copolymers. 1, *Macromolecules*, 26, 4928, 1993.

30. Doi, M., Harden, J.L., and Ohta, T., Anomalous rheological behavior of ordered phases of block copolymers. 2, *Macromolecules*, 26, 4935, 1993.

31. Boyer, D. and Vinals, J., Grain boundary pinning and glassy dynamics in stripe phases, *Phys. Rev. E*, 65, 046119, 2002.

32. Huang, Z.-F. and Vinals, J., Shear-induced grain boundary motion for lamellar phases in the weakly nonlinear regime, *Phys. Rev. E*, 69, 041504, 2004.

33. Giupponi, G., Harting, J., and Coveney, P.V., Emergence of rheological properties in lattice Boltzmann simulations of gyroid mesophases, *Europhys. Lett.*, 73, 533, 2006.

34. Khachaturyan, A.G., *Theory of Structural Transformations in Solids*, Wiley-Interscience, New York, 1983.

35. Sakurai, S., Umeda, H., Taie, K., and Nomura, S., Kinetics of morphological transition in polystyrene-block-polybutadiene-block-polystyrene triblock copolymer melt, *J. Chem. Phys.*, 105, 8902, 1996.

36. Sakurai, S., Umeda, H., Furukawa, C., Irie, H., Nomura, S., Lee, H.H., and Kim, J.K., Thermally induced morphological transition from lamella to gyroid in a binary blend of diblock copolymers, *J. Chem. Phys.*, 108, 4333, 1998.

37. Hamley, I.W., Fairclough, J.P.A., Ryan, A.J., Mai, S.-M., and Booth, C., Lamellar-to-gyroid transition in a poly(oxyethylene)-poly(oxybutylene) diblock copolymer melt, *Phys. Chem. Chem. Phys.*, 1, 2097, 1999.

38. Funari, S.S. and Rapp, G., A continuous topological change during phase transitions in amphiphile/water systems, *Proc. Nat. Acad. Sci.*, 96, 7756, 1999.

39. Floudas, G., Ulrich, R., and Wiesner, U., Microphase separation in poly(isoprene-b-ethylene oxide) diblock copolymer melts. I. Phase state and kinetics of the order-to-order transitions, *J. Chem. Phys.*, 110, 652, 1999.

40. Wang, C.-Y. and Lodge, T.P., Unexpected intermediate state for the cylinder-to-gyroid transition in a block copolymer solution, *Macromol. Rapid Commun.*, 23, 49, 2002.

41. Jeong, U., Lee, H.H., Yang, H., Kim, J.K., Okamoto, S., Aida, S., and Sakurai, S., Kinetics and mechanism of morphological transition from lamella to cylinder microdomain in polystyrene-block-poly(ethylene-co-but-1-ene)-block-polystyrene triblock copolymer, *Macromolecules*, 36, 1685, 2003.

42. Hajduk, D.A., Ho, R.-M., Hillmyer, M.A., Bates, F.S., and Almdal, K., Transition mechanisms for complex ordered phases in block copolymer melts, *J. Phys. Chem. B*, 102, 1356, 1998.

43. Kimishima, K., Koga, T., and Hashimoto, T., Order-order phase transition between spherical and cylindrical microdomain structures of block copolymer. I. Mechanism of the transition, *Macromolecules*, 33, 968, 2000.

44. Krishnamoorti, R., Silva, A.S., Modi, M.A., and Hammouda, B., Small-angle neutron scattering study of a cylinder-to-sphere order-order transition in block copolymers, *Macromolecules*, 33, 3803, 2000.

45. Imai, M., Saeki, A., Teramoto, T., Kawaguchi, A., and Nakaya, K., Kinetic pathway of lamellar → gyroid transition: Pretransition and transient states, *J. Chem. Phys.*, 115, 10525, 2001.

46. Squires, A.M., Templer, R.H., Seddon, J.M., Woenkhaus, J., Winter, R., Narayanan, T., and Finet, S., Kinetics and mechanism of the interconversion of inverse bicontinuous cubic mesophases, *Phys. Rev. E*, 72, 011502, 2005.

47. Qi, S. and Wang, Z.-G., Kinetics of phase transitions in weakly segregated block copolymers: Pseudostable and transient states, *Phys. Rev. E*, 55, 1682, 1997.

48. Kaneda, M., Tsubakiyama, T., Carlsson, A., Sakamoto, Y., Ohsuna, T., Terasaki, O., Joo, S.H., and Ryoo, R., Structural study of mesoporous MCM-48 and carbon networks synthesized in the spaces of MCM-48 by electron crystallography, *J. Phys. Chem. B*, 106, 1256, 2002.

49. Landh, T., From entangled membranes to eclectic morphologies: cubic membranes as subcellular space organizers, *FEBS Lett.*, 369, 13, 1995.

50. Sackmann, E., in *Physics of Biological Systems: From Molecules to Species*, Flyvbjerg, H., Hertz, J., Jensen, M.H., Mouritsen, O.G., and Sneppen, K., Eds., Springer-Verlag, Heidelberg, 1997.

51. *Biomathematics: Mathematics of Biostructures and Biodynamics*, Andersson, S., Larsson, K., Larsson, M., and Jacob, M., Eds., Elsevier, Amsterdam, 1999.

52. Edwards, S.F., The statistical mechanics of polymers with excluded volume, *Proc. Phys. Soc. London*, 85, 613, 1965.
53. Oono, Y., Statistical physics of polymer solutions—conformation-space renormalization-group approach, *Adv. Chem. Phys.*, 61, 301, 1985.
54. Onuki, A., *Phase Transition Dynamics*, Cambridge University Press, Cambridge, 2002.
55. Leibler, L., Theory of microphase separation in block copolymers, *Macromolecules*, 13, 1602, 1980.
56. Ohta, T. and Kawasaki, K., Equilibrium morphology of block copolymer melts, *Macromolecules*, 19, 2621, 1986.
57. Comment on the free energy functional of block copolymer melts in the strong segregation limit, *Macromolecules*, 23, 2413, 1990.
58. Oono, Y. and Ohta, T., The coarse grained approach in polymer physics, in *Stealing the Gold: A Celebration of the Pioneering Physics of Sam Edwards*, Goldbart, P.M., Goldenfeld, N., and Sherrington, D., eds., Oxford University Press, Oxford, 2004, chapter 10.
59. de Gennes, P.G. and Prost, J., *The Physics of Liquid Crystals*, Oxford University Press, Oxford, 1993.
60. Seul, M. and Andelman, D., Domain shapes and patterns: The phenomenology of modulated phases, *Science*, 267, 476, 1995.
61. Hashimoto, T., Shibayama, M., and Kawai, H., Ordered structure in block polymer solutions. 4. Scaling rules on size of fluctuations with block molecular weight, concentration, and temperature in segregation and homogeneous regimes, *Macromolecules*, 16, 1093, 1983.
62. Semenov, A.N., Contribution to the theory of microphase layering in block-copolymer melts, *Sov. Phys. JETP*, 61, 733, 1985.
63. Helfand, E., Theory of inhomogeneous polymers: Fundamentals of the Gaussian random-walk model, *J. Chem. Phys.*, 62, 999, 1975.
64. Matsen, M.W. and Bates, F.S., Origins of complex self-assembly in block copolymers, *Macromolecules*, 29, 7641, 1996.
65. Unifying weak- and strong-segregation block copolymer theories, *Macromolecules*, 29, 1091, 1996.
66. Kawakatsu, T., *Statistical Physics of Polymers*, Springer-Verlag, Heidelberg, 2004.
67. Fraaije, J.G.E.M., Dynamic density functional theory for microphase separation kinetics of block copolymer melts, *J. Chem. Phys.*, 99, 9202, 1993.
68. Hasegawa, R. and Doi, M., Adsorption dynamics. Extension of self-consistent field theory to dynamical problems, *Macromolecules*, 30, 3086, 1997.
69. Matsen, M.W., The standard Gaussian model for block copolymer melts, *J. Phys. Condens. Matter*, 14, R21, 2002.
70. Uneyama, T. and Doi, M., Density functional theory for block copolymer melts and blends, *Macromolecules*, 38, 196, 2005.
71. Bahiana, M. and Oono, Y., Cell dynamical system approach to block copolymers, *Phys. Rev. A*, 41, 6763, 1990.
72. Kawasaki, K. and Sekimoto, K., Morphology dynamics of block copolymer systems, *Physica A*, 148, 361, 1988.
73. Maurits, N.M. and Fraaije, J.G.E.M., Mesoscopic dynamics of copolymer melts: From density dynamics to external potential dynamics using nonlocal kinetic coupling, *J. Chem. Phys.*, 107, 5879, 1997.
74. Groot, R.D., Madden, T.J., and Tildesley, D.J., On the role of hydrodynamic interactions in block copolymer microphase separation, *J. Chem. Phys.*, 110, 9739, 1999.

75. Kawasaki, K., Kinetic equations and time correlation functions of critical fluctuations, *Ann. Phys.*, 61, 1, 1970.

76. Pincus, P., Dynamics of fluctuations and spinodal decomposition in polymer blends. II, *J. Chem. Phys.*, 75, 1996, 1981.

77. Doi, M. and Edwards, S.F., *The Theory of Polymer Dynamics*, Clarendon, Oxford, 1986.

78. Nonomura, M., Yamada, K., and Ohta, T., Formation and stability of double gyroid in microphase-separated diblock copolymers, *J. Phys. Condens. Matter*, 15, L423, 2003.

79. Yamada, K., Nonomura, M., and Ohta, T., Kinetics of morphological transitions in microphase-separated diblock copolymers, *Macromolecules*, 37, 5762, 2004.

80. Aksimetiev, A., Fialkowski, M., and Holyst, R., Morphology of surfaces in mesoscopic polymers, surfactants, electrons, or reaction-diffusion systems: Methods, simulations, and measurements, *Adv. Chem. Phys.*, 121, 141, 2002.

81. Yamada, K., Nonomura, M., and Ohta, T., Fddd structure in AB-type diblock copolymers, *J. Phys. Condens. Matter*, 18, L421, 2006.

82. Milner, S.T. and Olmsted, P.D., Analytical weak-segregation theory of bicontinuous phases in diblock copolymers, *J. Phys. II France*, 7, 249, 1997.

83. Tyler, C.A. and Morse, D.C., Orthorhombic Fddd network in triblock and diblock copolymer melts, *Phys. Rev. Lett.*, 94, 208302, 2005.

84. Hashimoto, T., Koizumi, S., Hasegawa, H., Izumitani, T., and Hyde, S.T., Observation of "mesh" and "strut" structures in block copolymer/homopolymer mixtures, *Macromolecules*, 25, 1433, 1992.

85. Hajduk, D.A., Takenouchi, H., Hillmyer, M.A., Bates, F.S., Vigild, M.E., and Almdal, K., Stability of the perforated layer (PL) phase in diblock copolymer melts, *Macromolecules*, 30, 3788, 1997.

86. Loo, Y.-L., Register, R.A., Adamson, D.H., and Ryan, A.J., A highly regular hexagonally perforated lamellar structure in a quiescent diblock copolymer, *Macromolecules*, 38, 4947, 2005.

87. Wang, H., Nieh, M.P., Hobbie, E.K., Glinka, C.J., and Katsaras, J., Kinetic pathway of the bilayered-micelle to perforated-lamellae transition, *Phys. Rev. E*, 67, 060902, 2003.

88. Lai, C., Loo, Y.-L., Register, R.A., and Adamson, D.H., Dynamics of a thermoreversible transition between cylindrical and hexagonally perforated lamellar mesophases, *Macromolecules*, 38, 7098, 2005.

89. Nonomura, M. and Ohta, T., Kinetics of morphological transitions between mesophases, *J. Phys. Condens. Matter*, 13, 9089, 2001.

90. Bailey, T.S., Hardy, C.M., Epps, T.H., and Bates, F.S., A noncubic triply periodic network morphology in poly(isoprene-b-styrene-b-ethylene oxide) triblock copolymers, *Macromolecules*, 35, 7007, 2002.

91. Takenaka, M., Wakada, T., Akasaka, S., Nishitsuji, S., Saijo, K., Shimizu, H., and Hasegawa, H., Orthorhombic Fddd network in diblock copolymer melts, *Macromolecules*, 40, 4399, 2007.

92. Takenaka, M., unpublished.

93. Kawasaki, K. and Ohta, T., Phase Hamiltonian in periodically modulated systems, *Physica A*, 139, 223, 1986.

94. Wang, Z.-G., Response and instabilities of the lamellar phase of diblock copolymers under uniaxial stress, *J. Chem. Phys.*, 100, 2298, 1994.

95. Yamada, K. and Ohta, T., Interface between lamellar and gyroid structures in diblock copolymer melts, *J. Phys. Soc. Jpn.*, 76, 084801, 2007.

96. Yamada, K. and Ohta, T., Elastic theory of microphase-separated interconnected structures, *Europhys. Lett.* 73, 614, 2006.

97. Tyler, C.A. and Morse, D.C., Linear elasticity of cubic phases in block copolymer melts by self-consistent field theory, *Macromolecules*, 36, 3764, 2003.

98. Thompson, R.B., Rasmussen, K.O., and Lookman, T., Elastic moduli of multiblock copolymers in the lamellar phase, *J. Chem. Phys.*, 120, 3990, 2004.

99. Oyamatsu, K., Nuclear shapes in the inner crust of a neutron star, *Nucl. Phys. A*, 561, 431, 1993.

100. Watanabe, G., Sato, K., Yasuoka, K., and Ebisuzaki, T., Microscopic study of slablike and rodlike nuclei: Quantum molecular dynamics approach, *Phys. Rev. C*, 66, 012801, 2002.

101. Honda, T. and Kawakatsu, T., Epitaxial transition from gyroid to cylinder in a diblock copolymer melt, *Macromolecules*, 39, 2340, 2006.

102. Ly, D.Q., Honda, T., Kawakatsu, T., and Zvelindovsky, A.V., unpublished.

103. Sevink, G.J.A. and Zvelindovsky, A.V., Self-assembly of complex vesicles, *Macromolecules*, 38, 7502, 2005.

104. Uneyama, T., unpublished.

105. Masum, S.M., Li, S.J., Awad, T.S., and Yamazaki, M., Effect of positively charged short peptides on stability of cubic phases of monoolein/dioleoylphosphatidic acid mixtures, *Langmuir*, 21, 5290, 2005.

106. Helfrich, W., Elastic properties of lipid bilayers: theory and possible experiments, *Z. Naturforsch. C*, 28, 693, 1973.

107. Seifert, U., Configurations of fluid membranes and vesicles, *Adv. Phys.*, 46, 13, 1997.

8 Dynamics of Phase Transitions in Solids

Akira Onuki, Akihiko Minami,
and Akira Furukawa

CONTENTS

8.1 INTRODUCTION

In Chapter 7, we discussed the kinetics of phase separation in soft-matter systems. In this chapter, we turn our attention to phase separation in "hard" systems, for example, solids. A variety of spatially modulated domain structures emerge in phase-ordering solids [1,2]. For example, in alloys, a difference arises in the lattice constants of the two phases, which is called the *lattice misfit*. In structural phase transitions, anisotropically deformed domains of the low-temperature phase emerge in the high-temperature phase. In these cases, elastic strains are induced that radically influence the phase transition behavior. Such strain or elastic effects have long been observed in phase transitions in solids. We need to understand the physical processes involved and make some predictions, which would be of great technological importance. In numerical studies of phase-ordering phenomena in solids, use has been made of a time-dependent Ginzburg–Landau (TDGL) model or a phase-field model, in which the order parameters are coupled to the elastic field in the coarse-grained free-energy functional [1–8]. Such approaches are powerful in understanding the mesoscopic dynamics of domain structures. Some relevant factors in theories and simulations of these processes are (i) elastic anisotropy [1,4], (ii) elastic inhomogeneity (the composition-dependence of the elastic moduli) [2,5,6], and (iii) the simultaneous presence of phase separation and order-disorder phase transitions [1,8].

We should also note that most previous studies have treated the elastic field under the *coherent* condition, in which the lattice planes are continuous through the interfaces without dislocations. However, in real alloys in multi-phase states, the *incoherent* situation is frequently encountered [9], where dislocations appear around the interface regions, leading to some loss of coherence. In Figure 8.1, we schematically illustrate the coherent and completely incoherent interface conditions.

We first show some typical experimental observations. Figure 8.2 shows the evolution of Ni_3Al domains (γ'-precipitates) from spheres to cuboids [10,11]. We can see that initially spherical domains change their shapes into cuboids with facets in the {100} planes as they grow. Figure 8.3 displays the evolution of Ni_4Mo domains in a Ni-Mo alloy [12]. Here, harder cuboids with a larger shear modulus C_{44} are wrapped by a softer matrix, and the coarsening almost stops in the final snapshot (d). (Ni_4Mo represents a Ni-base ordered structure.) The tendency of domain pinning has been observed for large volume fractions of precipitates or in spinodal decomposition.

Furthermore, the application of stretching or compression in the [100] direction in cubic solids is known to produce cylindrical or lamellar domains in the late stages with large domain sizes. Figure 8.4 shows examples of the morphology of γ'-precipitates under uniaxially applied stress in a Ni-Al alloy [13]. Both lamellar and cylindrical domain structures have been observed in a number of experiments, depending on

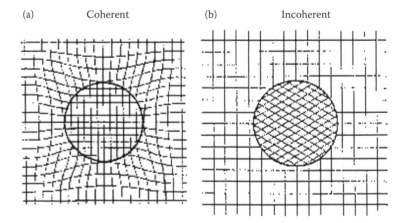

(a)   Coherent   (b)   Incoherent

FIGURE 8.1 (a) Coherent interface condition and (b) completely incoherent interface condition for a two-phase state with a lattice misfit. In real alloys, the coherency loss can also be partial.

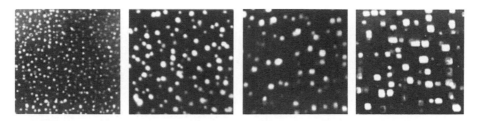

FIGURE 8.2 Transmission electron microscope images of γ'-precipitates in a Ni-13.8 wt.% Al alloy aged at 1023 K for 15 min, 2 h, 4 h, and 8 h (from left to right) [10,11]. Their shapes change into cuboids. (From Maheshwari, A. and Ardell, A.J., *Phys. Rev. Lett.*, 70, 2305, 1993 and Zhu, J.Z. et al., *Acta Mater.*, 52, 2837, 2004. With permission.)

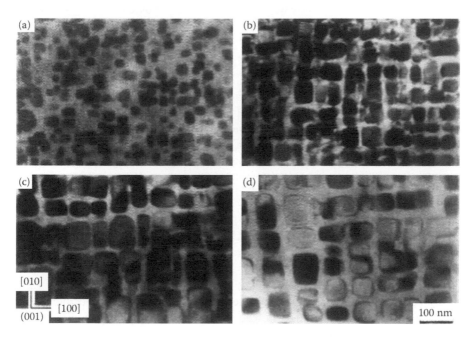

FIGURE 8.3 Evolution of Ni_4Mo precipitates in a Ni-16.3 at.% Mo alloy with isothermal aging for (a) 12.8 ks, (b) 864 ks, (c) 2.6 Ms, and (d) 5.2 Ms (From Miyazaki, T. and Doi, M., *Mater. Sci. Eng. A*, 110, 175, 1989. With permission.) The coarsening is nearly pinned in (d).

whether the deformation is stretching or compression. In these three examples, phase separation and an order-disorder transition occur simultaneously, as will be discussed in Section 8.3.

The organization of this chapter is as follows. In Section 8.2, we discuss elastic effects in phase separation separation by setting up the free energy and the dynamical equation coupled to elasticity in the coherent condition. In Section 8.3, we explain Ginzburg–Landau (GL) theories of order-disorder phase transitions in BCC and FCC cubic solids without and with elasticity. In Section 8.4, we present a phase-field model

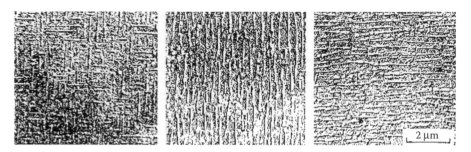

FIGURE 8.4   Replica microphotographs from the (100) side in a Ni-15 at.% Al alloy with no external stress (left frame), with tension (center frame), and with compression (right frame) (From Miyazaki, J. et al., *J. Mater. Sci.*, 14, 1827, 1979. With permission.) The stress was 147 MPa in the vertical direction. There were few effects of stress on the domain shapes in the early stages.

of phase transitions coupled to nonlinear elasticity, which can describe the birth and growth of dislocations in two-phase states. There are many important problems in the context of phase transitions coupled to elasticity. In a variety of structural phase transitions, we can see many unusual effects [14–16]. For example, in martensitic transitions, unique transitions, unique anisotropic domains can appear in cubic-to-tetragonal transformations. In ferroelectric solids, the electric polarization is the order parameter coupled to elasticity, and the domain dynamics is of great technological importance. In soft materials like gels and polymer solutions, phase separation is greatly influenced by elasticity on mesoscopic scales [2,17,18] (also see Chapter 7).

8.2   PHASE SEPARATION IN SOLIDS: COHERENT CASE

8.2.1   FREE-ENERGY FUNCTIONAL

We consider a binary alloy composed of A and B components (see Section 1.2.1). Assuming there are no vacancies and interstitials, we write the compositions of the two components as $c_A = c$ and $c_B = 1 - c$ with $0 \le c \le 1$. In this section, we neglect order-disorder phase transitions of the atomic configuration in each unit cell—then the composition c is the sole order parameter in the system. We assume cubic elastic symmetry and take the x, y, and z axes as being parallel to the crystal principal axes. The elastic displacement vector $\vec{u} = (u_x, u_y, u_z)$ is coupled to c in the free-energy functional

$$F = \int d\vec{r} f(c, \vec{u}). \tag{8.1}$$

We set $\vec{u} = 0$ in a reference one-phase crystal state without defects, where $c = $ const. We assume the free-energy density $f(c, \vec{u})$ has the form

$$f(c, \vec{u}) = f_0(c) + \frac{1}{2} C_0 |\vec{\nabla} c|^2 + \alpha c e_1 + f_{el}(\vec{u}, c). \tag{8.2}$$

The chemical part $f_0(c)$ is assumed to be of the Bragg–Williams form introduced in Section 1.2.2:

$$v_0 f_0(c) = k_B T[c \ln c + (1 - c) \ln(1 - c)] + 2k_B T_{c0} c(1 - c). \tag{8.3}$$

Here, v_0 is the volume of a unit cell in the reference state, and T_{c0} is the mean-field critical temperature in the absence of the coupling to elasticity. The lattice constant in the reference state is given by $v_0^{1/d}$ in d dimensions. The above free energy may be used in the strong-segregation case, where the composition difference Δc between the two phases is not small. The simpler Landau expansion follows in the weak-segregation case (cf. Equation 1.9):

$$f_0(c) = \frac{1}{2} a_0 (T - T_{c0}) \psi^2 + \frac{1}{4} u_0 \psi^4, \tag{8.4}$$

which is the expansion with respect to the composition deviation from the critical value c_c,

$$\psi = c - c_c. \tag{8.5}$$

Here, $c_c = 1/2$, $a_0 = 2k_B/v_0$, and $u_0 = k_B T/(6v_0)$ for the Bragg–Williams free energy in Equation 8.3. The second term in Equation 8.2 is the usual gradient term, accounting for an increase of the free energy due to inhomogeneity of $c(\vec{r})$. The coefficient C_0 is a positive constant of microscopic order ($\sim k_B T_0 v_0^{2/d}$).

The third term in Equation 8.2 is a coupling between the composition and the *volume change* or the *dilation strain*,

$$e_1 = \vec{\nabla} \cdot \vec{u} = \epsilon_{xx} + \epsilon_{yy} + \epsilon_{zz}, \tag{8.6}$$

where we define the symmetrized strain tensor $\epsilon_{ij} \equiv (\nabla_i u_j + \nabla_j u_i)/2$ with $\nabla_i = \partial/\partial x_i$. This coupling arises from a difference in the atomic sizes of the two components and has the following consequences.

(i) If the system is homogeneous in a one-phase state, the volume change measured from the state at $c = c_c$ is given by $e_1 = -\alpha(c - c_c)/K$ in the stress-free boundary condition, where K is the bulk modulus (its definition is provided below). If K is independent of c, the lattice constant a depends on c as [3]

$$\eta \equiv \frac{d \ln a}{dc} = -\frac{\alpha}{dK}. \tag{8.7}$$

This linear composition dependence, called the *Vegard law*, has been observed in many binary alloys in the stress-free boundary condition in one-phase states.

(ii) In two-phase states, the normalized lattice misfit $\Delta a/a$ is of order $\eta \Delta c$ in the coherent case, with Δc being the composition difference.

(iii) In real alloys, $|\eta|$ is at most of order 0.1 in crystal states. It should be noted that binary alloys can be in a crystal state without segregation only for small atomic size differences. For larger size differences, grain boundaries emerge enclosing small crystalline regions with different crystal axes (polycrystal), and eventually defects are proliferated resulting in highly disordered, amorphous solids (i.e., glasses) [19].

Finally, we consider the elastic free energy $f_{el}(\vec{u}, c)$ in cubic solids. In $d = 3$, the x, y, and z axes are taken along the principal crystal axes, and the strain components (other than e_1) used in the literature are introduced as [2]

$$e_2 = \epsilon_{xx} - \epsilon_{yy}, \quad e_3 = \frac{1}{\sqrt{3}}(2\epsilon_{zz} - \epsilon_{xx} - \epsilon_{yy}),$$

$$e_4 = 2\epsilon_{xy}, \quad e_5 = 2\epsilon_{yz}, \quad e_6 = 2\epsilon_{zx}. \tag{8.8}$$

We call e_2 and e_3 the tetragonal strains and e_3, e_4, and e_6 the shear strains. For small strains, we may use the following standard form bilinear in the strains [20],

$$f_{el}(\vec{u}, c) = \frac{K}{2}e_1^2 + \frac{\mu_2}{2}(e_2^2 + e_3^2) + \frac{\mu_3}{2}(e_4^2 + e_5^2 + e_6^2). \tag{8.9}$$

The first term arises from dilation, and the second and third terms arise from anisotropic strains. In terms of the usual elastic moduli C_{11}, C_{12}, and C_{44} [20], we have

$$\mu_2 = \frac{1}{2}(C_{11} - C_{12}), \quad \mu_3 = C_{44}, \tag{8.10}$$

where μ_2 is often written as C' in the literature. The case of isotropic elasticity is obtained for $\mu_2 = \mu_3 = \mu$, where we obtain the standard expression [20]

$$f_{el} = \frac{1}{2}Ke_1^2 + \mu \sum_{ij} \left(\epsilon_{ij} - \frac{1}{d}\delta_{ij}e_1 \right)^2, \tag{8.11}$$

which is invariant with respect to space rotations. For $\mu_2 \neq \mu_3$, anisotropic domains appear in phase separation [1,4,21]. The degree of cubic elastic anisotropy is often represented by

$$\xi_a = 2\frac{\mu_2}{\mu_3} - 2 = \frac{C_{11} - C_{12}}{C_{44}} - 2. \tag{8.12}$$

The composition-dependence of the elastic moduli, called the *elastic inhomogeneity*, gives rise to various important effects even when the dependence is weak [22–25]. For simplicity, we assume that the moduli $\mu_2 = \mu_2(c)$ and $\mu_3 = \mu_3(c)$ depend linearly on c as

$$\mu_2 = \mu_{20} + \mu_{21}(c - c_c),$$

$$\mu_3 = \mu_{30} + \mu_{31}(c - c_c), \tag{8.13}$$

where μ_{20}, μ_{30}, μ_{21}, and μ_{31} are constants independent of c. If $\mu_{21} > 0$ and $\mu_{31} > 0$, the regions with larger (smaller) c are elastically harder (softer) than those with smaller (larger) c. When μ_{21} and μ_{31} have different signs, we encounter some intriguing domain structures [2,23,24]. The elastic inhomogeneity should be weak when the elastic moduli of pure metals A and B are nearly the same, but it should be strong for large differences in the elastic moduli.

8.2.2 Elastic Stress, Elastic Equilibrium, and Dynamical Equation

We define the symmetric elastic stress tensor σ_{ij}, which satisfies

$$\frac{\delta F}{\delta u_i} = -\sum_j \nabla_j \sigma_{ij}, \tag{8.14}$$

at fixed c. In the linear-elasticity regime, σ_{ij} is given by

$$\begin{aligned} \sigma_{ii} &= (C_{11} - C_{12})\epsilon_{ii} + C_{12}e_1 + \alpha\psi, \\ \sigma_{ij} &= 2C_{44}\epsilon_{ij}, \qquad i \neq j. \end{aligned} \tag{8.15}$$

The diagonal component is rewritten as

$$\sigma_{ii} = C'\left(\epsilon_{ii} - \frac{e_1}{d}\right) + Ke_1 + \alpha\psi, \tag{8.16}$$

so that

$$K = C_{12} + \frac{C'}{d}. \tag{8.17}$$

The composition deviation changes the diagonal part by $\alpha\psi$ in the current model.

On the slow timescales of the composition field, the elastic field instantaneously relaxes and adjusts to a given composition field. Therefore, most previous theories of phase separation in solids have assumed a *mechanical equilibrium* condition in the evolution of the composition [1–8]:

$$\frac{\delta F}{\delta u_i} = -\sum_j \nabla_j \sigma_{ij} = 0. \tag{8.18}$$

Thus, $\vec{u}$ is determined as a functional of c. In general, we may apply an average homogeneous strain to a solid as

$$\langle \nabla_j u_i \rangle = A_{ij}, \tag{8.19}$$

where $\langle \cdots \rangle$ represents the space average in the solid. The displacement may be divided into an average and the deviation as

$$u_i = \sum_j A_{ij}x_j + \delta u_i, \tag{8.20}$$

where the tensor A_{ij} represents the affine deformation of the system. For simplicity, we fix the solid shape imposing clamped or periodic boundary conditions on $\delta \vec{u}$, where $\delta \vec{u}$ vanishes on the boundaries in the clamped case. Notice that the space average of $\sum_{ij} \sigma_{ij} \nabla_i \delta u_j$ vanishes in these cases. Because

$$f_{el} = \frac{1}{2} \sum_{ij} \sigma_{ij} \nabla_i u_j - \frac{1}{2} \alpha e_1 \psi \qquad (8.21)$$

holds even in the presence of elastic inhomogeneity, the total free energy is written as

$$F = \int d\vec{r} \left(f_0 + \frac{1}{2} C_0 |\vec{\nabla}\psi|^2 + \frac{1}{2} \alpha \psi e_1 + \frac{1}{2} \sum_{ij} \sigma_{ij} A_{ij} \right), \qquad (8.22)$$

under the mechanical equilibrium condition. The last term in the brackets is important only in the presence of elastic inhomogeneity. (If the elastic moduli are constants, σ_{ij}'s are linear in the strains or the composition.)

We next present the dynamical equation. In Section 1.5.1, the composition field is assumed to obey the Cahn–Hilliard (CH) diffusive equation,

$$\frac{\partial}{\partial t} c(\vec{r}, t) = \vec{\nabla} \cdot \left[\lambda_0 c(1 - c) \vec{\nabla} \mu \right]. \qquad (8.23)$$

Here, $D = \lambda_0 c(1 - c)$ is the composition-dependent kinetic coefficient, and $\mu = \delta F / \delta c$ is the chemical potential difference $\mu_A - \mu_B$ between the two components (in the presence of coupling to elasticity here) [2]. From Equations 8.2 and 8.3, we obtain

$$\mu = f_0' - C_0 \nabla^2 c + \alpha e_1 + \frac{\mu_{21}}{2}(e_2^2 + e_3^2) + \frac{\mu_{31}}{2}(e_4^2 + e_5^2 + e_6^2), \qquad (8.24)$$

where $f_0' = \partial f_0 / \partial c$. Here, we have

$$\frac{v_0 f_0'}{k_B} = T \ln \left(\frac{c}{1 - c} \right) + 2 T_0 (1 - 2c) \qquad (8.25)$$

for the Bragg–Williams form in Equation 8.3. The first two terms in Equation 8.24 are the usual terms in GL theory, as seen in previous chapters. The third term αe_1 is due to the lattice misfit, and the last two terms arise from the elastic inhomogeneity. The composition-dependence of the kinetic coefficient is important in the dilute limit of $c \rightarrow 0$ or $c \rightarrow 1$, where Equation 8.23 becomes the diffusion equation $\partial c / \partial t = D_0 \nabla_c^2$ if the coupling to the elastic field is neglected. The diffusion constant in the dilute limit is given by

$$D_0 = \frac{\lambda_0 k_B T}{v_0}. \qquad (8.26)$$

For the weak-segregation case, we neglect the composition-dependence of the kinetic coefficient and use the Landau expansion in Equation 8.4 to obtain (cf. Equation 1.177)

$$\frac{\partial}{\partial t}\psi(\vec{r},t) = \lambda_c \nabla^2 \mu, \tag{8.27}$$

where $\lambda_c = \lambda_0/4$ and $f_0' = a_0(T - T_{c0})\psi + u_0\psi^3$. We are neglecting the random source terms (divergence of the random diffusion current) in Equations 8.23 and 8.27 [2,26]. As mentioned in Section 1.5.1, the conserved order parameter model (without coupling to an elastic field) was introduced in the context of critical dynamics and has been called *Model B* [26].

8.2.3 BILINEAR INTERACTIONS IN CUBIC SOLIDS WITH HOMOGENEOUS ELASTIC MODULI

In cubic solids, elimination of the elastic field yields an anisotropic bilinear interaction term in the composition fluctuations, even for homogeneous elastic moduli [1,4,21]. This can give rise to cuboidal domains in simulations and can explain observations such as the one in Figure 8.2. However, its inclusion in simulations does not essentially alter the Lifshitz–Slyozov (LS) growth law $R(t) \propto t^{1/3}$ for the domain size in phase-separating systems [21].

We assume the mechanical equilibrium condition in Equation 8.18 and impose the fixed-volume condition without applied anisotropic strain. It is then convenient to consider the Fourier-transformed variables:

$$u_j(\vec{k}) = \sum_{\vec{r}} u_j(\vec{r})e^{i\vec{k}\cdot\vec{r}},$$

$$\psi_{\vec{k}} = \sum_{\vec{r}} \psi(\vec{r})e^{i\vec{k}\cdot\vec{r}}. \tag{8.28}$$

After some algebra, the mechanical equilibrium condition gives

$$u_j(\vec{k}) = \frac{ik_j}{C_{44}\left[1 + \varphi_0(\hat{k})\right](k^2 + \xi_a k_j^2)}\,\alpha\psi_{\vec{k}}. \tag{8.29}$$

The function $\varphi_0(\hat{k})$ depends on the direction $\hat{k} = k^{-1}\vec{k}$ of the wave vector $\vec{k}$ as

$$\varphi_0(\hat{k}) = \left(1 + \frac{C_{12}}{C_{44}}\right)\sum_j \frac{1}{1 + \xi_a \hat{k}_j^2}\,\hat{k}_j^2. \tag{8.30}$$

From Equation 8.22, the elastic part of F (denoted as F_{cub}) becomes a bilinear cubic interaction,

$$F_{\text{cub}} = \int d\vec{r}\,\frac{1}{2}\alpha e_1 \psi(\vec{r}) = \int \frac{d\vec{k}}{(2\pi)^d}\frac{1}{2}\tau_{\text{el}}(\hat{k})|\psi_{\vec{k}}|^2. \tag{8.31}$$

The coefficient $\tau_{el}(\hat{k})$ is negative and angle-dependent as [1,2]

$$
\tau_{el}(\hat{k}) = -\frac{\alpha^2}{C_{12} + C_{44}}\left[1 - \frac{1}{1 + \varphi_0(\hat{k})}\right]
$$

$$
= -\frac{\alpha^2(1 + 2\gamma_1 + 3\gamma_2)}{C_{11} + (C_{11} + C_{12})\gamma_1 + (C_{11} + 2C_{12} + C_{44})\gamma_2}, \tag{8.32}
$$

where the second line is the expression in $d = 3$ with

$$
\gamma_1 = \xi_a\left(\hat{k}_x^2\hat{k}_y^2 + \hat{k}_y^2\hat{k}_z^2 + \hat{k}_z^2\hat{k}_x^2\right),
$$

$$
\gamma_2 = \xi_a^2\left(\hat{k}_x\hat{k}_y\hat{k}_z\right)^2. \tag{8.33}
$$

Here, $\tau_{el}(\hat{k})$ is equal to $\tau_{el}[100] = -\alpha^2/C_{11}$ in the [100]-direction, and to $\tau_{el}[111] = -\alpha^2/L$ with $L = K + (2 - 2/d)C_{44} = C_{11} - (1 - 1/d)\xi_a C_{44}$ in the [111]-direction. However, its dependence on the angle is complicated, in general. We consider the case of small cubic anisotropy ξ_a to obtain the expansion,

$$
\tau_{el}(\hat{k}) = -\frac{\alpha^2}{C_{11}} + \tau_{cub}\left(\hat{k}_x^2\hat{k}_y^2 + \hat{k}_y^2\hat{k}_z^2 + \hat{k}_z^2\hat{k}_x^2\right) + O\left(\xi_a^2\right), \tag{8.34}
$$

where $\tau_{cub} = -2\alpha^2 C_{44}\xi_a/C_{11}^2$. The second (anisotropic) term in Equation 8.34 becomes $\tau_{cub}\hat{k}_x^2\hat{k}_y^2$ in $d = 2$. In earlier simulations, the above expansion has been used because of its simplicity.

The directions of the wave vector that minimize $\tau_{el}(\hat{k})$ [and maximize $\varphi_0(\hat{k})$] are called *elastically soft* directions. For most cubic crystals, ξ_a is negative (or $\mu_2 < \mu_3$) and the soft directions are $[\pm 1\,0\,0], [0\pm 1\,0]$, and $[0\,0\pm 1]$. On the other hand, if $\xi_a > 0$ (or $\mu_2 > \mu_3$), the softest directions are $[\pm 1\pm 1\pm 1]$. These results are obvious from the expansion in Equation 8.34 as $\tau_{cub} \propto \xi_a$. As the temperature is lowered, the composition fluctuations varying in the soft directions trigger spinodal decomposition, whereas those varying in the other directions remain stable. Below the spinodal, phase separation proceeds with the interface planes tending to be perpendicular to one of the soft directions. Thus, for $\xi_a < 0$, the critical temperature is shifted as

$$
T_c = T_{c0} + \frac{\alpha^2}{a_0 C_{11}}. \tag{8.35}
$$

Since Cahn's original papers [3,4], this T_c has been called the *coherent critical temperature*. Including this shift, we may introduce the coherent free-energy density,

$$
f_c = f_0 - a_0\left(T_c - \frac{1}{2}T_{c0}\right)\psi^2. \tag{8.36}
$$

In the weak-segregation case, we have $f_c = \tau\psi^2/2 + u_0\psi^4/4$ with

$$
\tau = a_0(T - T_c). \tag{8.37}
$$

We rewrite the CH equation in the weak-segregation case (Equation 8.27) by eliminating the elastic field. In terms of τ in Equation 8.37, it becomes

$$\frac{\partial}{\partial t}\psi(\vec{r},t) = \lambda_c \nabla^2 \left(\tau - C_0\nabla^2 + u_0\psi^2\right)\psi + \lambda_c I, \tag{8.38}$$

where I arises from the elastic interaction. In $d = 2$, we use Equation 8.34 to obtain

$$I = \nabla^2\left(\frac{\delta F_{cub}}{\delta\psi}\right) = \tau_{cub}\nabla_x^2\nabla_y^2 w. \tag{8.39}$$

Here, w is a potential determined by the Laplace equation,

$$\nabla^2 w = \delta\psi, \quad \text{or} \quad w = \frac{1}{\nabla^2}\delta\psi, \tag{8.40}$$

where $\delta\psi = \psi - \langle\psi\rangle$. Here, $1/\nabla^2$ is the inverse operator of ∇^2, so the Fourier component of w is related to that of ψ as $w_{\vec{k}} = -k^{-2}\psi_{\vec{k}}$. We integrated the dynamical equation 8.38 in the case $\tau < 0$ on a 128×128 lattice with periodic boundary conditions [21]. We measure space in units of $\xi_b = (C_0/|\tau|)^{1/2}$ and time in units of $t_0 = \xi_b^4/(\lambda_c C_0)$. In Figure 8.5, we display domain structures after quenching at $t = 0$ with $\tau_{cub}/|\tau| = 0.675$. The volume fraction of one component is 0.5 in the frames on the left and 0.3 in the frames on the right. The softest directions are [01] and [10], so the domains are rectangular stripes aligned in [10] or [01] without anisotropic external stress. The lengths of the shorter sides have a sharp peak at a length scale $R(t) \sim t^a$ with $a = 0.2$ to 0.3, and the lengths of the longer sides are broadly distributed. The growth law is not very different from the usual LS law $R(t) \sim t^{1/3}$, despite the highly anisotropic shapes of domains [21].

8.2.4 Interactions Arising from Elastic Inhomogeneity

In the presence of elastic inhomogeneity, we obtain a third-order interaction term in the composition fluctuations after elimination of the elastic field [5,6]. This gives rise to asymmetric domain shapes between soft and hard regions, leading to pinning of the coarsening. In addition, a dipolar interaction also arises under applied anisotropic strain [5,6].

In the case of isotropic elasticity, the shear modulus depends on $\psi = c - c_c$ as

$$\mu(c) = \mu_2(c) = \mu_3(c) = \mu_0 + \mu_1\psi, \tag{8.41}$$

and the bulk modulus is assumed to be a constant independent of c. We assume $\mu_1 > 0$: then, the phase with larger ψ is harder than that with smaller ψ. Hereafter, we calculate $\vec{u}$ by treating μ_1 as a small expansion parameter. This expansion is justified for $|\mu_1\psi| \ll L_0$, where L_0 is the elastic moduli for the longitudinal displacements (for longitudinal sounds),

$$L_0 = K + \left(2 - \frac{2}{d}\right)\mu_0. \tag{8.42}$$

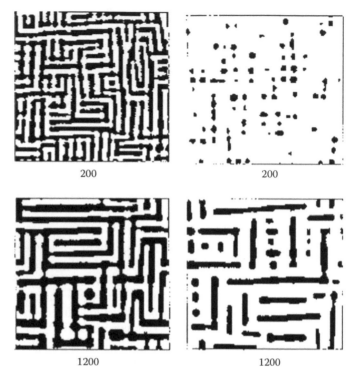

200   200

1200   1200

FIGURE 8.5 Evolution patterns in cubic alloys with $\tau_{\mathrm{cub}}/|\tau| = 0.625$, without external stress and elastic inhomogeneity. The volume fraction of one phase is 0.5 (left) and 0.3 (right). The numbers represent the time after quenching in units of $t_0 = \xi_b^4/\lambda_c C_0$. (From Nishimori, H. and Onuki, A., *Phys. Rev. B*, 42, 980, 1990. With permission.)

The mechanical equilibrium condition is rewritten as

$$\alpha \nabla_i \psi + (L_0 - \mu_0)\nabla_i e_1 + \mu_0 \nabla^2 u_i + 2\mu_1 \sum_j \nabla_j \left[\psi \left(\epsilon_{ij} - \frac{1}{d}\delta_{ij} e_1 \right) \right] = 0. \quad (8.43)$$

Taking the divergence of the above vector equation yields

$$\nabla^2 (L_0 e_1 + \alpha \psi) + 2\mu_1 \sum_{ij} \nabla_i \nabla_j \left[\psi \left(\epsilon_{ij} - \frac{1}{d}\delta_{ij} e_1 \right) \right] = 0. \quad (8.44)$$

The zeroth order solution for $\mu_1 = 0$ is calculated as

$$\delta \vec{u}^{(0)} = -\frac{\alpha}{L_0}\vec{\nabla} w, \quad (8.45)$$

where w is defined by Equation 8.40. The above form also follows from Equation 8.29 in the isotropic limit. Therefore, the dilation strain is expanded as

$$e_1 = -\frac{\alpha}{L_0}\psi + \frac{2\alpha\mu_1}{L_0^2 \nabla^2} \sum_{ij} \nabla_i \nabla_j \left[\psi \left(\nabla_i \nabla_j w - \frac{1}{d}\delta_{ij}\nabla^2 w \right) \right] + O(\mu_1^2). \quad (8.46)$$

Substituting the above results in the free energy in Equation 8.22, we may eliminate $\vec{u}$ in terms of ψ to obtain the free energy $F = F\{\psi\}$. To first order in μ_1, we find

$$F = \int d\vec{r}\left[f_c(\psi) + \frac{1}{2}C_0|\vec{\nabla}\psi|^2\right] + F_{\text{inh}} + F_{\text{ex}}, \qquad (8.47)$$

where the constant terms and the term linear in ψ are not written explicitly. We explain the three contributions in Equation 8.47.

(i) The free-energy density $f_c(\psi)$ is given by Equation 8.36 as C_{11} becomes L_0 in the isotropic case [3]. In his original paper [3], Cahn claimed a downward shift because his definition of the shift was $(\Delta T)_c \equiv T_c - T_{c0} - \alpha^2/(Ka_0) = -3\alpha^2\mu_0/(4KL_0a_0) < 0$ in the isotropic elasticity.

(ii) The second term F_{inh} is the third-order interaction due to the elastic inhomogeneity and is expressed as

$$F_{\text{inh}} = g_E \int d\vec{r} \, \psi\hat{Q}, \qquad (8.48)$$

where $g_E = \mu_1\alpha^2/L_0^2$. The quantity $\hat{Q}$ is written in terms of w in Equation 8.40 as

$$\hat{Q} = \sum_{ij}\left(\nabla_i\nabla_j w - \frac{1}{d}\delta_{ij}\nabla^2 w\right)^2, \qquad (8.49)$$

which represents the degree of anisotropic deformations.

(iii) From the last term on the right-hand side of Equation 8.22, the applied strain A_{ij} in Equation 8.19 yields a dipolar interaction [5,6],

$$F_{\text{ex}} = -g_{\text{ex}} \int d\vec{r} \sum_{ij} A_{ij}(\nabla_i\psi)(\nabla_j w)$$

$$= g_{\text{ex}} \sum_{\vec{k}} \sum_{ij} A_{ij}\hat{k}_i\hat{k}_j|\psi_{\vec{k}}|^2, \qquad (8.50)$$

where $g_{\text{ex}} = -2\mu_1\alpha/L_0$ ($\sim \mu_1\eta$ with η being defined by Equation 8.7). The second line is the expression in terms of the Fourier components $\psi_{\vec{k}}$. Because $g_{\text{ex}} \propto \mu_1$, the composition fluctuations can be influenced by externally applied strain only in the presence of elastic inhomogeneity. We may apply a uniaxial deformation, for which the strain tensor $\{A_{ij}\}$ is expressed as $A_{zz} = \lambda_{\parallel}$, $A_{jj} = \lambda_{\perp}$ ($j \neq z$), and $A_{ij} = 0$ ($i \neq j$). Then, F_{ex} becomes of the same form as the dipolar interaction in uniaxial ferromagnets [27,28],

$$F_{\text{ex}} = g_{\text{ex}}(\lambda_{\parallel} - \lambda_{\perp})\sum_{\vec{k}}\left(\hat{k}_z^2 - \frac{1}{d}\right)|\psi_{\vec{k}}|^2. \qquad (8.51)$$

For positive $g_{ex}(\lambda_\parallel - \lambda_\perp)$, the interface normal tends to be perpendicular to the z-axis. If this quantity is negative, the interface normal tends to be parallel to the z-axis. Thus, the observation in Figure 8.4 is in accord with our expression for F_{ex}. Furthermore, in the experiment, the effect of applied stress appeared only in late-stage phase separation [13], which is the case for $|g_{ex}(\lambda_\parallel - \lambda_\perp)| < |\tau|$ in our theory.

The elastic inhomogeneity leads to unique domain morphologies. Next, we present two-dimensional simulation results in the case of isotropic elasticity with F_{inh} but without F_{ex}. The dynamical equation is of the form in Equation 8.38 with

$$I = \nabla^2\left(\frac{\delta F_{inh}}{\delta\psi}\right) = g_E\nabla^2\hat{Q} + g_E\sum_{ij}\nabla_i\nabla_j[\psi(2\nabla_i\nabla_j w - \delta_{ij}\nabla^2 w)]. \qquad (8.52)$$

This equation was integrated for $\tau < 0$ on a 128×128 lattice with periodic boundary conditions [22]. To characterize the strength of the inhomogeneity, we introduce the dimensionless parameter,

$$g_E^* = \frac{g_E}{\sqrt{|\tau|u_0}}$$

$$= \frac{\mu_1\alpha^2}{L_0^2\sqrt{|\tau|u_0}} \sim \frac{v_0\mu_1\eta^2}{k_B\sqrt{T(T_c - T)}}, \qquad (8.53)$$

where η is defined by Equation 8.7. In Figure 8.6, we display the evolution of domains at $g_E^* = 0.07$ in $d = 2$, where the volume fraction of the soft region is $\phi_s = 0.7, 0.5$, and 0.3. Space and time are measured in units of $\xi_b = (C_0/|\tau|)^{1/2}$ and $t_0 = \xi_b^4/(\lambda_c C_0)$. We can see significant shape changes of domains from circles. In Figures 8.6b and c, once the soft regions form networks enclosing hard droplet-like domains, the domain coarsening is nearly stopped. In accordance with the observation in Figure 8.3, the soft phase tends to be percolated even if it is the minority phase. The two-phase states are driven into metastable states because of asymmetric elastic deformations in the soft and hard regions. This picture becomes evident in Figure 8.7, where the degree of anisotropic deformations $\hat{Q}$ is displayed in a pinned state with $\phi_s = 0.5$ and $g_E^* = 0.05$. This plot shows *mountains*, characteristic of local elastic energy barriers preventing further coarsening. In two-phase states, the surface energy ($\sim \sigma R^{d-1}$, σ being the surface tension) and the elastic inhomogeneity free energy [$\sim g_E(\Delta c)^3 R^d$] per domain are of the same order. When the pinning sets in, these free energies should be balanced, leading to the characteristic length,

$$R_E = \frac{\sigma}{g_E(\Delta c)^3}. \qquad (8.54)$$

Furthermore, we may introduce crossover values of the reduced temperature and the average order parameter $M = \langle\psi\rangle$ as

$$\tau_E = \frac{g_E^2}{u_0}, \quad M_E = \frac{g_E}{u_0}. \qquad (8.55)$$

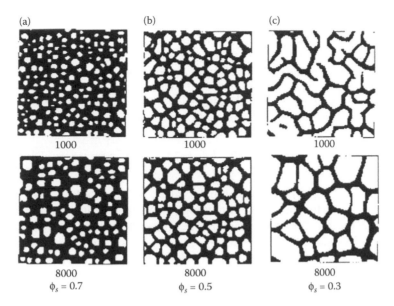

FIGURE 8.6 Evolution patterns at $g_E^* = 0.07$ for (a) $\phi_s = 0.7$, (b) $\phi_s = 0.5$, and (c) $\phi_s = 0.3$, with elastic inhomogeneity in isotropic elastic theory. The numbers denote the times after quenching at $t = 0$. The black regions represent soft domains, and the white regions represent hard domains. (From Onuki, A. and Nishimori, H., *Phys. Rev. B*, 43, 13649, 1991. With permission.)

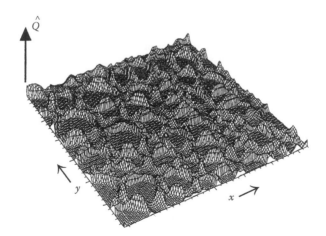

FIGURE 8.7 "Mountain structure" of the degree of anisotropic elastic deformations $\hat{Q}$ for $\phi_s = 0.5$ and $g_E^* = 0.05$ in isotropic elasticity theory. The soft regions are mostly uniaxially deformed, whereas the hard regions are only isotropically dilated and $\hat{Q} \simeq 0$. (From Onuki, A. and Nishimori, H., *Phys. Rev. B*, 43, 13649, 1991. With permission.)

Then, $g_E^* = (\tau_E/|\tau|)^{1/2}$ in Equation 8.53. In mean-field theory, we have $\Delta c \sim (|\tau|/u_0)^{1/2}$ and $\sigma \sim C_0(\Delta c)^2/\xi_b$, so that $R_E \sim (C_0/\tau_E)^{1/2} \sim \xi_b/g_E^*$. We recognize that $R_E \gg \xi_b$ for $g_E^* \ll 1$ or for $|\tau| \gg \tau_E$. It was also found that the LS growth law $R(t) \sim t^{1/3}$ is obeyed for $R < R_E$ or in the time-regime $t < t_0$ $(\xi_b/R_E)^3$ [2,25]. The pinned two-phase states realized in Figure 8.6 are metastable or glassy if the free-energy barrier, written as ΔF_E, is much larger than the thermal energy $k_B T$. The barrier (per domain) is estimated as

$$\Delta F_E \sim \sigma R_E^{d-1} = k_B T A_\sigma \left(\frac{R_E}{\xi_b}\right)^{d-1}, \qquad (8.56)$$

where $A_\sigma \equiv \sigma \xi_b^{d-1}/(k_B T)$. In mean-field theory, A_σ is of order $C_0^2 \xi_b^{d-4}/(k_B T u_0)$ (while $A_\sigma \sim 0.1$ in the asymptotic critical region). We notice that the condition $A_\sigma \gtrsim 1$ is equivalent to the Ginzburg criterion that ensures the validity of mean-field theory [2], so we find $\Delta F_E \gg k_B T$ for $g_E^* \ll 1$. Figure 8.8 shows the phase diagram in the (M, τ)-plane, where $M = \langle \psi \rangle$. We show the instability points of one-phase states (*) and those of two-phase glassy states ($\times$). A first-order phase transition, occurs at the transition points (+), where the free-energy values in the two states coincide. There is no continuous transition, and the critical point becomes nonexistent with elastic inhomogeneity. The *coherent critical point* is meaningful only for weak elastic inhomogeneity. For $|\tau| \lesssim \tau_E$ (or for $g_E^* \gtrsim 1$), and $|M| \lesssim M_E$, the barrier in the glassy two-phase states is weakened, which suggests the appearance of periodic two-phase states in equilibrium. Note that the simulations yielding Figure 8.8 were performed without a random diffusion current, that is, no noise.

8.2.5 EFFECT OF $F_{cub} + F_{inh}$

Next, we examine the domain morphology in the presence of F_{cub} and F_{inh} without applied anisotropic stress [29]. In Figure 8.9, we display the evolution and pinning of two-phase patterns for $\phi_s = 0.5$ (upper frames) and 0.3 (lower frames) in the presence of F_{cub} and F_{inh}, where we set $\tau_{cub}/|\tau| = 0.675$ and $g_E^* = 0.07$. Here, the role of elastic cubic anisotropy is to orient the interfaces in the preferred directions. The patterns obtained closely resemble those in Figure 8.3 [12], observed in a Ni-base alloy at high volume fractions of the hard ordered phase, where the component with smaller C_{44} forms a network. We may also calculate the phase diagram in the presence of F_{cub}, which is similar to that in Figure 8.8 [25].

8.3 ORDER-DISORDER PHASE TRANSITIONS AND PHASE SEPARATION IN BINARY ALLOYS

In real alloys, it is common that an order-disorder phase transition and phase separation occur simultaneously. Figure 8.10 illustrates representative intermetallic ordered phases. In this section, we consider body-centered-cubic (BCC) and face-centered-cubic (FCC) binary alloys, neglecting interstitials and vacancies. In particular, we show a number of experiments on Ni-base alloys. The Ni-atom can easily form solid solutions, intermetallic phases, or chemical compounds with other elements. This affinity stems from the partially-filled 3d shell of Ni.

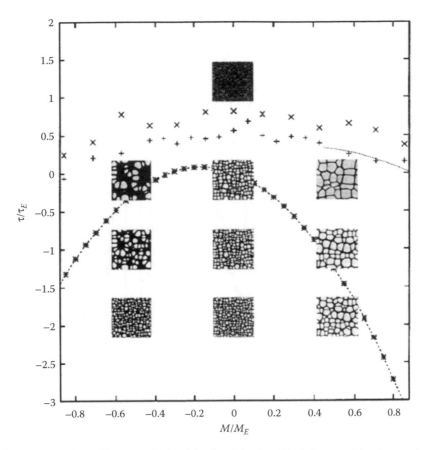

FIGURE 8.8 Phase diagram calculated in $d = 2$ in the (M, τ) (composition-temperature) plane with elastic inhomogeneity, and in isotropic elasticity theory. The data points are as follows: +, first-order transition points; *, instability points of one-phase states; ×, instability points of pinned two-phase states. The dotted line is a theoretical first-order transition line. The domain patterns in pinned states are also shown, with the soft regions marked black. (From Onuki, A. and Furukawa, A., *Phys. Rev. Lett.*, 86, 452, 2001. With permission.)

8.3.1 GINZBURG–LANDAU THEORY OF BCC ALLOYS

Let us consider a binary (AB) alloy forming a BCC lattice such as Fe-Be and Cu-Zn. The lattice may be divided into two sublattices, as shown in Figure 8.10. The compositions of A atoms on the two sublattice sites are written as

$$c_1 = c + \frac{1}{2}\eta, \quad c_2 = c - \frac{1}{2}\eta, \tag{8.57}$$

where η is the order parameter of the order-disorder phase transition (which should not be confused with η in Equation 8.7). The compositions of B atoms are $1 - c_1$ and $1 - c_2$ on the two sublattice sites. We may assume $0 < c \le 1/2$ without loss of generality—then $|\eta| \le 2c$. The lattice structure in the ordered phase ($\eta \ne 0$) is

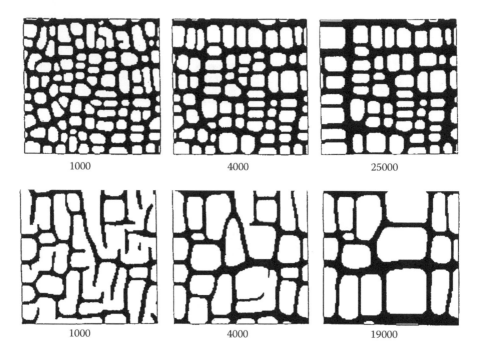

<table>
<tr><td>1000</td><td>4000</td><td>25000</td></tr>
</table>

<table>
<tr><td>1000</td><td>4000</td><td>19000</td></tr>
</table>

FIGURE 8.9 Evolution patterns in the presence of elastic inhomogeneity in a cubic solid with $\tau_{cub} = 0.675$, $g_E = 0.07$, and $g_{cub} = 0$ at $\phi_s = 0.5$. The snapshots resemble those in Figure 8.3. The domain growth is pinned at the final times. (From Nishimori, H. and Onuki, A., *J. Phys. Soc. Jpn.*, 60, 1208, 1991. With permission.)

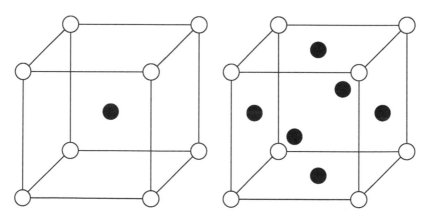

FIGURE 8.10 The left frame shows the $L1_0$ structure on a BCC lattice—examples are Fe-Be and Cu-Zn. The right frame shows the $L1_2$ structure on a FCC lattice—examples are Al_3Li, Ni_3Cr, etc. (Al: ●; Li: ○). For $Al + Li$, domains of the $L1_2$ structure (γ'-precipitates) appear in an Al-rich metastable, disordered phase (γ-matrix), as T is lowered or the Li composition is increased.

called L1$_0$ or B2. The system is invariant with respect to a change of the sign of η because the two sublattices are symmetrical, so the free energy is an even function of η. We assume there are short-range interactions between the nearest-neighbor and next-nearest-neighbor pairs. The chemical part of the free-energy density is then of the following form [1,30,31]:

$$v_0 f_0(c, \eta) = \frac{1}{2} k_B T \sum_{k=1,2} [c_k \ln c_k + (1 - c_k) \ln(1 - c_k)] - w_0 c^2 - w_1 \eta^2, \quad (8.58)$$

where the linear term in c is not written explicitly, and f_0 is of the form in Equation 8.3 for $\eta = 0$. The two parameters w_0 and w_1 are combinations of the pair interaction energies—phase separation is favored by increasing w_0 and structural ordering by increasing w_1. For the moment, we neglect the coupling to elasticity and discuss the phase behavior determined by $f_0(c, \eta)$. The phase diagrams are very complicated, depending on the ratio w_0/w_1, as illustrated in Figure 8.11 [30]. The equation that determines η follows from $\partial f_0/\partial \eta = 0$ (at fixed c) as

$$\ln \left[\frac{(c + \eta/2)(1 - c + \eta/2)}{(c - \eta/2)(1 - c - \eta/2)} \right] = \frac{8w_1}{k_B T} \eta. \quad (8.59)$$

Here, η can be nonvanishing only for $w_1 > 0$. The solution $\eta = \eta(c)$, which gives the minimum of $f_0(c, \eta)$ at each c, needs to be calculated. Then $f_0(c, \eta(c))$ becomes a function of c only, and determines the phase behavior presented in Figure 8.11. Clearly, it is much more complicated than the Bragg–Williams form in Equation 8.3.

In particular, if η is small, we may expand f_0 with respect to η^2 as

$$f_0 = \frac{k_B T}{v_0} [c \ln c + (1 - c) \ln(1 - c)] - \frac{w_0}{v_0} c^2 + \frac{1}{2} \tau(c) \eta^2 + \frac{1}{4} \bar{u}_0(c) \eta^4 + \cdots,$$

$$(8.60)$$

where $v_0 \tau(c) = k_B T/[4c(1 - c)] - 2w_1$ and $v_0 \bar{u}_0(c) = k_B T[c^{-3} + (1 - c)^{-3}]$. The disordered phase becomes unstable for $\tau(c) < 0$ or for $k_B T < 8c(1 - c)w_1$. If the deviation $m = c - \langle c \rangle$ is small, we may approximate f_0 in the standard GL form [2,26],

$$f_0(m, \eta) = \frac{1}{2\chi_0} m^2 + \gamma_0 m \eta^2 + \frac{1}{2} \tau \eta^2 + \frac{1}{4} \bar{u}_0 \eta^4 + \frac{1}{6} v_0 \eta^6. \quad (8.61)$$

Here, χ_0, γ_0, $\bar{u}_0$, and v_0 are appropriate coefficients dependent on T and $\langle c \rangle$. The critical line is determined by $\tau = 0$ and $u_0 = \bar{u}_0 - 2\gamma_0 \chi_0 > 0$. In equilibrium, we may set $m = -\chi_0 \gamma_0 (\eta^2 - \langle \eta^2 \rangle)$. Then, the coefficient of the quartic term in f_0 is decreased to $u_0 = \bar{u}_0 - 2\gamma_0 \chi_0$. As a result, the tricritical condition $u_0 = 0$ can be realized if w_1

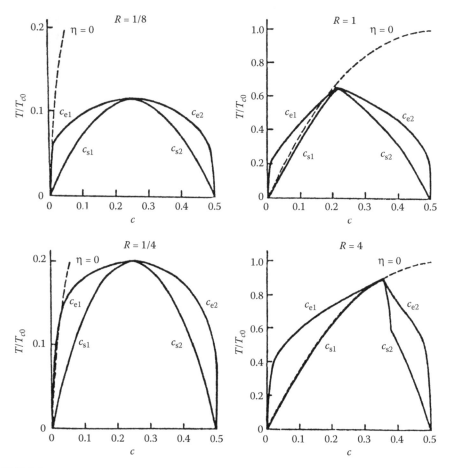

FIGURE 8.11 Phase diagrams of BCC alloys for $w_1 > 0$, obtained from Equation 8.58 without coupling to elasticity [30]. The parameter R is defined by $w_0/w_1 = 4(R - 1)/(R + 1)$. The temperature is scaled by $k_B T_{c0} = 2w_1$. The lines c_{e1} and c_{e2} denote the solubility (coexistence) lines, and c_{s1} and c_{s2} denote the spinodal lines. The instability line $T/T_{c0} = 4c(1 - c)$ against ordering is also shown (broken line). (From Ino, H., *Acta Metall.*, 26, 827, 1978. With permission.)

is larger than $w_0/4$ and $-w_0/2$. The tricritical temperature and composition are

$$T_t = \frac{8w_1(2w_1 + w_0)}{3k_B(4w_1 + w_0)},$$

$$c_t = \frac{1}{2} - \sqrt{\frac{1}{12} \cdot \frac{4w_1 - w_0}{4w_1 + w_0}}. \tag{8.62}$$

To simulate phase separation in BCC alloys including elastic effects, Sagui et al. [8] set up the following free-energy density:

$$f = f_0(m, \eta) + \frac{C_0}{2}|\vec{\nabla}\eta|^2 + (\alpha_1 m + \alpha_2 \eta^2)e_1 + f_{el}(\vec{u}, m, \eta). \tag{8.63}$$

The first chemical term is of the tricritical form in Equation 8.61, and the third term represents the coupling to the dilation strain. Sagui et al. studied the case of isotropic elasticity with inhomogeneous shear modulus,

$$\mu = \mu_0 + \mu_1 m + \mu_2 \eta^2. \tag{8.64}$$

After elimination of the elastic field, we obtain the elastic inhomogeneity interaction,

$$F_{\text{inh}} = \frac{1}{L_0^2} \int d\vec{r} (\mu_1 m + \mu_2 \eta^2) \sum_{ij} \left(\nabla_i \nabla_j w - \frac{1}{d} \delta_{ij} \nabla^2 w \right)^2, \tag{8.65}$$

correct to first order in μ_1 and μ_2. The quantity L_0 is the modulus in Equation 8.42, and w is determined by

$$\nabla^2 w = \alpha_1 m + \alpha_2 (\eta^2 - \langle \eta^2 \rangle). \tag{8.66}$$

The dynamical equations for m and η are given by

$$\frac{\partial}{\partial t} m(\vec{r}, t) = \lambda_0 \nabla^2 \left(\frac{\delta F}{\delta m} \right), \tag{8.67}$$

$$\frac{\partial}{\partial t} \eta(\vec{r}, t) = -L_0 \frac{\delta F}{\delta \eta}, \tag{8.68}$$

which are called the *Model C* equations in critical dynamics [26]. In Figure 8.12, the average composition is common in the upper and lower panels. In the upper panel, the ordered regions (white or gray) are harder and take droplet shapes, and the disordered regions (black) are percolated. They closely resemble the evolution patterns in Figure 8.6 of Model B coupled to isotropic elasticity, except that there are two variants of the ordered phase. Here we do not see antiphase boundaries (interfaces between the two variants). In the lower panel, the disordered regions (black) are harder—they form wetting layers in an early stage due to the nature of the Model C quench, but they tend to become droplet-like in late stages because they are hard. We can see the appearance of antiphase boundaries in the lower panel because the two variants touch after the shape changes of the disordered domains. We make some further remarks. (i) In the simulation, coarsening was observed to slow down considerably compared with the case without elasticity [2], but in the lower panel the asymptotic growth behavior remains unclear. (ii) The anisotropy arising from cubic elasticity is neglected in the simulation—this would give rise to closer resemblance of simulated patterns and real morphologies. (iii) Near the tricritical point, the elastic inhomogeneity interaction is marginal. In fact, R_E in Equation 8.54 is proportional to the correlation length ξ_b on the first-order transition line. Thus, for sufficiently small μ_1 and μ_2 in Equation 8.64, $R_E \gg \xi_b$ and the elastic inhomogeneity remains weak as the tricritical point is approached.

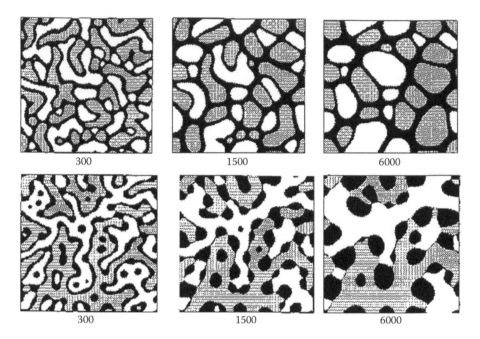

FIGURE 8.12 Evolution patterns in Model C when the ordered phase is the hard phase (upper frames) or the soft phase (lower frames). The snapshots correspond with $t = 300, 1500$, and 6000 (in appropriate units). (From Sagui, C., *Phys. Rev. E*, 50, 4865, 1994. With permission.)

8.3.2 GINZBURG–LANDAU THEORY OF FCC ALLOYS

We next consider a binary alloy (such as Al + Li or a Ni-base alloy) having a FCC lattice. As illustrated in Figure 8.10, the composition of A atoms on the corner sites (denoted by the subscript 1) and those on the face sites (denoted by the subscripts 2, 3, 4) are expressed as [1,32,33]

$$c_1 = c + \eta_1 + \eta_2 + \eta_3, \qquad c_2 = c + \eta_1 - \eta_2 - \eta_3,$$
$$c_3 = c + \eta_2 - \eta_3 - \eta_1, \qquad c_4 = c + \eta_3 - \eta_1 - \eta_2. \qquad (8.69)$$

Here, c is the average composition of A atoms, and $\vec{\eta} = (\eta_1, \eta_2, \eta_3)$ constitutes a three-component order parameter. Therefore, there can be a variety of ordered phases, as will be discussed below. The compositions of B atoms are given by $1 - c_k$, assuming there are no interstitials and vacancies. We consider the nearest-neighbor and next-nearest-neighbor pair interactions and obtain the following expression:

$$v_0 f_0 = \frac{k_B T}{4} \sum_{k=1}^{4} [c_k \ln c_k + (1 - c_k) \ln(1 - c_k)] - w_0 c^2 - w_1 \sum_{k=1}^{3} \eta_k^2, \qquad (8.70)$$

where w_0 and w_1 are appropriate combinations of the interaction energies. Because two atoms at the sites 1 and 2 (corner-face) and those at the sites 2 and 3 (face-face) are equally separated, they interact with the same potential. Thus, the nearest-neighbor interaction energy becomes proportional to $c_1(c_2 + c_3 + c_4) + (c_2c_3 + c_3c_4 + c_4c_2)$, leading to the last term of Equation 8.70. The Landau expansion of f_0 in powers of η_k becomes

$$\frac{v_0 f_0(c, \vec{\eta})}{k_B T} = c \ln c + (1 - c) \ln(1 - c) - \frac{w_0}{k_B T} c^2 + \left[\frac{1}{2c(1-c)} - \frac{w_1}{k_B T} \right] \sum_{k=1}^{3} \eta_k^2$$

$$+ \frac{2c-1}{2c^2(1-c)^2} \eta_1 \eta_2 \eta_3 + \frac{1}{12} \left[\frac{1}{c^3} + \frac{1}{(1-c)^3} \right]$$

$$\times \left[\sum_{k=1}^{3} \eta_k^4 + 6(\eta_1^2 \eta_2^2 + \eta_2^2 \eta_3^2 + \eta_3^2 \eta_1^2) \right] + \cdots . \tag{8.71}$$

The free energy is isotropic up to the second-order terms. The instability curve at homogeneous c and $\eta_k = 0$ ($k = 1, 2, 3$) is given by $T = 2w_1 c(1 - c)$ for $w_1 > 0$, below which small fluctuations of η_k grow. More phenomenologically, we may write the Landau expansion up to sixth-order terms from the symmetry requirements of the FCC structure as [33,34]

$$\frac{v_0 f_0(c, \vec{\eta})}{k_B T} = \frac{f_0(c)}{k_B T} + a_2 \sum_{k=1}^{3} \eta_k^2 + \left(a_3 + a_5 \sum_{k=1}^{3} \eta_k^2 \right) \eta_1 \eta_2 \eta_3$$

$$+ \left(a_{41} + a_{62} \sum_{k=1}^{3} \eta_k^2 \right) \sum_{k=1}^{3} \eta_k^4 + a_{42} \left(\eta_1^2 \eta_2^2 + \eta_2^2 \eta_3^2 + \eta_3^2 \eta_1^2 \right)$$

$$+ a_{61} \sum_{k=1}^{3} \eta_k^6 + a_{63} \eta_1^2 \eta_2^2 \eta_3^2 + \cdots , \tag{8.72}$$

where $f_0(c)$ depends only on c, and the coefficients $a_2, a_3, \ldots$ are functions of c and T.

(i) We consider the $L1_0$ phase first. Depending on the values of the coefficients at each c in Equation 8.71 or Equation 8.72, ordered states with the form $\vec{\eta} = (\pm\eta, 0, 0)$ can be stable, where we have $c_1 = c_2 = c + \eta_1$ and $c_3 = c_4 = c - \eta_1$. Then, the free energy in Equation 8.70 assumes the same form as that in Equation 8.58, leading to an $L1_0$ structure. Equivalently, we may set $\vec{\eta} = (0 \pm \eta, 0)$ or $(0, 0, \pm\eta)$. In $d = 3$, there are six variants with the $L1_0$ structure that emerge in phase-ordering processes. In real FCC crystals, such atomic displacements in a preferred direction cause a cubic-to-tetragonal change of the lattice structure, if the elasticity is coupled.

(ii) The $L1_2$ structure in Figure 8.10 is considered next. It is realized for the isotropic ordering $\eta_1 = \eta_2 = \eta_3 = \eta$, so $c = \eta = 1/4$ for a perfect $L1_2$

crystal. Here, the free-energy density becomes [35,36]

$$v_0 f_0(c, \eta) = \frac{k_B T}{4} [(c + 3\eta) \ln(c + 3\eta) + (1 - c - 3\eta) \ln(1 - c - 3\eta)$$

$$+ 3(c - \eta) \ln(c - \eta) + 3(1 - c + \eta) \ln(1 - c + \eta)]$$

$$- w_0 c^2 - 3 w_1 \eta^2. \tag{8.73}$$

Equivalently, we may set $\vec{\eta} = (\eta, -\eta, -\eta), (-\eta, \eta, -\eta)$, or $(-\eta, -\eta, \eta)$ from the FCC symmetry [34]. Note that f_0 is invariant with respect to the transformation $(\eta_1, \eta_2, \eta_3) \rightarrow (-\eta_1, -\eta_2, \eta_3)$, and so on. Thus there are four equivalent ordered variants.

Further including the gradient free energy, Braun et al. [33] investigated interfaces such as the one between L1$_0$ and L1$_2$ phases and antiphase boundaries between two variants of the L1$_2$ phase. Khachaturyan et al. [36] examined the consequences of the mean-field free energy in Equation 8.73 for Al + Li. As in the BCC case, ordering can first take place without appreciable change of large-scale composition fluctuations, and the resultant order can then induce spinodal decomposition for relatively deep quenching. In Al + Li, elastic effects are suppressed because of very small lattice mismatch.

Wang et al. [34] studied phase-separation dynamics of the γ'-γ alloy in $d = 2$ in the presence of cubic elasticity (but with homogeneous elastic moduli). The free-energy density has the form:

$$f(c, \vec{\eta}, \vec{u}) = f_0(c, \vec{\eta}) + \frac{C_0}{2} |\vec{\nabla} c|^2 + \frac{D_0}{2} |\nabla \vec{\eta}|^2 + \alpha c e_1 + f_{el}(\vec{u}). \tag{8.74}$$

With the cubic interaction F_{cub} in Equation 8.31 following from the stress-balance condition in Equation 8.18, they integrated the diffusive Equation 8.23 for c and the relaxation-type equations,

$$\frac{\partial}{\partial t} \eta_i(\vec{r}, t) = -L_0 \frac{\delta F}{\delta \eta_i}. \tag{8.75}$$

Wang et al. observed the appearance of γ'-precipitates and subsequent shape changes into rectangles in the late stages, in accordance with Figure 8.2 [10]. In a recent 3-d simulation shown in Figure 8.13, Zhu et al. [11] followed the evolution of γ'-precipitates in a relatively early stage, using a similar model where the free-energy density is more complicated than Equation 8.74. The parameters were chosen to reproduce experiments in an Al-Ni alloy, where the elastic inhomogeneity effect is relatively weak and was neglected. As future work, it is of great interest to perform 3-d simulations with both cubic anisotropy and elastic inhomogeneity.

8.4   NONLINEAR ELASTICITY MODEL: INCOHERENT CASE

In multi-phase states of alloys, dislocations are produced around the interfaces when the lattice constants or the crystalline structures of the different phases are not close.

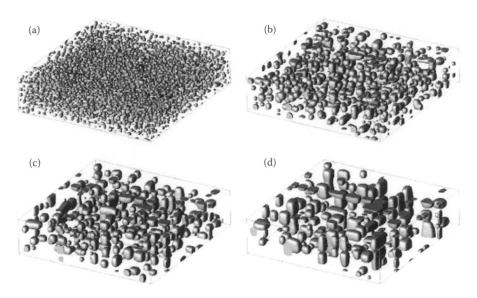

FIGURE 8.13 Evolution of γ'-precipitates obtained from 3-d simulations. The cuboidal shapes are due to cubic elasticity. The parameters are chosen to describe a Ni-13.8 at% Al alloy aged at 1023 K; (a) $t = 15$ min; (b) $t = 2$ h; (c) $t = 4$ h; (d) $t = 8$ h. The computational domain size is $160 \times 640 \times 640$ nm^3. (From Zhu, J.Z. et al., *Acta Mater.*, 52, 2837, 2004. With permission.)

The left panel of Figure 8.14 shows dislocation lines around a large γ'-precipitate in an Al-Sc alloy [37]. In this experiment, the transition from the coherent to the semi-coherent (partially incoherent) interface condition was observed when the precipitate radii exceeded 15–40 nm. Moreover, dislocations are proliferated in plastic deformations. The dislocations thus generated decisively affect the macroscopic mechanical properties. The right panel of Figure 8.14 shows a dislocation network produced in a softer matrix under stretching in a Ni-base γ'-γ alloy [38]. In such systems with harder precipitates, dislocations appear in the interface regions and grow preferentially into the softer matrix. The former occurs because considerable strains preexist around the interfaces, whereas the latter occurs because the elastic energy of the dislocation core increases with increasing elastic moduli. The dislocations do not penetrate into the precipitates in the presence of elastic inhomogeneity, so γ'-γ alloys such as Ni-Al, Ti-Al, or Al-Li have been used as *precipitation-hardened* materials against plastic deformations in technology [9].

The coupling between the composition and dislocations is also noteworthy, both in one-phase and multi-phase states. It is known that dilational changes around an edge dislocation give rise to local solute enrichment or depletion in the presence of the size difference among different atomic species. This results in the so-called *Cottrell atmosphere* [39], as was observed using 3-d atomic probe techniques [40]. On the other hand, a composition change around a screw dislocation is due to the composition-dependence of the elastic moduli [41]. As a result, heterogeneous nucleation is favored around dislocations, as has been observed in a number of experiments [42,43]. Some authors have numerically studied phase separation in the presence of dislocations in

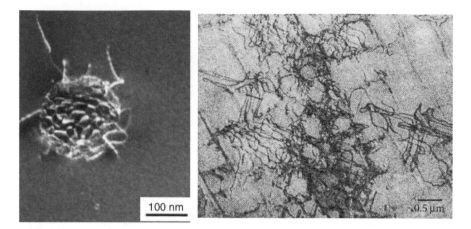

FIGURE 8.14 Images of dislocations surrounding a γ'-precipitate in a γ-matrix (left) (From Iwamura, S. and Miura, Y., *Acta Mater.*, 52, 591, 2004. With permission.) and dislocations formed under stretching in a γ-matrix (right) (From Pollock, T.M. and Argon, A.S., *Acta Mater.*, 40, 1, 1992. With permission.)

$d = 2$ [44–46]. We also mention a recent 3-d simulation study on the effect of solutes on dislocation motion [47].

Dislocation dynamics in $d = 3$ is already very complicated in one-phase states without coupling to the composition. Hence, large-scale molecular dynamics simulations have been performed to understand the dynamics [48,49]. As more phenomenological approaches, we mention phase-field theories of dislocation dynamics [50–52], which allow more efficient simulations on mesoscopic scales. In this section, we will discuss our phase-field approach to study dislocation formation and phase separation around dislocations in $d = 3$. We shall see that the elastic inhomogeneity is crucial in dislocation dynamics in two-phase states.

8.4.1 Nonlinear Elastic Energy

We consider Model B coupled with nonlinear elasticity via the Vegard law, neglecting the order-disorder phase transition. For a cubic crystal, we propose a nonlinear form for the elastic energy density f_{el} in Equation 8.2. We allow the strains $e_2, \ldots, e_6$ to take large values, but e_1 is assumed to be small. We then use the following form:

$$f_{el} = \frac{1}{2} K e_1^2 + \Phi(c, e_2, e_3) + \Psi(c, e_4, e_5, e_6). \tag{8.76}$$

The bulk modulus K is assumed to be independent of c. The second term $\Phi(c, e_2, e_3)$ represents an increase of the elastic energy due to stretching or compression along the principal axes, and the third term $\Psi(c, e_4, e_5, e_6)$ represents that due to shear deformations. In setting up Φ and Ψ, we require (i) crystal cubic symmetry, (ii) crystal periodicity, and (iii) compatibility with linear elasticity theory.

We note that the elastic energy should be invariant with respect to a $\pi/2$-rotation around each axis. With respect to a $\pi/2$-rotation around the z-axis, the strain components in the new reference frame $\{e_j'\}$ are expressed in terms of the original strains $\{e_j\}$ as

$$e_2' = -e_2, \quad e_3' = e_3, \quad e_4' = -e_4, \quad e_5' = -e_6, \quad e_6' = e_5. \tag{8.77}$$

However, for a $\pi/2$-rotation around the x-axis, the transformations are more complex as

$$e_2' = \frac{1}{2}e_2 - \frac{\sqrt{3}}{2}e_3, \quad e_3' = -\frac{\sqrt{3}}{2}e_2 - \frac{1}{2}e_3,$$
$$e_4' = e_6, \quad e_5' = -e_5, \quad e_6' = -e_4. \tag{8.78}$$

For a $\pi/2$-rotation around the y-axis, the corresponding transformations are

$$e_2' = \frac{1}{2}e_2 + \frac{\sqrt{3}}{2}e_3, \quad e_3' = \frac{\sqrt{3}}{2}e_2 - \frac{1}{2}e_3,$$
$$e_4' = -e_5, \quad e_5' = e_4, \quad e_6' = -e_6. \tag{8.79}$$

Notice that (e_2, e_3) and (e_4, e_5, e_6) are not mixed in these transformations, so we have assumed the form in Equation 8.76 for the elastic energy. Second, to obtain slip planes, we should assume that Φ and Ψ are periodic functions of the strains. For example, slips appear on the (1 0 0) and (0 1 0) planes under shear deformations, where e_4 is nonvanishing, and we require that Ψ is a periodic function of e_4. Here, the period of Ψ with respect to e_4, and that of Φ with respect to e_2, are both chosen to be 1. Third, for small strains, the standard form of f_{el} in the linear-elasticity regime should be obtained as Equation 8.9. The simplest forms of Φ and Ψ satisfying the above requirements are

$$\Phi(c, e_2, e_3) = \frac{\mu_2}{8\pi^2}\left[3 - \cos(2\pi e_+) - \cos(2\pi e_-) - \cos\left(\frac{4\pi e_3}{\sqrt{3}}\right)\right], \tag{8.80}$$

$$\Psi(c, e_4, e_5, e_6) = \frac{\mu_3}{4\pi^2}[3 - \cos(2\pi e_4) - \cos(2\pi e_5) - \cos(2\pi e_6)], \tag{8.81}$$

where $e_\pm \equiv e_2 \pm e_3/\sqrt{3}$ in Φ. Here, $\Phi \geq 0$ and $\Psi \geq 0$. The moduli $\mu_2 = \mu_2(c)$ and $\mu_3 = \mu_3(c)$ can depend on c as in Equation 8.13. The elastic inhomogeneity thus introduced is crucial in dislocation dynamics, as stated at the beginning of this section. As a generalization of Equation 8.15, the symmetric elastic stress tensor σ_{ij} can be obtained as follows. The diagonal components are

$$\sigma_{xx} = Ke_1 + \alpha c + \mu_2(f_+ + 2f_- - f_3),$$
$$\sigma_{yy} = Ke_1 + \alpha c - \mu_2(2f_+ + f_- + f_3), \tag{8.82}$$
$$\sigma_{zz} = Ke_1 + \alpha c + \mu_2(f_+ - f_- + 2f_3),$$

where $f_\pm = \sin(2\pi e_\pm)/(6\pi)$ and $f_3 = \sin(4\pi e_3/\sqrt{3})/(6\pi)$. The off-diagonal components are expressed as

$$\sigma_{xy} = \frac{\mu_3}{2\pi} \sin(2\pi e_4),$$

$$\sigma_{yz} = \frac{\mu_3}{2\pi} \sin(2\pi e_5), \qquad (8.83)$$

$$\sigma_{zx} = \frac{\mu_3}{2\pi} \sin(2\pi e_6).$$

For small strains, we reproduce the stress tensor in Equation 8.15 linear in the strains.

8.4.2 DISLOCATION DYNAMICS AND RHEOLOGY

The timescale of the birth and growth of dislocations is much faster than that of diffusion. To describe these processes, we should integrate the momentum equation for the lattice velocity field $\vec{v} = \partial \vec{u}/\partial t$. In this section, we assume the following form for this equation:

$$\rho_0 \frac{\partial}{\partial t} \vec{v}(\vec{r}, t) = \vec{\nabla} \cdot \overset{\leftrightarrow}{\sigma} + \eta_0 \nabla^2 \vec{v} + \left(\zeta_0 + \frac{\eta_0}{3}\right) \vec{\nabla}\vec{\nabla} \cdot \vec{v}, \qquad (8.84)$$

where the mass density ρ_0 is assumed to be a constant. We introduce the shear viscosity η_0 and the bulk viscosity ζ_0 [20], which give rise to damping of $\vec{v}$ and $\vec{u}$. In Equation 8.84, the nonlinearity is only in the force density $\vec{\nabla} \cdot \overset{\leftrightarrow}{\sigma}$. As usual, the composition obeys the CH equation 8.23. In nonlinear elasticity theory, the chemical potential difference is written as

$$\mu(\vec{r}, t) = \frac{k_B}{v_0} \left[T \ln\left(\frac{c}{1-c}\right) + 2T_0(1 - 2c) \right]$$

$$- C_0 \nabla^2 c + \alpha e_1 + \frac{\mu_{21}}{\mu_2} \Phi + \frac{\mu_{31}}{\mu_3} \Psi. \qquad (8.85)$$

The elastic inhomogeneity is assumed to be of the form in Equation 8.13. On the scale of the lattice constant $a = v_0^{1/d}$, there are three characteristic times. The *acoustic time* is defined by

$$\tau_0 = \left(\frac{\rho_0}{\mu_{20}}\right)^{1/2} a, \qquad (8.86)$$

where $(\mu_{20}/\rho_0)^{1/2}$ is the transverse sound velocity. The *viscous damping* time is

$$\tau_v = \frac{\rho_0 a^2}{\eta_0}. \qquad (8.87)$$

Finally, the *composition diffusion* time is

$$\tau_D = D_0^{-1} a^2, \qquad (8.88)$$

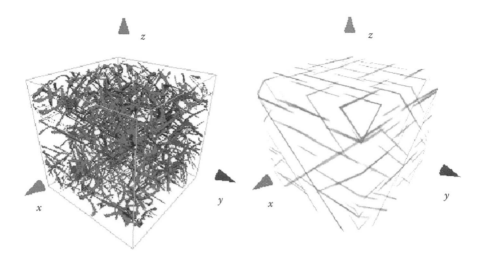

FIGURE 8.15 Dislocation loops (left) and slip planes (right) created by uniaxial stretching along the z-axis at $\epsilon = \dot\epsilon t = 0.281$ with $\dot\epsilon = 10^{-5}\tau_0^{-1}$ in a one-phase state. (From Minami, A. and Onuki, A., *Acta Mater.*, 55, 2375, 2007. With permission.)

where D_0 is given by Equation 8.26. The ratio of the relaxation times of $\vec{u}$ and c is then given by $\tau_v/\tau_D = D_0\rho_0/\eta_0$, where we assume $\zeta_0 \sim \eta_0$. In most real alloys, this ratio is very small, and the two timescales are much separated.

We performed 3-d simulations on a $128 \times 128 \times 128$ cubic lattice with periodic boundary conditions. The system length was $L = 128a$. We set $\mu_{30}/\mu_{20} = 1.1$ and $\mu_{21} = \mu_{31} = 1.2\mu_{20}$; then, the cubic anisotropy is rather weak but the elastic inhomogeneity is strong. In the first example, we apply uniaxial stretching along the z-axis for $t > 0$, where the average strains are expressed as

$$\langle \epsilon_{zz} \rangle = -2\langle \epsilon_{xx} \rangle = -2\langle \epsilon_{yy} \rangle = \frac{1}{\sqrt{3}}\dot\epsilon t, \qquad (8.89)$$

in terms of the applied strain rate $\dot\epsilon$. We then obtain $\langle e_3 \rangle = \epsilon$ with $\epsilon = \dot\epsilon t$. In previous simulations in $d = 3$ [48–50], *dislocation tangles* have been observed under stretching—these are also observed in our model. To demonstrate this, we first show results in the one-phase state in Figure 8.15. Here, $\dot\epsilon = 10^{-5}\tau_0^{-1}$ and dislocations emerge for $\epsilon \gtrsim 0.26$. In the left panel, string-like regions represent dislocation loops, where the elastic energy $\Phi(e_2, e_3) > 0.025\mu_{20}$. In the right panel, we display e_3 on the surface of the simulation cell, which is relatively large in slip planes. We can see four kinds of slips due to the cubic elastic anisotropy, whose normals make an angle of $\pi/4$ with respect to the z- and are perpendicular to the x- or y-axis.

Next, we consider the two-phase case. In Figure 8.16, we prepare a hard spherical domain at the center of the cell. At $\epsilon \sim 0.1$, dislocation loops are created in the interface region upon stretching. Subsequently, the loops preferentially glide into the soft region. Their ends are trapped at the surface in an early stage. As each dislocation becomes extended, the two ends on the surface are connected and detached from it, resulting in a new loop with a kink at the connected point. Remarkably,

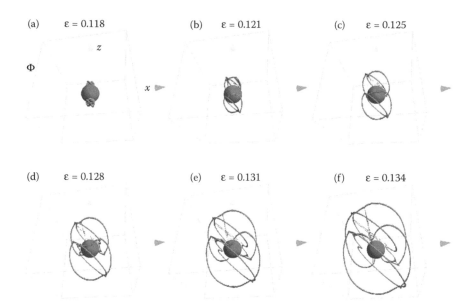

FIGURE 8.16 Dislocation formation around a single hard precipitate under uniaxial stretching for various values of $\epsilon = \dot{\epsilon}t$ with $\dot{\epsilon} = 10^{-4}\tau_0^{-1}$. Here, $c = 1/2$ on the domain surface and $\Phi = 0.03\mu_{20}$ on the dislocation lines. (From Minami, A. and Onuki, A., *Acta Mater.*, 55, 2375, 2007. With permission.)

a hard precipitate can act as a dislocation source. This is analogous to the *Frank–Read process* [53], where dislocation multiplication occurs around a preexisting dislocation.

In Figure 8.17, we prepare a number of coherent cuboidal hard domains within a soft matrix at $\langle c \rangle = 1/2$. In this case, the soft region is percolated and encloses hard cuboidal domains, owing to the elastic inhomogeneity. We display the evolution of dislocation loops, using the form of Φ under stretching in Equation 8.89. We again find the birth of dislocations at the interface regions and subsequent network growth in the soft matrix, in agreement with the experiment shown in Figure 8.14 [38].

We also examine the mechanical response in two-phase states. In Figure 8.18, we display the space average of the tetragonal stress component for the domain configuration in Figure 8.17:

$$\sigma_3 = \frac{1}{\sqrt{3}}(2\sigma_{zz} - \sigma_{xx} - \sigma_{yy}). \tag{8.90}$$

Here, cyclic stretching is applied with $|\dot{\epsilon}(t)| = 10^{-4}\tau_0^{-1}$. In the early stage in the first cycle, $\langle \sigma_3 \rangle$ increases linearly with increasing $\epsilon = \dot{\epsilon}t$. The average stress exhibits an overshoot upon dislocation appearance at $\epsilon \sim 0.11$. However, the first overshoot disappears from the second cycle. The curves of the second and third cycles are not much different, where there remains a residual stress ($\sim -0.2\mu_{20}$) at zero strain. It is a general trend that the peak height of the stress decreases with an increase in the number of structural defects.

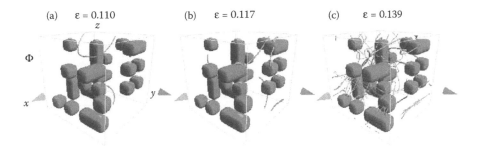

FIGURE 8.17 Dislocation formation in the soft region outside cuboidal domains under uni-axial stretching for various values of $\epsilon = \dot{\epsilon} t$ with $\dot{\epsilon} = 10^{-4} \tau_0^{-1}$. Here, $c = 1/2$ on the surfaces of the cuboids and $\Phi = 0.053$ on the lines representing dislocations. (From Minami, A. and Onuki, A., *Acta Mater.*, 55, 2375, 2007. With permission.)

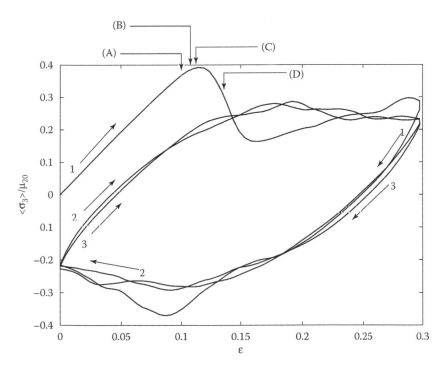

FIGURE 8.18 Stress-strain curve under cyclic stretching in the first three cycles at $|\dot{\epsilon}| = 10^{-4} \tau_0^{-1}$. The points A, B, and C correspond with the snapshots in Figure 8.17. (From Minami, A. and Onuki, A., *Acta Mater.*, 55, 2375, 2007. With permission.)

8.4.3 PHASE SEPARATION AROUND DISLOCATIONS

For incoherent phase separation, we need to describe dislocation formation and domain coarsening. The former takes place on the short timescale of τ_0 in

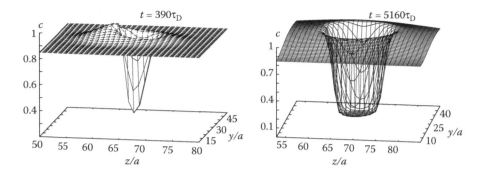

FIGURE 8.19 Composition profiles around an edge dislocation at $t = 390\tau_D$ (left) and $t = 5160\tau_D$ (right) with $\langle c \rangle = 0.85$. The profiles remain two-dimensional along the dislocation line.

Equation 8.86, and the latter occurs on timescales much longer than J_D in Equation 8.88. We recently integrated Equations 8.23 and 8.84 in $d = 2$ on long timescales by setting $\tau_0/\tau_D = 10^{-3}$ [46]. (In $d = 3$, however, this task is not easy at present.) In this section, therefore, we initially prepare dislocations and follow phase separation around them. New dislocations do not appear, so we may calculate $\vec{u}$ assuming the mechanical equilibrium condition in Equation 8.18 and integrate the diffusive equation 8.23. In this section, the system is quenched from a high temperature to $0.5T_c$ without applied anisotropic stress.

In our first example, we consider the case of heterogeneous nucleation around edge dislocations [45,52]. We start with a pair of parallel edge dislocations along the x-axis with the average composition $\langle c \rangle = 0.85$. In Figure 8.19, we show the composition profiles around one of the dislocations at $t = 390\tau_D$ and $5160\tau_D$. In the chemical potential difference μ in Equation 8.24, the term αe_1 is relevant in the initial stage, as it is of the form $Cz/(y^2 + z^2)$ in linear elasticity theory [20]. Here, C is a constant, and (y, z) represents the position in the (y, z)-plane. In the early stages, the accumulation of the soft component is much stronger than that of the hard component for the current off-critical composition. In the late stage, however, the dislocations

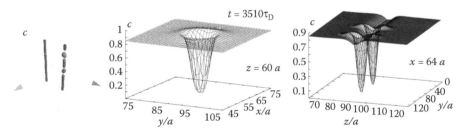

FIGURE 8.20 Heterogeneous nucleation around a pair of parallel screw dislocations at $3510\tau_D$ with $\langle c \rangle = 0.85$. The surface with $c = 0.825$ is shown on the left. We also show the cross-sectional profiles for the left dislocation at $z = 60a$ (middle) and at $x = 64a$ (right). (From Minami, A. and Onuki, A., *Acta Mater.*, 55, 2375, 2007. With permission.)

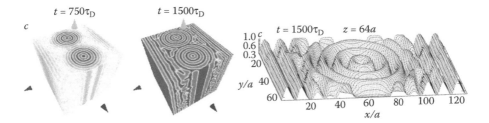

FIGURE 8.21 Spinodal decomposition around a pair of parallel screw dislocations at $t = 750\tau_D$ (left) and $t = 1500\tau_D$ (middle) with $\langle c \rangle = 0.5$. Phase separation starts around the dislocation lines. The cross-sectional profile of c is also shown at $z = L/2$ (right). (From Minami, A. and Onuki, A., *Acta Mater.*, 55, 2375, 2007. With permission.)

are wrapped by cylindrical domains of the new phase with its radius growing in time. In accordance with this result, mesoscopic cylindrical domains have been observed around dislocation lines [42,43]. For a screw dislocation line, the displacement is along the line, and we need to perform 3-d simulations. Here, the accumulation of the soft component around a defect decreases the local elastic moduli and lowers the elastic energy associated with the defect [41]. In Figures 8.20 and 8.21, we start with a pair of screw dislocations along the z-axis. In Figure 8.20, where $\langle c \rangle = 0.85$, we show the composition profile in a late stage at $t = 3510\tau_D$. In the chemical potential difference μ in Equation 8.24, the last term arising from the elastic inhomogeneity serves to attract the soft component to the dislocations. In our simulation [52], the composition was nearly homogeneous along the z-axis in the early stages. However, droplet-like soft regions suddenly appeared around the dislocation line in the late stage due to surface tension. In Figure 8.21, we set $\langle c \rangle = 0.5$ and then the bulk region is unstable. Here, the dislocations trigger and affect spinodal decomposition. The composition pattern is concentric near the dislocation cores and tends to be lamellar far from them.

REFERENCES

1. Khachaturyan, A.G., *Theory of Structural Transformations in Solids*, Wiley-Interscience, New York, 1983.
2. Onuki, A., *Phase Transition Dynamics*, Cambridge University Press, Cambridge, 2002.
3. Cahn, J.W., On spinodal decomposition, *Acta Metall.*, 9, 795, 1961.
4. Cahn, J.W., Hardening by spinodal decomposition, *Acta Metall.*, 11, 1275, 1963.
5. Onuki, A., Ginzburg–Landau approach to elastic effects in the phase separation of solids, *J. Phys. Soc. Jpn.*, 58, 3065, 1989.
6. Onuki, A., Long-range interactions through elastic fields in phase-separating solids, *J. Phys. Soc. Jpn.*, 58, 3069, 1989.
7. Fratzl, P., Penrose, O., and Lebowitz, J.L., Modeling of phase separation in alloys with coherent elastic misfit, *J. Stat. Phys.*, 95, 1429, 1999.
8. Sagui, C., Somoza, A.M., and Desai, R.C., Spinodal decomposition in an order-disorder phase transition with elastic fields, *Phys. Rev. E*, 50, 4865, 1994.
9. Cahn, R.W. and Haasen, P., Eds., *Physical Metallurgy*, North-Holland, Amsterdam, 1996.

10. Maheshwari, A. and Ardell, A.J., Morphological evolution of coherent misfitting precipitates in anisotropic elastic media, *Phys. Rev. Lett.*, 70, 2305, 1993.
11. Zhu, J.Z., Wang, T., Ardell, A.J., Zhou, S.H., Liu, Z.K., and Chen, L.Q., Three-dimensional phase-field simulations of coarsening kinetics of γ' particles in binary Ni-Al alloys, *Acta Mater.*, 52, 2837, 2004.
12. Miyazaki, T. and Doi, M., Shape bifurcations in the coarsening of precipitates in elastically constrained systems, *Mater. Sci. Eng. A*, 110, 175, 1989.
13. Miyazaki, J., Nakamura, K., and Mori, H., Experimental and theoretical investigations on morphological changes of γ' precipitates in Ni-Al single crystals during uniaxial stress-annealing, *J. Mater. Sci.*, 14, 1827, 1979.
14. Müller, K.A. and Thomas, H., *Structural Phase Transitions I*, Springer-Verlag, Heidelberg, 1981.
15. Cowley, R.A., Structural phase transitions I. Landau theory, *Adv. Phys.*, 29, 1, 1980.
16. Bruce, A.D., Structural phase transitions. II. Static critical behaviour, *Adv. Phys.*, 29, 111, 1980.
17. Onuki, A. and Puri, S., Spinodal decomposition in gels, *Phys. Rev. E*, 59, R1331, 1999.
18. Imaeda, T., Furukawa, A., and Onuki, A., Viscoelastic phase separation in shear flow, *Phys. Rev. E*, 70, 051503, 2004.
19. Hamanaka, T. and Onuki, A., Transitions among crystal, glass, and liquid in a binary mixture with changing particle-size ratio and temperature, *Phys. Rev. E*, 74, 011506, 2006.
20. Landau, L.D. and Lifshitz, E.M., *Theory of Elasticity*, Pergamon, New York, 1973.
21. Nishimori, H. and Onuki, A., Pattern formation in phase-separating alloys with cubic symmetry, *Phys. Rev. B*, 42, 980, 1990.
22. Onuki, A. and Nishimori, H., Anomalously slow domain growth due to a modulus inhomogeneity in phase-separating alloys, *Phys. Rev. B*, 43, 13649, 1991.
23. Orlikowski, D., Sagui, C., Somoza, A.M., and Roland, C., Large-scale simulations of phase separation of elastically coherent binary alloy systems, *Phys. Rev. B*, 59, 8646, 1999.
24. Orlikowski, D., Sagui, C., Somoza, A.M., and Roland, C., Two- and three-dimensional simulations of the phase separation of elastically coherent binary alloys subject to external stresses, *Phys. Rev. B*, 62, 3160, 2000.
25. Onuki, A. and Furukawa, A., Phase transitions of binary alloys with elastic inhomogeneity, *Phys. Rev. Lett.*, 86, 452, 2001.
26. Hohenberg, P.C. and Halperin, B.I., Theory of dynamic critical phenomena, *Rev. Mod. Phys.*, 49, 435, 1977.
27. Aharony, A., Dependence of universal critical behavior on symmetry and range of interaction, in *Phase Transitions and Critical Phenomena*, Vol. 6, Domb, C. and Green, M.S., Eds., Academic Press, London, 1976, 357.
28. Garel, T. and Doniach, S., Phase transitions with spontaneous modulation—the dipolar Ising ferromagnet, *Phys. Rev. B*, 26, 325, 1982.
29. Nishimori, H. and Onuki, A., Freezing of domain growth in cubic solids with elastic misfit, *J. Phys. Soc. Jpn.*, 60, 1208, 1991.
30. Ino, H., A pairwise interaction model for decomposition and ordering processes in BCC binary alloys and its application to the Fe-Be system, *Acta Metall.*, 26, 827, 1978.
31. Kubo, H. and Wayman, C.M., A theoretical basis for spinodal decomposition in ordered alloys, *Acta Metall.*, 28, 395, 1979.
32. Landau, L.D. and Lifshitz, E.M., *Statistical Physics*, Pergamon Press, Oxford, 1980.

33. Braun, J., Cahn, J.W., McFadden, G.B., and Wheeler, A.A., Anisotropy of interfaces in an ordered alloy: a multiple-order-parameter model, *Philos. Trans. Roy. Soc. London A*, 355, 1787, 1997.

34. Wang, Y., Banerjee, D., Su, C.C., and Khachaturyan, A.G., Field kinetic model and computer simulation of precipitation of $L1_2$ ordered intermetallics from FCC solid solution, *Acta Mater.*, 46, 2983, 1998.

35. Soffa, W.A. and Laughlin, D.E., Decomposition and ordering processes involving thermodynamically first-order order-disorder transformations, *Acta Metall.*, 37, 3019, 1989.

36. Khachaturyan, A.G., Lindsey, T.F., and Morris, Jr., J.W., Theoretical investigation of the precipitation of γ' in Al-Li, *Metall. Tran. A*, 19, 249, 1988.

37. Iwamura, S. and Miura, Y., Loss in coherency and coarsening behavior of Al_3Sc precipitates, *Acta Mater.*, 52, 591, 2004.

38. Pollock, T.M. and Argon, A.S., Creep resistance of CMSX-3 nickel base superalloy single crystals, *Acta Mater.*, 40, 1, 1992.

39. Cottrell, A.H., *Dislocations and Plastic Flow in Crystals*, Clarendon Press, Oxford, 1953.

40. Blavette, D., Cadel, E., Fraczkiewicz, A., and Menand, A., Three-dimensional atomic-scale imaging of impurity segregation to line defects, *Science*, 286, 2317, 1999.

41. Fleisher, R.L., Solution hardening, *Acta Metall.*, 9, 996, 1961.

42. Dash, W.C., Copper precipitation on dislocations in silicon, *J. Appl. Phys.*, 27, 1193, 1956.

43. Matsushita, K. and Koda, S., Precipitation and generation of dislocations in an Al-Mg alloy, *J. Phys. Soc. Jpn.*, 20, 251, 1965.

44. Léonard, F. and Desai, R.C., Spinodal decomposition and dislocation lines in thin films and bulk materials, *Phys. Rev. B*, 58, 8277, 1998.

45. Hu, S.Y. and Chen, L.Q., Solute segregation and coherent nucleation and growth near a dislocation—a phase-field model integrating defect and phase microstructures, *Acta Mater.*, 49, 463, 2001.

46. Minami, A. and Onuki, A., Dislocation formation in two-phase alloys, *Phys. Rev. B*, 70, 184114, 2004.

47. Hu, S.Y., Li, Y.L., Zheng, Y.X., and Chen, L.Q., Effect of solutes on dislocation motion—a phase-field simulation, *Int. J. Plasticity*, 20, 403, 2004.

48. Bulatov, V.V., Abraham, F.F., Kubin, L.P., Devincre, B., and Yip, S., Connecting atomistic and mesoscale simulations of crystal plasticity, *Nature*, 391, 669, 1998.

49. Buehler, M.J., Hartmaier, A., Gao, H., Duchaineau, M., and Abraham, F.F., Atomic plasticity: description and analysis of a one-billion atom simulation of ductile materials failure, *Comput. Methods Appl. Mech. Eng.*, 193, 5257, 2004.

50. Wang, Y.U., Jin, Y.M., Cuitiño, A.M., and Khachaturyan, A.G., Nanoscale phase field microelasticity theory of dislocations: model and 3D simulations, *Acta Mater.*, 49, 1847, 2001.

51. Minami, A. and Onuki, A., Dislocation formation and plastic flow in binary alloys in three dimensions, *Phys. Rev. B*, 72, 100101, 2005.

52. Minami, A. and Onuki, A., Nonlinear elasticity theory of dislocation formation and composition change in binary alloys in three dimensions, *Acta Mater.*, 55, 2375, 2007.

53. Frank, F.C. and Read, W.T., Multiplication processes for slow moving dislocations, *Phys. Rev.*, 79, 722, 1950.

Index

323

T - #0361 - 071024 - C8 - 234/156/16 - PB - 9780367385859 - Gloss Lamination